Decision Control, Management, and Support in Adaptive and Complex Systems:

Quantitative Models

Yuri P. Pavlov
Bulgarian Academy of Sciences, Bulgaria

Rumen D. Andreev
Bulgarian Academy of Sciences, Bulgaria

Managing Director: Lindsay Johnston
Editorial Director: Joel Gamon
Book Production Manager: Jennifer Yoder
Publishing Systems Analyst: Adrienne Freeland
Development Editor: Myla Merkel
Assistant Acquisitions Editor: Kayla Wolfe
Typesetter: Christina Henning
Cover Design: Jason Mull

Published in the United States of America by
Information Science Reference (an imprint of IGI Global)
701 E. Chocolate Avenue
Hershey PA 17033
Tel: 717-533-8845
Fax: 717-533-8661
E-mail: cust@igi-global.com
Web site: http://www.igi-global.com

Library of Congress Cataloging-in-Publication Data

Pavlov, Yuri P., 1952-
Decision control, management, and support in adaptive and complex systems: quantitative models / by Yuri P. Pavlov and Rumen D. Andreev.
p. cm.
Includes bibliographical references and index.
Summary: "This book presents an application and demonstration of a new mathematical technique for descriptions of complex systems"--Provided by publisher.
ISBN 978-1-4666-2967-7 (hardcover) -- ISBN 978-1-4666-2968-4 (ebook) -- ISBN 978-1-4666-2969-1 (print & perpetual access) 1. Decision making--Mathematical models. 2. Decision support systems--Mathematical models. 3. Problem solving--Mathematical models. I. Andreev, R. D. (Rumen D.) II. Title.
HD30.23P384 2013
302.3'5--dc23

2012037377

British Cataloguing in Publication Data
A Cataloguing in Publication record for this book is available from the British Library.

All work contributed to this book is new, previously-unpublished material. The views expressed in this book are those of the authors, but not necessarily of the publisher.

Table of Contents

Foreword

This book, written by leading Bulgarian specialists in decision-making, presents a new approach to the adaptive mathematical modeling of complex systems, in which human knowledge is decisive to achieve the best result. As it is widely appreciated now, the human thinking and preferences have a quantitative nature that makes the decision-making problems in complex systems to be considered as quantitative due to the difficulties or even impossibility for qualitative description. The methods presented in the book are appropriate for decision-making when scarce information is available, especially in the initial development phase of complex problems and situations. The main direction of the presentation is toward the branch of model-driven decision-making, value-driven design, and value-based management as crucial aspects of model-driven design.

The subject area of the book addresses theory of measurement (scaling), utility theory and Bayesian approach in decision-making theory, stochastic approximation and potential function method, machine learning and optimal control theory. The representation of complex systems using utility function as objective function needs mathematical tools for evaluation of this function on the basis of qualitative human information. The utility evaluation is a stochastic procedure developed in a machine learning approach and permits development of information and decision support of systems based on mathematically grounded solutions. This approach together with the classical application of control theory ensures control solutions with mathematical exactness in regard to complex systems with human participation and permits a new point of view for development of information systems for individual utility assessment. Some important ways of applications are given in practical cases of model-driven design and optimal control in biotechnology, value-based management in bird prediction, and e-learning personalization.

The book reflects the author's opinion that the mathematical proofs are necessary to be provided in details for readers/scholars with mathematical background and sufficiently serious mathematical experience. In that way they could study in-depth outlined principles and methods and eventually further develop some of them.

The book has a monographic character and includes theoretical results as well as practical applications of scientific research carried out by the authors in the Bulgarian Academy of Sciences during the previous three decades. It provides a general mathematical paradigm as well as many numerical methods suitable for implementation by practitioners constructing value models in new applications in different important areas such as sociology, medicine, agriculture, control design, and economics. The references contain carefully chosen fundamental bibliographic sources that may be consulted in case of deeper investigation. In this way, the book could be useful for researchers and scholars working in the fields where the quantitative decision-making is applied. Reading of the book requires some basic mathematic knowledge about sets, relations, and operations over them. The practitioners could omit the proofs of the theorems and could follow only the examples of applications of the mathematical methods. They could follow, based on these examples, the ways of development of value-driven modeling and models in the different chapters. That is why, as the authors stress, the book is not a simple collection of value-driven models in different areas of human activity, but a guidance on how to design value-driven models and how to perform value-driven decision-making in complex processes. The book can supplement upper-level courses and postgraduate courses, and it can serve decision makers as a valuable resource for designing complex systems.

Petko H. Petkov
Technical University of Sofia, Bulgaria

Petko Petkov *received M.S. and Ph.D. degrees in Control Engineering from the Technical University of Sofia in 1971 and 1979. Since 1973, he has been a faculty member in the Department of Systems and Control at the Technical University of Sofia. He has been a Professor of Control Theory in the same department since 1995. His research interests include numerical methods and software for control systems analysis and design, perturbation linear algebra and control, robust control systems design, and implementation of digital control algorithms. He coauthored Computational Methods for Linear Control Systems (Prentice Hall, 1991), Perturbation Theory for Matrix Equations (North-Holland, 2003), and Robust Control Design with MATLAB (Springer, 2005, 2012). Dr. Petkov is a member of the American Mathematical Society and a corresponding member of the Bulgarian Academy of Sciences.*

Preface

The book has a monographic nature and contains scientific results in the field of contemporary approaches to adaptive decision-making and decision. Its purpose is to present an application and demonstration of a new mathematical technique and possibilities for mathematical descriptions of complex systems, in which human speculative knowledge is decisive for the final solution. The aspiration for quantity measurements, estimations, and prognosis at all phases of decision-making and problem solving is natural. However, this task is carried out with very scarce initial information, especially in the initial development phase in complex problems and situations. In the initial stage of a decision process, the heuristic of the investigator is very important, because in most of the cases there is a lack of measurements or even clear scales under which to implement these measurements and computations. This stage is often outside of the strict logic and mathematics and is close to the art, in the widest sense of the word, to choose the right decision among a great number of circumstances and often without associative examples of similar activity. The correct assessment of the degree of informativity and usability of this type of knowledge requires careful analysis of the terms measurement, formalization, and admissible mathematical operations under the respective scale, which do not distort the initial empirical information.

In the book, we propose approaches and methods for measurement and analytical presentations of empirical and scientific knowledge expressed as preferences. The problem of human preference and their implementation in information systems is certainly one of the important topics in decision-making for the forthcoming years, as witnessed by some current activity in this area, from the standpoint of utility theory. According to social-cognitive theories, people's strategies are guided both by internal expectations about their own capabilities of getting results and by external feedback. Internal human expectations and assessments are generally expressed by qualitative preferences. Following the ideas of Professor Ralph Keeney, the main assumption in each management or control decision is that the values of the subject making the decision are the locomotive force, and as such, they are the main moment in supporting the decisions. The values are the guiding force in supporting the decisions and due to this are determining the formation of the decisions. In information systems in general and especially in the expert systems, the values are implicitly and heuristically included.

The productive merger of the mathematical exactness with the empirical uncertainty in the human notions is the main challenge here because people's preferences contain uncertainty due to the qualitative type of the empirical information and the qualitative nature of the human notions. This uncertainty is of subjective and probability nature. The uncertainty of the subjective preferences could be eliminated, as is typical in the stochastic approximation procedures and *machine-learning* based on the stochastic programming.

Due to multidisciplinary nature of the cognitive process and to the multidisciplinary nature of the fields of applications, our choice of scientific methods is oriented toward the utilization of the stochastic programming, the theory of measurement, the utility theory, and on some flexible techniques for extrapolation and prognosis based on the multilinker extrapolation and pseudo inverse matrices. In this manner, we have posed the decision-making problem as a problem of constructing value and utility functions based on stochastic recurrent procedures as machine learning, which can later be used in optimization problems. Following from this choice, the orientation of the book is toward the branch of model-driven decision-making. Model-driven decision-making emphasizes access to and manipulation of a statistical, financial, optimization, or simulation model. Model-driven decision-making uses data and parameters provided by users to assist decision makers in analyzing a situation.

While the role of knowledge management for decision support is well acknowledged, there is a socio-technical gap between existing theory and actual practice in real-life decision-making. In recent years, bridging this gap has been a challenge in many areas of research. This is a great divide between the social aspects aimed to be supported and those that are actually supported. In decision support, this challenge has raised several important questions concerned with the account and encapsulation of social aspects of managerial decision-making as well as with the representation of certain human cognitive aspects, such as intuition or insights within computational systems. In this manner the problem of constructing analytically valuable utility functions with flexible mathematical techniques is now a problem and a challenge. Model-driven value-driven design can be defined as a software development paradigm, in which required human value considerations are importantly and equally engineered into best practices, activities, and management. Value-driven design creates an environment that enables design optimization by providing designers with an objective function. The objective value function inputs all the important attributes of the system being designed and outputs a score. The purpose is to enable the assessment of a value for every design option so that options can be rationally compared and a choice taken. At the whole system level, the objective function that performs this assessment of value is called a value model.

Value-Based Management (VBM) concepts are prevalent in theory and practice, since value creation is commonly considered the paramount business goal. However, VBM mainly applies data-driven concepts to support decision-making, disregarding model-driven approaches. Validate mathematical preferences value evaluation could be the first step in realization of a human-adapted design process and decision-making in VBM. The objective is to avoid the contradictions in human decisions and to permit mathematical calculations in these fields and rational management and control decisions. The analytical description of the expert's preferences as value or utility function will allow mathematically the inclusion of the decision maker in the description of the complex system "technologist or manager-process." Value-based decision enables the assessment of a value for every design option so that options can be rationally compared and a choice taken. At the whole system level, the objective function that performs this assessment of value is called a value model. The rational and logical description of value-based decisions requires basic analytical description and representation of the DM's preferences. The mathematical description on such a socio-technical fundamental level requires basic mathematical terms as well, as equivalent of this description with respect to a given real object. In the last aspect of equivalency of the mathematical descriptions, we enter the theory of measurements and scaling.

The book has two aspects of orientation. In the first aspect, it suggests a common mathematical conception and appropriate numeric methods for development of model-driven decision-making based on value models. This approach realizes the concepts of value modeling in decision-making on the basis of the achievements of the utility theory, stochastic programming, and multilinker extrapolation. In that aspect, the book provides the general mathematical paradigm and lots of numerical methods the practitioners for development of value-driven models in new applications in sociology, medicine, agriculture, control design in economics. It will be of interest to advanced students and professionals working in the subject of decision theory, control and management of complex system, developers of information systems, as well as to economists and other social scientists. The proposed methods are unique and permit incorporation of machine learning with mathematical exactness in the decision-making and decision support practices.

In the second aspect, whenever deemed necessary, the mathematical proofs are provided in detail so that the reader with mathematical experience and mathematical background can study in depth the principles and methods outlined in this book and to eventually develop some of them. The reason is that the used stochastic programming approach, the theory of potential function method, is not well-known to the wide scientific community.

The practitioners could omit the proofs of the theorems and could follow only the examples of applications of the mathematical methods. They could follow, based on these examples, the ways of development of value-driven modeling and models in the different chapters. In this manner, the book is not a simple enumeration of value-driven models in different areas of human activities, but a guidance of how to design value-driven models and value-driven decision-making in complex processes.

Chapter one presents a conceptual framework of decision-making domain. Decision-making is the major component of a problem-solving system based on the principles of decision theory. The important factors that influence decision-making activity are decision environment and decision situation (decision context) that is interpreted usually as a complex adaptive system. Decision support is based on the idea that many important classes of decision situations can be supported by providing decision makers with a decision support environment. The construction of such environments concerns the development of Decision-Making Support Systems (DMSS). The main strategies in their development are determined in the framework of a generalized decision-making process. The development of DMSSs depends on the accepted implementation method, architectural representation of these systems, implementation approaches, and used information, communication, and computer technologies.

Chapter two has an introductory nature for the principles and some concepts from probability theory. We consider the concept probability in two aspects: objective probability as frequency of occurrence and in the aspect of human expectation. Special attention is devoted to the concept "generalized gradient" because we discuss recurrent procedures based on it, and theorems and proofs regarding the convergence of the recurrent procedures are provided. Presenting strict mathematical proofs, we give the mathematically prepared reader an opportunity to enter in the mechanisms of the proposed approaches and methods for evaluating subjective knowledge, empirical, and verbally expressed as preferences.

Chapter three has an introductory nature with respect to the decision-making theory. It begins with description of the notion measurement and scale. The concepts of "value" and "utility" are interpreted from the point of view of the theory of measurement (scaling). Special attention is paid to the concept "subjective probability." The chapter ends with an example for evaluation of subjective probabilities and examples of utility functions that have methodological generality.

Chapter four takes into consideration the main results and theorems for the existence of value and utility function. Our choice of theoretical results and theorems is conditioned methodologically by the respect to the utility theory's use and application in practical problems. This section concludes with three examples of the application of utility theory in practice. The chapter requires certain persistence and mathematical inclination with clear definitions and description of the used mathematical concepts.

Chapter five considers in detail a direction of the stochastic programming used in the stochastic procedure for analytical construction of value and utility functions, and subjective probabilities. The question of special interest is an important topic from the so-called "potential function method." This theory is relatively unknown in the scientific literature, in particular in the field of decision-making. Because of this, we have provided the used theoretical results with detailed proofs.

Chapter six presents the main theorems and methods for analytical representation of utility functions and a theorem for the convergence of the proposed recurrent procedures. The use of the proposed stochastic procedures for utility function construction is described from the viewpoint of Kahneman and Tversky and their Prospect Theory. Their understanding is based on the techniques given in chapters two, four, and five at the level of applicability of the discussed mathematical notions and methods.

Chapter seven takes into consideration stochastic procedures for the evaluation of subjective probability. They are based on the theoretical results outlined in chapter five. The described stochastic procedures are used for the decomposition of the multiattribute utility function to functions of fewer variables. This decomposition is fundamental for the use of utility theory in practice and in the modeling of more complex and difficult for formalization processes.

Chapter eight is devoted to the theoretical description of the Multilinear Extrapolation (MLE) approach of Rastrigin. This method is effective and suitable for making prognosis and evaluations in complex systems, including as analytical mathematical relationships. It allows adaptation of the prognoses with easy to implement iterative procedures and easy generation of training samples with applied significance. Such adaptation is also allowed by stochastic recurrent procedures. This makes the methods convenient as a basis for the development of adaptive, complex systems with human participation in the final decision.

Chapters nine, ten, and eleven are devoted to examples of application of the previously presented approaches and methods. Each of them may be read independently of the others according to the interests and orientation of the reader. Chapter nine shows the applicability of the proposed approaches and methods for complete description of the complex system "human-biotechnologist, bio-process." It presents a complete solution of a practically significant problem, which has not been formally described. Chapter ten regards the teaching/learning process as decision-making process where the teacher plays the role of decision maker. Determining the competence of a learner to use learning resources in one way or another is done on the basis of measuring his/her preferences. This chapter demonstrates the capabilities of the new approach to describe complex, socially determined concepts and processes with the formal methods of the decision-making theory. The next chapter is a demonstration of modeling and managing complex process from stockbreeding. It includes an evaluation of the scale by which the initial information is obtained, factor analysis for eliminating the overlapping, formal model based on MLE, and the utility theory for description of the processes of the considered field of stockbreeding.

Chapter twelve together with chapter one describes the structure and methods for the development of DMSS (on the base of data mining and knowledge discovery) using decision-aiding tools in the data mining process. It gives a brief description of the decision support system, which allows the construction of value and utility functions of the individual user. It describes a prototype of an information decision support system for the individual's utility evaluation as proposed by Professor Ralph Keeney. Mathematically it is backed up by the methods from the preceding chapters. The prototype of the information system is developed on the basis of the mathematical methods from chapters five, six, and seven.

Acknowledgment

The authors are deeply grateful to scientists and colleagues for their useful comments and recommendations and for the fruitful discussions in seminars, conferences, and scientific meetings in the Bulgarian Academy of Sciences and in the Sofia University "Kliment Ohridski" during the last decade. We are also grateful to the Chairmen of the Departments in the Bulgarian Academy of Sciences who initiated us in the field of the theory of Decision-Making. The authors are deeply grateful to Peter Vassilev for his professional help in the mathematical English translation. Our managing development editor at IGI Global, Myla Merkel, deserves a special mention and appreciation. We are deeply grateful to the IGI Global Editorial Division for the possibility to write this book.

Yuri Pavlov
Bulgarian Academy of Sciences, Bulgaria

Rumen Andreev
Bulgarian Academy of Sciences, Bulgaria

August 2012

Chapter 1
Decision Support Fundamentals

ABSTRACT

The chapter presents a conceptual framework of decision-making domain. Decision-making is the major component of problem-solving systems based on the principles of decision theory. The important factors that influence decision-making activities are decision environment and decision situation (decision context). Decision-making is observed from several viewpoints that determine different cognitive frameworks, within which the decision-making is interpreted and evaluated. It is considered in respect of the following major points of view: psychological viewpoint, viewpoints determined by decision's attributes, decision environment's attributes, and decision situation's features. The term situation is used to mean any issue, problem, condition, opportunity, or bounded system that has to be changed, improved, transformed, etc. Usually, the situation is interpreted to mean a complex adaptive system. Decision support is based on the idea that many important classes of decision situations can be supported by providing decision-makers with a decision support environment.

Rational approaches to decision-making are classified in one of the following categories: descriptive, normative, or prescriptive. The main normative models that are presented concern value functions, expected utility and subjective expected utility. Several alternative normative frameworks that have appeared in the literature of the last thirty years particularly for attacking the problem of conflicting objectives—the analytic hierarchical process, multi-criteria decision-making movement, outranking—are described. A Framework to align decision support-driven initiatives with the decision-making vision is given in the chapter. It divides the objective-oriented systems that determine the structure of decision-making domain from the strategic actions in this domain that have to determine the decision-making process. The latter serves as the basis for defining the main objectives, which have to be achieved in the

DOI: 10.4018/978-1-4666-2967-7.ch001

development of Decision-Making Support Systems (DMSS). A classification scheme of the main categories of such systems is suggested. The development of DMSSs depends on the accepted implementation method, architectural representation of these systems, implementation approaches and used information, communication and computer technologies. They guarantee not only the capabilities of decision-making support systems, but its characteristics as well.

1. DECISION-MAKING AND DECISION SUPPORT

Ackoff (1981) presents four ways to cope with difficulties (problems) of the material word—*absolution*, *resolution*, *solution,* or *dissolution*.

- **Absolution:** The problem remains outstanding in the hope of self-disappearing;
- **Resolving:** This approach requires to select a course of action that yields an outcome (artifact) that is good enough for a purpose or meets the needs of the specific context, in which the artifact will be used (the "best-fit" artifact), i.e. the result satisfies;
- **Solving:** This is to select a course of action that is believed to yield the best possible outcome;
- **Dissolving:** It is a way to remove the difficulties through changing the nature and/or the environment of the entity which brings the problem, i.e. the problem is eliminated by its reducing.

The resolution differs from the solution in the kind of causality, on which they are based. In resolving the selection of an option causes a satisfactory result, i.e. the relation of cause and effect can be described as efficient. In solving the selection of a course of action produces an optimal result—the causality optimizes. It is not necessary to contrast efficiency with optimization, since the sufficient and necessary condition for realization of a sustainable choice is guaranteed by its efficiency and optimality. The problem resolution and problem solution are two aspects of an important theme of cognitive psychology that is known as problem solving.

The purpose of decision-making is to make deliberately opting for one choice from two or more, proactively to optimize a situation or outcome and not let it happen by default. The main attributes of decision-making are:

- Cognitive process that focuses on the choice phase of a problem-solving activity;
- Selection of an option over others, which could include no action;
- **Risky Choice:** Under conditions that are uncertain;
- Decision in order to reach a specified goal or outcome.

Human activity in decision-making terms is subject of active research that is performed in the decision-making domain from several perspectives. From a problem solving perspective, the decision-making process is the most important component of a problem solving activity. From a cognitive perspective, it is necessary to examine decision-making in its interactions with the environment. According to distributed cognition that is a branch of cognitive science, knowledge, and cognition are not confined to an individual, rather they are distributed across objects, individuals, artifacts, and tools in the environment. The goal is to describe how distributed units are coordinated by analyzing the interactions between individuals, the representational media used, and the environment within which the activity takes place. Distributed Cognition is a useful descriptive framework that describes decision-making activity in informational and computational terms and suitable for analyzing situations that involve problem-solving (Perry, 2003). Since the purpose of decision-making is to make rational choices, it is worth noting that from a normative perspective the analysis of individual decisions is concerned with the logic of decision-making and rationality it leads to.

All aspects of decision-making domain are presented in Figure 1. The decision-making activity bases on the knowledge that is systematized in *the decision theory* and is regarded as subsystem of the *problem-solving framework*. The used decision theory outlines the decision maker's behaviour. *The decision environment* concerns decision-making style and decision context. *The situation* describes the context in which the problem is set. It identifies the features that affect and are affected by the eventual outcome of the decision-making process. It is imperative that the decision-makers have a detailed knowledge of the situation in which the decision is being taken, all the possible or likely alternatives that can be taken, and the outcomes of each of these alternatives. Anticipating the characteristics of the decision-taking situation is difficult. Any representation of the situation is an approximation of reality at some level, and the key is to identify the features that are instrumental in determining the situation in correspondence with the main objective of the decision maker.

Figure 1. Decision-making domain

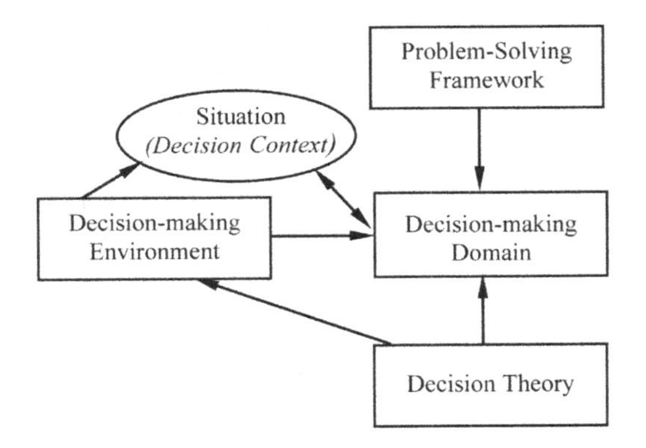

We can define a decision as it has "hard" characteristics, if the situation is uncertain, inherently complex with many different issues and there are several goals that must be achieved but one or more is blocked and compromises are needed. The difficulties that a situation gives birth to decision-making could increase, when the decision environment involves two or more people in making a decision. Different perspectives can lead to different conclusions and may disagree about the assumptions, probable outcomes and even the decision. The decision context is crucial to understanding the real nature of the decision to be made and the needs it addresses. It is impossible to take decisions in isolation from the situation, i.e. the circumstances or context cannot be disregarded.

Problem solving is a central topic of cognitive psychology concentrated on study of cognition. The cognition covers higher mental processes including the ways of knowing and understanding of world, information processing, decision taking and make assessment, knowledge representation to the others. The problem solving passes through the following basic stages—preparation for alternatives generation, identification of the possible alternatives (options), determination of all the possible consequences of these alternatives (outcomes) and evaluation of all the possible consequences of each option (Greeno, 1978). Alternatives refer to the identified choices available to the decision-taker and outcomes are the predicted results of choosing each of the alternatives. Essentially, the decision-making process is the act of differentiating between the alternatives given the situation, in order to arrive at an optimal or satisfactory outcome. The preparation for alternatives generation concerns problem formulation that is a result of problem analysis. The other two stages coincide with decision-making. Problem analysis must be done first then the information gathered in that process is used by decision-making.

1.1. Problem Solving

From a work in cognitive psychology (Courtney & Paradice, 1993), *problem analysis* is considered as a process that has three phases—*problem identification*, *problem definition,* and *problem structuring*. Problem identification occurs when the need for a decision is perceived by a decision maker. When characterized as a problem situation, this need is typically perceived in the context of some perceived deviation from expectations. A problem situation is not necessarily a bad situation, since an opportunity could also create the need for a decision. Problem definition involves determining the relevant properties of the situation. Problem structuring examines the problem situation to determine a strategy for addressing the situation. The problems can differ from one another with respect to their structure. There are two extreme kinds of problems—structured and unstructured. A problem that is represented by formal structure is regarding as hard problem or well-defined problem. The other kind of problems is known as ill-defined problems. In order to reduce uncertainly and increase structure in an ill-defined problem the problem solver has to transform it into a well defined one.

Smith (1988) classifies the existing conceptualizations of problem structure by four notions: goal state, problem space, knowledge, and process:

- The first conceptualization relates the degree of the problem structure to the clarity of the major objective state of the problem; If the goal is adequately specified, the problem is structured;
- The second conceptualization relates the degree of the problem structure to its representability, if the characteristics of the problem can be easily measured, and the relationships between them can be formulated explicitly and quantitatively, then the problem is structured;
- According to the third conceptualization, the problem structure is related to the solver's knowledge: By this notion, the degree of the problem structure is regarded as person dependent rather than of a natural kind;
- Finally, the degree of the problem structure can be seen in light of the solution process: If an effective solution procedure is known and regularly used, then the problem is structured—on the other hand, if no solution strategy can be found, the problem is unstructured.

Problems may also not be structured or unstructured in their entirety but only in terms of some stages in the solution procedure. Since the problem structure is rather a continuous than a dichotomous concept, there is always something between the two extremes, namely, semistructured problems. Semistructured problems may exist between the extremes according to any of the conceptualizations or they may be structured according to one conceptualization but unstructured according to another.

There is no unanimous viewpoint on the presentation of a problem analysis process. Kepner and Tregoe (1981) present a five-step process for problem analysis:

1. **Definition of the Problem:** A brief statement of "what is broken" or "why we know that something is broken."
2. **Description of the Problem in Four Dimensions:** Identity, "what"; location, "where"; timing, "when"; magnitude, "how much, how many".
3. **Extraction of Key Information in all Four Dimensions:** What *is* known and what *could be* wrong.
4. **Testing for the most Probable Cause:** Throw out the unlikely and examine the feasibility of the most likely cause.
5. Verification of the True Cause.

The determination of a set of alternatives for solving a problem is the culmination in this cognitive process, since the lack of options, especially when none of the options seems to address the real needs recognized in a decision context, causes a major difficulty in achievement of a good solution. The successful generation of alternatives depends on two factors:

- Adequate problem analysis that identifies the proper situation and what may be done to address the issues;
- Presence of creativity; Generation of ideas for achievement of objectives is really creative work.

The performance of an informal activity as alternatives generation could be structured in some degree, if in solution of this problem it is used a cognitive scheme known as "inductive reasoning" (heuristic), i.e. the performance bases on inherent heuristic. Heuristics are basic empiric rules that often ensure the achievement of successful results, but cannot guarantee them. There are some general methods of heuristics that are usually used (Groner, Groner, & Bischof, 1983):

1. **Representativeness:** It supports rapid assessment based on stereotypical classification. If a newly proposed product has similar characteristics to an earlier success or failure, then it will be similarly classified. Such thinking may not be an unreasonable starting point but may lead to unnecessarily rapid conclusions when resource is available to gather further information.
2. **Availability:** Frequently occurring events are easy to recall, and it is natural to associate a high probability with them. However, peripheral events can modify information that is available from memory, causing inappropriate probabilities to be assumed. For instance, peculiar events often receive high publicity, giving the impression that they are more common than they are in actuality. When statistically considering association between events, many would use.
3. **Anchoring and Adjustment:** This is an example of adjusting away from an initial start point. Once more, the heuristic can be useful, but, if an initial value or viewpoint is unreasonable, then the adjustment is often insufficient to overcome the anchor.
4. **Approaches to Increase the Number of Alternatives:** Looking for alternatives in the scientific, firm and patent literature; using of competent experts; modification of existing alternatives; combination of present alternatives; inclusion of alternatives that are antipode of already suggested; other approaches.
5. **Optimal Conditions for Creative Work:** It is necessary to know the factors that help for achievement of desired result in variants generation of a team.
6. **Goal Seeking:** It bases on an inverted view on alternatives generation where a preferred or optimal solution is known, and it is looking for options, with which help the achievement of the optimal or preferred solution is possible.
7. **Forward-Orientation of Variants Identification Process:** The problem solution is presented through transitional stages and subgoals measured by substitution through quantitative criteria. The variants for achievement of a subgoal are a part of the set of variants giving problem solution.

There are many other non-heuristic ways of facilitating and stimulating the generation of sufficient options:

- **Brainstorming:** Where ideas are collected without criticism and then considered;
- **Bringing in Diverse Groups of People with Different Perspectives and Experience:** The team assembled included unusual members such as psychologists;

- **Insight:** It is a surprise enlightenment for the relationships among different elements describing a problem that are considered as independent—the focus is not on "step by step" process for variants generation;
- **Morphological Analysis:** It bases on identification of all independent variables that represented a situation (problem, system), determination of their values and generation alternatives that present all possible combination of variables and their values;
- **Serious Games:** The players perform specific role in a virtual reality that represent a real situation. During the game, the actors present different variations for achievement of an objective.

Evaluation of possible consequences of each option and their comparison with each other is a central problem in problem solving and any decision process. It requires a complete knowledge of all future events and their probabilities. Difficulties are frequently encountered when identifying all alternatives and outcomes owing to a lack of complete knowledge of the situation and a lack of perfect foresight. Since to provide perfect knowledge is impossible, uncertainty is inherent in decision taking. It could arise from several sources, among which are:

- The subjectivity of decision-takers. Individuals may interpret outcomes in different ways;
- Incomplete information about the problem, alternatives, and/or outcomes;
- Poorly differentiated alternatives.

The question is "which option has the most acceptable set of consequences, given my current context and my desired goals?" It is not about whether to take a risk or not. It is about how to take reasonable risks and how to evaluate the impact of those risks. Reasonableness is subjective and differs between people, organizations, and situations. In essence, the talk is about the potential negative value placed on the consequences of a course of action (or inaction).

It is necessary to make a clear distinction between risk and strict uncertainty. A classification of risk and uncertainty is due to Knight (1921):

- Decisions involving risk are those where the decision-takers can assign probabilities or probability distributions to describe the randomness they face;
- Decisions that are strictly uncertain are those for which probability distributions cannot be assigned to the possible outcomes of the decision.

This distinction implies that risky decisions are those for which the situation, alternatives, and/or outcomes can be anticipated, whereas strictly uncertain decisions are those for which the situation, alternatives and/or outcomes cannot be anticipated. At first glance, this appears to be a reasonable distinction between circumstances where the possibilities have a known degree of certainty and those where at least some of the possibilities are unknown. Under an assumption of certainty a specific outcome is assumed to follow from a specific course of action. Under an assumption of risk, one of several outcomes with knowable/estimable probabilities is assumed to follow from a specific course of action, and the laws of probability are assumed to hold. Under an assumption of strict uncertainty, one of several outcomes with unknowable probabilities is assumed to follow from a specific course of action (subjective probabilities).

A variant of analysis of the consequence of an option in condition of strict uncertainty is expected value analysis. Expected value is the measure of the gain (or loss) from a decision derived by multiplying the probability of an outcome by the value of that outcome. The decision choices that people face are presented either positively in terms of potential gains or negatively as losses.

While the presentation of alternatives may seem trivial and inconsequential, the valence (positive or negative) can have drastic effects on how risky people will be in their decisions—a reflection of the value that evaluates the alternatives.

1.2. Decision-Making Perspectives

There are two approaches to consider decision-making: analytic and holistic. In conformity with the analytic approach, the decision-making is broken down into distinct elements that are used later for synthesizing a decision-making process. The holistic approach, on the other hand, focuses on way of behaving (manner, style) of decision-making in the context of view of the decision-making problem. The latter could be observed from several viewpoints that determine different cognitive frameworks, within which the decision-making is interpreted and evaluated. The decision-making is considered in respect of the following major points of view: psychological viewpoint, viewpoints determined by decision's attributes, decision environment's attributes, and decision situation's features (Figure 2).

According to the psychological viewpoint, the cognitive decision-making is considered in a framework, in which the human consciousness is presented. It is determined by two types of psychological functions—rational and irrational. In the group of rational functions are included thinking and feeling. Intuition and perception are representatives of the other group. Usually, when the talk is about rational functions, it has in mind thinking. Intuition is usually defined as knowing or sensing something without the use of rational processes. Alternatively, it has been described as a perception of reality not known to consciousness, in which the intuition knows, but does not know how it knows. There are attempts to redefined intuition as rational process (Westcott, 1968), stating that it is a process in which an individual reaches a conclusion on the basis of less explicit information than is ordinarily required to reach that decision. In this aspect the utility function could be viewed as analytical representation of the empiric knowledge. Rational/intuitive orientation of decision-making is a concept that has yet to make a significant impact on mainstream decision-making research. It is due to fact that the rational decision mainly coincides with the analytical approach to decision-making that is basis for development of decision theory. The latter is foundation for performance of analytic decision-making.

Sauter classifies four problem solving/decision styles (Sauter, 1999): left-brain style, right-brain style, accommodating, and integrated styles. The *left-brain style* employs analytical and quantitative techniques and relies on rational and logical reasoning. In an effort to achieve predictability and minimize uncertainty, problems are explicitly defined, solution methods are determined, orderly information searches are conducted, and analysis

Figure 2. Holistic approach to decision-making

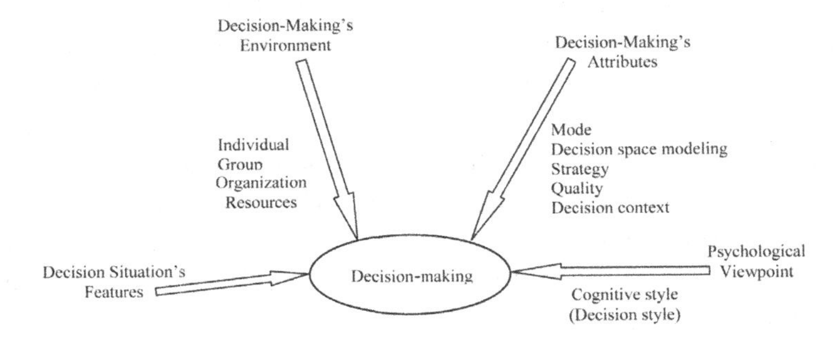

is increasingly refined. In direct contrast, *right-brain style* decision-making is based on intuitive techniques. *Accommodating* decision style is a non-dominant style. Lastly, in an *integrate style* decision makers are able to combine the left- and right-brain styles—they use analytical processes to filter information and intuition to contend with uncertainty and complexity.

The focus of a decision-making process is decision taking that involves a choice from possible actions or alternatives to satisfy one or several given objectives. The consideration of decision-making as a process-oriented problem requires its regarding from perspective of process-oriented viewpoint that is determined by the following attributes, which concern different aspects of a process:

- *Process presentation in operational context.* It is the way in which decision-making is done (mode of process performance);
- *The way of achieving of process results* or decision-making strategy;
- *Description of process environment* or modeling of decision space;
- *Evaluation of process carrying out* or decision-making quality.

In correspondence with the existing modes of decision-making performance, there are two general categories of decision-making: *one-stage (one-shot)* decision-making (single decision) and *multi-stage* decision-making. The latter is known also in the literature as "sequential decision-making" or "dynamic decision-making." A single decision is made based on the current knowledge about the situation and about the environment. Regrettably, life is not so simple that a decision stands alone all of the time. It is far more commonplace to find decisions occurring in sequence, in which each decision is contingent upon the previous decisions taken and will influence subsequent or future decisions. Single decisions should be optimal. However, optimality of a single decision requires consideration and measurements of both its immediate contribution and its impact on successive decisions. A relatively large number of decisions must be coordinated throughout the time horizon. The search for an optimal decision must anticipate a decision's future consequences.

Multi-stage decision-making presents Ackoff's idea of a system of decisions as a set of actions in which the outcome of each action depends on earlier actions and the interactions of those earlier actions, this dependence and interdependence being created by the results of the previous actions and the situation's response (Ackoff, 1998). Such a transformation usually requires a continuing process consisting in a series of decisions in a temporal order. At each decision time t a single decision dt is determined based on the current knowledge about resources and other relevant attributes of a particular system under consideration. This knowledge is denoted as the state st of the system at time t. A state st defines the set Dt of feasible decisions at time t. Any decision creates (discovers) new knowledge by modification of resource characteristics. Future decisions are subject to the results of such a modification as well as subject to the results of changing conditions due to additional events occurring independently of the decision maker's influence. The sequence of interdependent decisions is carried out for achievement of a superordinate goal. A single decision generates a distinct contribution to this goal. At the same time, it influences the conditions of the contributions of the following decisions. The sum of the single contributions of a sequence represents its total value.

Several strategies for decision-making have been documented—for example, optimizing and satisfying (Simon, 1986). *Optimizing* involves identifying as many different options as possible and choosing the best. How thoroughly this can be performed depends on the importance of the problem, the time available for solving it, availability of resources and knowledge, and the value or desirability of each outcome. *Satisfying*, on the other hand, centers around a process of goal adjustment and trade-offs whereby lower-level goals are substituted for maximized goals such that the first satisfactory, rather than the best, option is selected. Although perhaps ideal, optimized decision-making often proves to be impracticable and, in reality, satisfying is often used.

An important aspect of decision-making is the construction of suitable models of decision space. These models, known as decision models can be considered as ideas or abstractions whose primary function is to represent decision environment, in which a decision-making process is performed. They determine the thinking style of decision-making and are categorized as *mental models, abstract models,* and *mathematical models*. Mental models represent personal understanding of a situation - how the individual construes the world. Such mental constructs can be used to identify possible objectives and outcomes resulting from strategies that the decision-taker could choose. The *personal construct theory* captures how individuals make sense of the world from both similarity and difference (Kelly, 1955). Within this theory, personalized constructs are organized hierarchically, so that above and below each pole of a construct there is another construct. Although each person's construct hierarchy is different, to the extent to which two individuals have the same construct hierarchy, they construe the world in the same way. Two techniques have been used to elicit constructs from individuals and from groups, the *repertory grid technique* (Fransella, Bell, & Bannister, 2003), and the *cognitive mapping technique* (Eden & Ackermann, 2001). The latter of these has been widely used as a way into complex or messy problems, those situations in which alternatives and outcomes arc not initially obvious.

An abstract model can be useful in interrogating a decision situation so as to draw out key characteristics of concern. Two possible approaches are the use of *organizational metaphors* (Morgan, 2006) and the *Soft Systems Methodology* (SSM) (Chcckland, 1999). Morgan's organizational metaphors consist of eight different perspective lenses: the organization, organism, brain, culture, political system, flux and transformation, instrument of domination, and psychic prison. An initial diagnostic reading asks how a situation is like each metaphor, each bringing a different view and different theories to bear on the decision situation. The diagnostic reading can then be used to generate one or more storyline readings of a situation. A storyline reading has one metaphor as a dominant frame and others supporting. In the SSM approach, a decision situation is first entered and explored with tools such as rich pictures (using drawings and graphics). Having explored the situation, root definitions are developed of identified purposeful activity systems that do P by M to achieve R. Abstract models are then developed from these root definitions that is, models of the purposeful activity systems are made, not of what is actually done. The models can then be used to interrogate current practice. What is in practice but not in the model? What is in the model but not in practice? From this questioning a greater understanding of situation, alternatives, and outcomes can surface.

Mathematical models are used in order to identify the essence and develop an understanding of decision situation. This modeling guarantees the mathematical perspective of decision-making. It involves recognizing the key aspects of the decision context and the alternatives, and disregarding those aspects that have little or no influence over the situation at hand. This is often an iterative process. Mathematical models couch the terms of the situation in mathematical equations. These equations are then applied to current data in order to predict future values, are solved simultaneously in order to give an optimal solution. Statistical models are similar but primarily identify patterns and relationships within previous data from the system. The resulting information is then used to predict potential outcomes.

Decision quality is judged according to whether or not the decision (1) meets the objectives as thoroughly and completely as possible, (2) meets the objectives efficiently with concern for cost, energy, and side effects, and (3) takes into account valuable bi-products or indirect advantages. Decisions can be right (good) or wrong (bad). A *good* decision is logical, based on available information, and reflects context-sensitive values set for the problem solution (Beynon, Rasmequan, & Russ, 2002). A *bad* decision, on the other hand, is based on inadequate information and does not reflect intended values. The quality of a decision is not necessarily reflected in its outcome—a good decision can have either a good or a bad outcome.

Decisions can be qualified as ethical or moral. The key to ethical or moral decisions is that they are not between right and wrong but, more likely, between right and right depending on your perspective. Ethics or morals ask questions about the way of our action and consider the standards against which actions should be judged right or wrong. One person's decision is another's rejection. This can lead to disputes and disagreements since the basis for decisions is often subjective or emotional. As a result, there is no right or wrong answer because everybody will interpret the outcome according to his or her own stance. Different people will have different perspectives in different situations.

Decision context is determined by the activity, within which framework decision taking has place. Management is a classical framework for decision-making. There are three broad categories that encompass all managerial activities:

- **Operational Control:** Its objective is an efficient and effective execution of specific tasks;
- **Management Control:** It concerns efficient utilization and acquisition of resources in the accomplishment of organizational goals;
- **Strategic Planning:** The long-range goals and policies for resource allocation.

This categorization defines the existing managerial styles of decision-making.

As a cognitive process, the decision-making is carried out in some environment, which provides collection of resources at the time of decision that supports this process. The decision-making depends on the cognitive capacity of decision environment, which is characterized by the number of actors that take part in this process; their organization and their roles; a set of data and information sources, a set of models for knowledge discovery and tools for their providing; conditions that influence human's attention, stress and other psychological factors, which can limit the cognitive capacity of humans; computer background—a set of software and hardware technology. All these resources act as constraints on decision-making and determine the way of its performance. There are numerous tools for data mining, knowledge discovery, or information providing, but not all are appropriate to every situation and not every situation has a suitable tool. They can be placed in the following categories:

- Expert knowledge systems;
- Risk management tools;
- Calculation engines;
- Simulation models.

In dependence of the number of persons that take part in decision-making process we can defined two decision-making categories:

- **Independent Decision-Making:** Involves one decision-maker to reach a decision without the need for assistance from other persons. It is the exception because of the common need for collaboration with other experts;
- **Pooled Interdependent (Group) Decision-Making:** A collaborative decision-making process whereby all participants in this process work together on solution of a problem.

Much research has focused on the differences between individual and group decision-making. Rarely are important or critical decisions taken by one person. Usually several people are involved, whether through an organization like a hierarchical process or in a group, team or committee. Group dynamics are different from individual dynamics. Members of a group will have group objectives but also their own agenda—their own goals and characteristics. The group decision-making requires an analysis for determination who should be involved in decisions and to what extent. It sets out all stakeholders in the decision and then looks at whether they should be: responsible for decisions, accountable for decisions, consulted about decisions, informed about decisions.

There are several key factors that determine the relationships between members of a group (group characteristics): communication pattern in a group, group thinking, goals, and interests of group members and usage of environmental resources. In correspondence with various group characteristics there are various group decision-making styles. We considered three major communication patterns in a group that determine the following decision-making styles:

- **Autocratic Style:** The group has a leader that makes a decision with available information or seek specific information from the others members of the group before taking the decision;
- **Consultative Style:** A leader consults on either an individual basis or on a group basis, seeking ideas and suggestions with respect to the problem. After the consultation, the leader takes the decision;
- **Negotiation Style:** Discretion of the group is high and the leader facilitates the communications in the group for making suggestions and decision taking.

If the group thinking bases on advanced agreement about a rule-based formula, the decision-making style is called *formulaic*. In this case, the disputes in the group are refereed to arbitration that guarantees a decision taking in correspondence with standard rules. The other two group decision-making styles ensure *collective* or *cooperative* decision-making. In collective decision-making, the participants in this process have common goals and interests. When the members of a group use common decision resources during decision-making, this group decision-making is known as cooperative decision-making. Which approach is most appropriate depends upon the following features of the decision situation—available information to the leader, clarity of problem structure, and importance of decision acceptance to the choice of style.

According to the idea to view of the organization as a "role system," the decision-making environment could be regarding as a group of people with interacting roles. This viewpoint requires understanding how, through role-playing, an individual behaviour is presented in the framework of decision-making. Decision-taking (making) are slowed due to bottlenecks when there is lack of clarity with respect to which component of a decision-making is to be taken in which part of an organization. Distinctive bottlenecks are global versus local, function versus function and inside versus outside partners. Assigning specific decision-making roles remove the bottlenecks, since their identification fixes groups and group-individual relationships. The key observation is that ownership of a decision does not always rest with the actors who control an organizational network. Although being well connected is a desirable attribute, it does not necessarily mean actual decisional power. It appeared that in many cases decision owners rarely had the power to actually make the decision (thus commit the required resources to the implementation). Decision power to commit resources stays with the decision maker, while decision owners are the people who lead the decision from start to finish either because they have been assigned the supervision of a decision.

While ownership may change from decision to decision, the overall control of the decision-making process usually remains in the same hands. The differentiation of the decision maker and the decision owner as well as the shift from a rational to a not-so-rational view of behaviour had resulted in new approaches in decision-making science.

The consideration of decision maker as an actor-like player first appears in soft systems methodology (Checkland, 1981). It applies the idea of systems thinking to the imprecise and unstructured area of human activity. Soft systems or Human Activity Systems (HAS) are human activities purposefully carried out by people and their relationships within a highly structured organization (hard system). This methodology uses root definitions to identify the emergent properties of a system to be modeled in a HAS: customer, actor, transformation (the transformation carried out by the system), worldview, ownership, and environment. Checkland's methodology differentiates several roles in relation to problem content that concern decision-making:

- **Client:** He wants to know or do something about the situation and has the power to commission a problem-solving activity;
- **Decision Taker:** The decision taker can alter the activities and may commit to resource allocation;
- **Problem Owners:** They are people who are affected by a situation, experience it as a problem, and wish something were done about it; they may not necessarily be in the position of precisely defining the problem or identifying a solution.
- **Problem Solver (Independent):** He feels capable of solving the problem and is able to find ways to improve the situation.

Rogers and Blenko (2006) suggest a richer conception of the roles involved in organizational decision-making. They point toward tacit knowledge in the minds of various expert employees. They identify the following aspects of a decision maker that are presented by five significant roles: *Recommend, Agree, Perform, Input,* and *Decide* (RAPID). Within any decision, some of these roles may be located in different parts of the organization, but this should be clearly specified. For instance, recommending proposals, formally deciding, and agreeing to a recommendation might be globally taken, whereas inputting to a decision and performing may be locally taken. Vári and Vecsenyi's result is an even finer, more detailed differentiation of certain roles (Vári, 1984). They identify the following roles:

- **Decision Makers:** These are people who have the executive power to determine the use of outputs of the decision-making process.
- **Proposers:** Proposers make recommendations to decision makers.
- **Experts:** They provide input and knowledge to model and structure the problem.
- **Implementation-Oriented:** They participate in the execution of the selected and approved solution.
- **Clients:** In case decision support is sought, they are the ones who have asked for such help.
- **Decision Analysts or Consultants:** Consultants help with methodology and modeling choices as well as with decision-making procedures, or may facilitate the collaboration among all participants.

Compared to Soft System Methodology (SSM), the above set of roles is more focused on motivations and roles related to decision support, while the SSM approach is a full methodology and provides better guidelines for the handling of multiple interpretations of the complex and interrelated issues that exist in organizations.

It is necessary to present the possible mode of decision-making according to the complexity of the situation and its uncertainty. In low complex-

Figure 3. Situation dependent decision-making

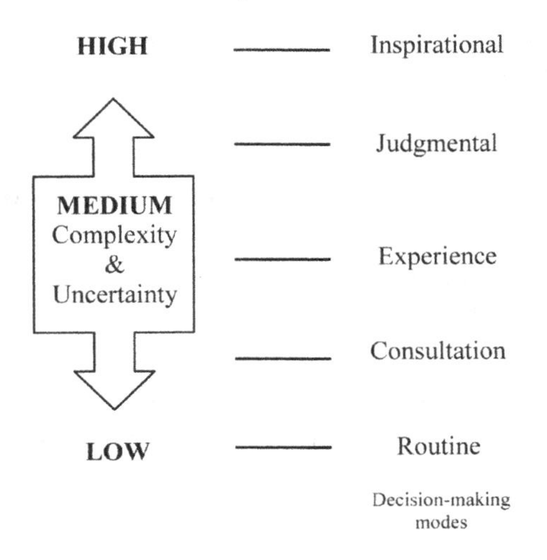

ity and uncertainty of a situation, the appropriate approach to decision-making is the use of programmed decision. Decisions are programmed to the extent that they are repetitive and routine, to the extent that a definite procedure has been worked out for handling them so that they do not have to be treated from scratch each time. Decisions on the other hand are non-programmed to the extent that they are novel, unstructured, and unusually consequential. Figure 3 presents decision-making modes appropriate to various levels of decision situation complexity and uncertainty.

At the extreme where it is uncertain then, often, inspiration is needed in order to see the optimum decision

1.3. Decision Situation

The term *situation* is used to mean any issue, problem, condition, opportunity, or bounded system that has to be changed, improved, transformed, etc. Usually, the situation is interpreted to mean a complex adaptive system. By a complex situation (problem) in a complex environment we mean one that is defined difficulty and may significantly change in response to some solution, has many interrelated causative forces, has no (or few) precedents and is often prone to surprise. In short, lack of enough knowledge and skills for description and determination of the main objective and solutions in synergy with the main objective. These complex situations may be within an organization, a part of the organization or in the organization's external environment. Such situations are called "messes," since they produce conditions where one knows that a problem exists but it is not clear what the problem is. A complex situation in a complex environment is regarding as a *Complex Adaptive Mess* (CAM) (Bennet & Bannet, 2008). CAM has the following properties: emergence, the butterfly effect, tipping points, feedback loops. *Emergence* is a global/local property of a complex system in disequilibrium that results from the interactions and relationships among its agents, and between the agents and their environment. These interactions may exert a strong influence within the system. The *butterfly effect* occurs when a very small change in one part of a complex adaptive system—which may initially go unrecognized—results in a huge or massive disruption, surprise, or turbulence. These results may be impossible, or extremely difficult, to predict. A *tipping point* occurs when a complex system changes slowly until it unpredictably hits a threshold, which creates a large-scale change throughout the system (bifurcation). *Feedback loops* can be self-reinforcing or damping, improving a situation or making it worse.

A perspective indicates *how* one looks at a problem (Linstone, 1999). It bases on the notion of three perspectives inherent in any complex problem situation: technical (T), personal (P), and organizational (O). The T, P, and O perspectives yield quite distinct pictures of a complex situation. The P perspective provides an immediate grasp of the essential aspects of a situation, but it also reflects the bias of the analyst that constructed it. The T perspective may be a parsimonious model of a situation, but it can also be criticized, as it reflects an analyst's biases in the choice of assumptions, variables, and simplifications. An O perspective may give some indication of how an organization may be expected to move in a general sense. However, in the case of a strong organizational leader, one can argue that the O perspective is biased as well, since the leader's P perspective becomes adopted as the O perspective. These biases act as filters, blocking some aspects of a situation from view and allowing other aspects to be recognized.

In preparing for making decisions that deal with complex adaptive messes, a number of broad cognitive competencies may prove helpful. They are not typically part of professional discipline training and education, and therefore may be unfamiliar to many decision-makers. It is used the term *integrative competencies,* since they provide connective tissue, thereby creating knowledge, skills, abilities, and behaviours that support and enhance other competencies. They also help decision-makers deal with larger, more complex aspects of a CAM, either integrating data, information, knowledge, or actions, or helping the decision-maker perceive and comprehend the complexity around them by clarifying events, patterns, and structures. In the extreme, the landscape of a complex situation is one with multiple and diverse connections with dynamic and interdependent relationships, events and processes. While there are certainly trends and patterns, they may well be entangled in such a way as to make them indiscernible, and compounded by time-delays, non-linearity and a myriad of feedback loops.

As with any informed decision-making process, we move into the complexity decision space with the best toolset and as deep an understanding of the situation as possible. That toolset may include experience, education, relationship networks, knowledge of past successes, and historic individual preferences, multiple frames of reference, cognitive insights, wellness (mental, emotional, and physical), and knowledge of related external and internal environmental pressures. The decision space in which the CAM (Complex Adaptive Mess) is to be considered—using relevant decision support processes such as the analytical hierarchy process, systems dynamic modeling, scenario development, etc., and information and technology systems—includes situation and decision characteristics, outcome scenarios, a potential solution set, resources, goals, limits, and a knowledge of political, sociological, and economic conditions, i.e., ontology.

When the traditional decision process is applied to a simple or complicated situation, the objective is to move the situation from the current state to some desired future state. As introduced above, when dealing with complex problems the decision process often requires a commitment to *embark on a journey toward an uncertain future*, creating a set of iterative actions whose consequences will cause a move from the current situation (A) *toward* a desired future situation (B). Since there is no direct cause and effect relationship that is traceable from the decision to the desired future state, the journey may require extensive preparation. The decision strategy must have the capacity and internal support mechanisms needed for an implementation journey that cannot be predetermined. Thus, in trying to understand or generate a decision strategy for a specific CAM, one not only needs rules, patterns and relationships but also the underlying theories, principles and guidelines (where available) to allow generalized knowledge creation from and application to the specific situation. This generalization may be quite challenging for complex problems, which may not repeat themselves within a semblance of coherence. However, before the mind can effectively observe, reflect, and interpret a situation in the external world, the frame of reference of the decision-maker must be recognized since that frame will define and limit what is sensed, interpreted, and understood. Multiple frames of reference serve as tools to observe and interpret the system from differing perspectives, providing the opportunity to find the best interpretations and explanations of the complex situation. To find a frame of reference applicable to complexity requires an appropriate language, a set of concepts and ways to characterize the situation.

There are several mechanisms for influencing complex situations:

1. **Ontology-Oriented Approach:** The ontology of the decision process represents the schema or set of characteristics and conditions surrounding the decision strategy that potentially have an important influence on the desired outcome. To the extent that these factors may be identified, they can then be prioritized, rated for significance, visualized through graphics and used to develop the optimum decision strategy.
2. **Structural Adaptation:** When the complex situation/problem lies within the decision-makers own organization, special considerations come into play. It may become necessary to view the problem as part of the organization in the sense that both "systems" are complex and interconnected. The organization may be part of the problem and any successful solution would include changes both inside the "problem" as well as in the surrounding organization. There are two coupled complex systems that are connected such that any change in one is likely to affect the other. Since such structural coupling or adaptation is common, the decision strategy may be to use this coupling to pull the problem situation along in the desired direction. Structural adaptation is a good way to influence complex organizations.
3. **Boundary Management:** It is a technique for influencing complex situations by controlling/influencing their boundary. Complex system behaviour is usually very sensitive to its boundary conditions because that is where the energy comes from that keeps it alive and in disequilibrium.
4. **Sense and Respond:** Another strategy to deal with CAMs. This is a testing approach where the problem is observed, then perturbed, and the response studied. This begins a learning process that helps the decision-maker better understand the behaviour of the CAM. Using a variety of sensing and perturbations provides the opportunity to dig into the nature of the situation/problem before taking strong action. Closely coupled to the sense and respond approach is that of *amplification*, used where the problem is very complex and the situation's response is unknown. This is the evolutionary approach where a variety of actions are tried to determine which ones succeed. The successful actions are then used over and over again in similar situations (the process of amplification) as long as they yield the desired results. When actions fail, they are discarded and new actions are tried
5. **Absorption:** It brings the complex situation into a larger complex system so that the two slowly intermix, thereby resolving the original problem by dissolving the problem system.
6. **Simplification:** It reduces our own uncertainty, makes decisions easier, and allows logical explanations of those decisions. Simplicity captivates the mind; complexity confuses and forces us to use intuition and judgment, both difficult to explain to others.

1.4. Decision Support

Decision support is based on the idea that many important classes of decision situations can be supported by providing decision-makers with a decision support environment. It is result of two forces, firstly the demand of the decision maker for a better support in order to perform the task he is in charge of more effectively and secondly, the opportunities provided by technology and the culture prevailing in the organization. This concept coincides with a type of task that here implies there is a decision to take with some necessary preparatory work. This idea is realized through Decision Support Systems (DSS). It serves as basis for construction of the conceptual framework of DSS that is embedded in a DSS development environment, which uses the available technology at the time of implementation of the development environment (Klein & Methlie, 1990).

Decision support systems are intended to aid the decision maker in difficult for formal description situations and problems or under the conditions of complex tasks, in which the final decision is determined by subject and structuring of his/hers goals and elimination of the errors generated by the uncertainty or the complexity of the solved problem is required. Their role in such problems is to help the decision-making subject to extract the necessary professional knowledge, including at empirical expert level. The aim is to account for the total initial information with its inherent uncertainty, if human assessment is considered and its inclusion in models or problems at the level of measurement scales or data acquisition. The solution of these problems allows for structuring and formal analysis, under which the contradictions or errors in the initial data are removed or there is diminishment in the uncertainty and inconsistencies in the subjective assessments and analysis on expert or management level. Hence, these DSS contain within formal models for analysis and synthesis of decisions in accordance with the measurement scales of the initial information and the nature of the considered problem. They, have a mandatory element, a subsystem for dialogue with the user, in which the initial data is input and an iterative change of the final goal of the considered problem is conducted in dialogue with the user. The task the decision maker performs and is expected to be supported can be broken down into a series of sub-tasks:

- *Recognition* of a problem or situation requiring design of a solution and a decision;
- Decision problem *diagnosis* and *structuring*;
- Generation of *alternatives* and their description (outcomes, uncertainty);
- *Choice of an alternative* (decision criteria, preferences);
- Decision *implementation and monitoring* (in particular for group decision).

The concept of decision support implies the belief that a decision can be improved. This hypothesis is based on the belief that the decision basis and process of analyzing decision situation can be improved. The improvements related to the decision basis—information concerning the description of decision situation, creation of interesting alternatives, computation and visualization of criteria, the reliability and pertinence of information associated with alternatives, and so forth. In simple cases, the decision maker just applies its preferences to select a course of action among several. In complex decision situations, the improvement of the decision analysis is related to the knowledge of decision maker and rigour of his reasoning. The support of reasoning is consequently also a way to improve the decision. For most decision support systems, the steps of problem recognition and structuring are bypassed, since their application is designed for a class of decisions, which is known in advance. For other situations, these steps may require analysis. The decision maker can be a single person or a group involved in the decision process. In the case of the more complex situation of a group, the commitment to the decision is made easier, if the group has been involved in the different decision steps and is truly convinced by the arguments used for selecting the alternative. In a group, due to possible conflicts of criteria and preferences, it is normal that different alternatives be preferred by different persons facing an identical decision situation.

The capabilities of a DSS for supporting of decision makers' tasks are presented by the following functions:

- Easy capture concerning information (variables, facts, documents…) relevant and useful for the study of the decision class;

- Management of data relevant to the decision class including analytical tools;
- Presentation of information needed for the analysis of the decision;
- Computing of criteria associated with alternatives;
- Easy and flexible interaction between the decisions makers and the DSSs;
- Communication among the decision makers allowing DSS application sharing in case of a decision involving distant DSS users in a group decision.

In DSS designed for a specific decision class within a given decision context that is designed to support repetitive decisions, the organization defines to a certain extent the method used to make a decision.

2. DECISION THEORY: RATIONAL DECISION-MAKING

Theories of decision-making as intentional, consequential action based on knowledge of alternatives and their consequences evaluated in terms of a consistent preference ordering are known as Rational Decision Theories. Rational approaches to decision-making can generally be classified in one of the following categories: descriptive, normative, or prescriptive (Bell, Raiffa, & Tversky, 1988). Although the definitions of these three are relatively dynamic within the research literature, they can be loosely described as follows:

1. **Descriptive Methods:** They analyze how decisions are taken and determine optimal choices based on what is or what has been done. Descriptive techniques analyze how real people take decisions, including how they perceive the situation, how they determine alternatives and the resulting outcomes, and how this is influenced by their biases. These techniques draw on methodologies from several different fields including psychology and statistics.
2. **Normative Methods:** Also known as strict mathematical theories based on the axiomatic approach. They determine what choice(s), in theory, should be taken on the base of strict mathematical derivations. Normative techniques are based on and demand the assumption that the decision-taker is rational and abstract from cognitive bias. The choice of optimal outcomes identified via normative decision-taking techniques represents rational selection from the alternatives. The process is transparent and is without bias.
3. **Prescriptive Methods:** They are related to normative methods in that they determine optimal choices in theory, but these choices are constrained by limitations of what can be done in reality. Prescriptive techniques relax the assumption of rationality of the decision-maker (taker), but, rather than account for bias, these techniques account for real-world constraints and indicate sub-optimal choice according to these constraints. Therefore, prescriptive techniques generate understanding of the choices that should be made given that rationality might not be a possible or optimal mode of operation of the decision-taker and given that some alternatives might not be possible. The prescriptive techniques generate understanding of the choices that the empiric subjective knowledge should be included in the definitive solution in assistance of the normative solution.

A normative model of decision-making is different from a descriptive model. Descriptive models are specifications of the response patterns of human beings and theoretical accounts of those observed response patterns in terms of psychological mechanisms. This type of model is the goal of most work in empirical psychology. In contrast, normative models embody mathematical standards for action and belief-standards that, if met, serve to optimize the accuracy of beliefs and the efficacy of actions. Value functions, expected utility, and subjective expected utility are normative models—they tell us how we should make decisions. We will see that a very large literature in cognitive psychology indicates that the descriptive model of how people actually behave deviates from the normative models.

The primarily mathematical techniques presented in this text fall into the normative category but from the position of the prescriptive approach. To assist people in thinking through complex decision problems, a quantitative framework is needed to encourage consistency, the essence of rationality. The primarily mathematical techniques in this book take in consideration the real word restriction as uncertainty in the DM's preferences and notions and the objective probabilities appearing in the real world events. They give analytical presentation of the DM's preferences; even these preferences describe empiric feelings and knowledge. When it is the choice between several different alternatives, we seek to satisfy ourselves that the choice we make is consistent with our own beliefs and values, and with the values of the organization that we serve. The practical role of decision theory is to provide a framework that assists people in achieving such consistency. Nevertheless, when implementing these approaches to support decision taking, the prescriptive approach should be taken where the optimal choices resulting from the decision analysis will be tempered by real-world restrictions.

2.1. Utility and Value

Cognitive psychologist Jonathan Baron (2008) has discussed how the term utility is a slippery word. It is used technically in decision science in ways that do not map exactly into anything in general discourse. The term as used in cognitive science does not refer to its primary dictionary definition of "usefulness": Instead in decision theory it means something closer to "goodness." The utility is not the same as pleasure. Instead, utility refers to the good that accrues when people achieve their goals and is thus more closely related to the notions of worth or desirability than it is to pleasure. Hastie and Dawes (2001) suggest that it is perhaps best to think of utility as an index of subjective value. Viewing utility as subjective value or utility provides a nice link to the most basic normative model of optimal decision-making: maximizing value or expected utility in the case of uncertainty and risk.

The concept of a numerical measure to describe the value of alternative choices has come to be referred to as utility theory, with the Utility function being the numerical measure itself. There are two types of utility:

- **Ordinal Value (Utility):** It is making the assumption that it is possible to say for any two bundles of objects, which is preferred;
- **Cardinal Value (Utility):** Making the assumption that a number is available representing the worth of any bundle of objects.

On the base of cardinal utility, it is developed the expected utility theory to prescribe how people should evaluate options about which they are uncertain. A separate development of ideas about utility was based on new procedures for measuring the utility of the older kind, most notably the theories of difference measurement and conjoint measurement. Although ordinal utility is all that is needed to direct choice between multi-attributed alternatives, it is a difficult task to attempt to elicit such a function by plotting indifference curves. The theory of conjoint measurement and other developments in measurement theory have spawned a revival of interest in cardinal utility, distinct from that of von Neumann and Morgenstern but applicable in the absence of uncertainty.

Decision analysts have made this distinction between two types of utility, which can be characterized as the distinction between value and risk preferences, ever since the subject first developed.

Value preferences are made between competing objectives or attributes when no uncertainty is present. Risk preference (as measured by the expected utility, mathematical expectation of utility functions discussed above) addresses the decision-maker's aversion to, indifference to, or desire for risk-taking.

Value Function

The idea of a value function is introduced to assist people faced with decision problems; we can characterize this problem as follows. A set of options has been well defined. Note that the terms alternatives, choices, decisions, or actions are used by some writers in place of the word options. The literature sometimes makes careful distinctions between them and sometimes, as here, uses them interchangeably. We assume that we know what the options are and, perhaps more importantly, that we know what the attributes (or factors) are which discriminate amongst them. We suppose that there are n such attributes, named X1, X2,..., Xn, and that it is possible unequivocally to provide a numerical score for each option with respect to each attribute. The set of scores with respect to the attributes for a particular option will be represented by a vector x = {x1, x2,. . ., xn}. The problem is now reduced to deciding which vector of scores is most attractive. If this could be made by a function this function could be said value function. We present the following aspects of value function:

1. **Definition and Alternatives:** Suppose that we can construct a real-valued function v(.), with the property that v(x) > v(y) if and only if we prefer the option whose vector of scores is x in situation A to another with score vector y in situation B; and with the further property that v(x) = v(y) if and only if we are indifferent between A and B. We shall refer to such a function as a value function. Then the choice of option is simple. We merely choose the option with the score vector x for which v(.) is largest. The difficulty will be in constructing this function v(.).
2. **Existence of the Value Function:** The existence of a value function to represent preferences is equivalent to there being a complete preference order over the options. Clearly if a value function exists it induces a complete ordering on the vectors x. Conversely, it is easy to see that if we do have a complete preference order, then we can construct a value function by merely placing the vectors in increasing preference order and attaching any increasing sequence of numbers to the vectors.
3. **Additive Value Functions:** The most often used form for a value function is the additive one, in most cases incorrectly:

$$v(x) = \sum_i v_i(x_i).$$

The simplicity of this form and the ability to work on each attribute separately make it an attractive candidate for our consideration, although it is obviously quite possible for a particular preference structure to lead to a non-additive value function. The essential condition is that of preference independence (Fishburn, 1970). A pair of attributes, X_1, X_2, is said to be preference independent of all the other attributes $\{X_i\ i = 3, \ldots, n\}$ if preferences between different combinations of levels of X_1, X_2, with the level of all other attributes being held at constant values, do not depend on what these constant values are.

Uncertainty

The problem of uncertainty is of particular importance, since it is very hard to discover any real decision problems where we are not at least slightly uncertain about the outcomes. Some years ago, most texts on decision analysis presented the subject as being almost entirely concerned with uncertainty, and generally, where there was only one attribute of interest, usually money. The subject has advanced and the analysis of conflicting objectives is now emerging as the more useful area in practice. Nonetheless, dealing with uncertainty is vital in many applications.

1. **Descriptions of Uncertainty:** The relative frequency theory holds that the probability of an event is the long-run frequency with which the event occurs in an infinite repetition of an experiment; it is thus seen as an objective property of the real world. Subjective probability present probability as the degree of belief which an individual has in a proposition; this is a property of the individual's subjective perception (or state of knowledge) of the real world. The former theory has dominated scientific thinking for many years, because it has the appearance of conforming to the empirical objectivity which science is sometimes claimed to need and possess.
2. **Theoretical Background to Subjective Probability:** Probability theory is an obvious construct for describing subjectively perceived uncertainty.
3. **Using Probabilities:** To use probabilities in a decision analysis it will be necessary to know some of the details of probability theory.

In particular and not only, the following concepts will be needed:

- Random variables, both discrete and continuous;
- Cumulative probabilities, probability density functions;
- Moments of distributions, means, and variances;
- Conditional, joint, and marginal probabilities;
- Independence;
- Distributions of functions of random variables.

Risk Preferences

Utility Theory

Now that we have seen how to describe our subjectively perceived uncertainties in numerical terms, we can move on to discussing how this formalism can be used to aid decision-makers faced with alternative courses of action, each involving uncertainty. We now seek some way to evaluate the uncertain options (or lotteries or gambles, as we shall refer to them) facing a decision-maker. We want to construct an evaluation method such that, once a problem has been specified, its decision tree drawn, and the uncertainties measured using probabilities, we will be able to compute a number as mathematical expectation of a utility function that represents the values of the possible outcomes (alternatives). This utility function has to be consistent with all our judgments and additive in regards to probabilities describing the uncertainty.

It was with the publication of von Neumann and Morgenstern's "The Theory of Games and Economic Behaviour" in 1944 that modern utility theory can be said to have begun. Their imaginative definition of utility led to a coherent normative theory for decision-making in the presence of uncertainty. The idea is to define the utility of an outcome as equal to the probability of winning a standard prize in a gamble such that the individual is indifferent between receiving the outcome for sure and accepting the gamble. This rather non-obvious definition needs fleshing out, but coupled with some fairly innocuous commonly accepted behavioural postulates, it leads to an operational utility theory, and to expected utility as a sensible measure for gambles.

Possible axiomatic for utility is:

1. The decision-maker can place all possible lotteries in a preference order. That is, for any two lotteries, A and B, he can state which he prefers, or whether he is indifferent between them; furthermore, if he prefers A to B and B to C, then he prefers A to C.
2. If the decision-maker prefers P1 to P2 and P2 to P3, where P1, P2, P3 are three prizes, then there is some probability p, between 0 and 1, such that he is indifferent between getting P2 for sure and a lottery giving him P1 with probability (p) and P3 with probability (1 - p).
3. If the decision-maker is indifferent between two prizes, P1 and P2, then he is also indifferent between two lotteries, the first giving P1, with probability p and a prize P3 with probability (1 - p), and the second giving P2 with probability p and P3 with probability (1 - p). This must hold whatever P3 and pare.
4. If two lotteries each lead only to the two prizes P1 and P2 and the decision-maker prefers P1 to P2 ($P1 \succ P2$), then he prefers the lottery with the higher chance of winning P1.

Suppose lottery A1, A2 and A3 gives a prize P1 with probability (p_i) and a prize P2 with probability ($1-p_i$), for i= 1, 2, 3; and suppose lottery B gives entry to lottery A2 with probability (q), and entry to lottery A3 with probability (1 - q); then the decision-maker is indifferent between lottery A and lottery B if, and only if:

$$p_1 = q\,p_2 + (1 - q)\,p_3$$

Implications of Utility Theory

Here we describe some of the ideas that have resulted from the concept of utility. The first is "certainty equivalence." One of the key ideas in utility theory is the substitution principle (Raiffa, 1968); this is the supposition that, in the analysis of a decision problem, we can always substitute a gamble for a certain consequence, and vice versa. In a particular instance, the sum of money which a decision-maker is prepared to swap for a gamble is his "certainty equivalent" of the gamble. Clearly, the utility of the certainty equivalent must be equal to the expected utility of the gamble; so folding back a decision tree could be interpreted as determining the certainty equivalent at each node in the tree (Raiffa, 1968; Keeney, 1993).

Risk aversion is associated with a convex utility function. A little thought should convince the reader that, if a utility function has a decreasing first derivative then the owner of that utility function will be risk-averse for any gamble he faces; that is, the certainty equivalent will always be less than the expected value. Although risk aversion is by far the most common attitude towards risk, there are some situations where the opposite preferences may be displayed. If the certainty equivalent exceeds the expected value, the owner of the utility function is displaying risk-seeking behaviour. If a utility function is concave (i.e. its first derivative is increasing), then the person whose risk preferences are modeled by the utility function will be risk seeking for all gambles he faces. If the utility function is neither convex nor concave over the whole range, his attitude towards risk will depend on which gamble he is considering. People who enjoy betting might be risk seeking for gambles involving small amounts but risk-averse for uncertain options involving larger stakes. Finally, if a utility function is linear, we say that its owner is risk-neutral; in this case, his certainty equivalent is equal to the expected value for any gamble that he faces, and so he need not compute utilities at all. It is sometimes useful to work with special analytic forms for the utility function. Since utility functions are elicited empirically, there is no reason why any special analytic form should emerge. However, in many other areas of the application of mathematics, empirical curves may often be well approximated by an analytic function, and by doing so analysis may be easier; and the same is true in utility theory. Perhaps the most often used form is the exponential:

$u(x) = a — bx\exp(-x)$.

This may not fit all situations, since it is bounded above and determined by only one parameter; but, apart from its advantage in statistical calculations, it has some interesting properties. If the decision-maker has an exponential utility function, then his certainty equivalent for increments in wealth for a gamble will not depend on his initial asset position, contrary to the general case. The methodology proposed in this book for the evaluation of individual subjective utility function is based on stochastic procedures and constructs polynomial approximation of utility independently of the type of the function (Aizerman, 1970; Pavlov, 2011). For this reason, the descriptions of the forms of the utility functions remain outside of the scope of the book.

2.2. Other Normative Approaches

Some of these methods are closely related to the decision theory we have described in this chapter, while others stand more on their own is described an approach to handling decision problems, which is based on reasonable behavioural postulates. There is no obligation for any decision-maker to follow this approach. Therefore, it is natural to make an attempt to describe several alternative normative frameworks that have appeared in the literature of the last thirty years, particularly for attacking the problem of conflicting objectives. Although there are numerous versions of these techniques, we will outline just five of the most prominent of them in this section, in each case giving references where the interested reader can study the ideas further.

The Analytic Hierarchy Process (AHP)

Theoretically, the AHP is based on the following axioms given by (Saaty & Vargas, 1984):

Axiom 1: The decision-maker can provide paired comparisons a_{ij} of two alternatives i and j corresponding to a criterion/sub-criterion on a ratio scale which is reciprocal, i.e. $a_{ji} = 1/a_{ij}$.
Axiom 2: The decision-maker never judges one alternative to be infinitely better than another corresponding to a criterion, i.e. $a_{ij} \neq \infty$.
Axiom 3: The decision problem can be formulated as a hierarchy.
Axiom 4: All criteria/sub-criteria which have some impact on the given problem, and all the relevant alternatives, are represented in the hierarchy in one go.

In nutshell, there are three major concepts behind the AHP, as follows:

- **The AHP is Analytic:** mathematical and logical reasoning for arriving at the decision is the strength of the AHP. It helps in analyzing the decision problem on a logical footing and assists in converting decision-makers' intuition and gut feelings into numbers which can be openly questioned by others and can also be explained to others.
- **The AHP Structures the Problem as a Hierarchy:** Hierarchic decomposition comes naturally to human beings. Reducing the complex problem into subproblems to be tackled one at a time is the fundamental way that human decision-makers have worked. Evidence from psychological studies suggests that human beings can compare 7 ± 2 things at a time. Hence, to deal with a large and complex decision-making problem it is essential to break it down as a hierarchy. The AHP allows that.

- **The AHP defines a process for decision-making:** Formal processes for decision-making are the need of the hour. Decisions, especially collective ones, need to evolve. A process is required that will incorporate the decision-maker's inputs, revisions and learning and communicate them to others so as to reach a collective decision. The AHP has been created to formalize the process and place it on a scientific footing. The AHP helps in aiding the natural decision-making process.

The methodology of the AHP can be explained in the following steps:

Step 1: The problem is decomposed into a hierarchy of goal, criteria, sub-criteria, and alternatives. This is the most creative and important part of decision-making. Structuring the decision problem as a hierarchy is fundamental to the process of the AHP. Hierarchy indicates a relationship between elements of one level with those of the level immediately below. This relationship percolates down to the lowest levels of the hierarchy, and in this manner, every element is connected to every other one, at least in an indirect manner. A hierarchy is a more orderly form of a network. An inverted tree structure is similar to a hierarchy. Saaty suggests that a useful way to structure the hierarchy is to work down from the goal as far as one can and then work up from the alternatives until the levels of the two processes are linked in such a way as to make comparisons possible.

Step 2: Data are collected from experts or decision-makers corresponding to the hierarchic structure, in the pair-wise comparison of alternatives on a qualitative scale as described below. Experts can rate the comparison as equal, marginally strong, strong, very strong, and extremely strong.

Step 3: The pair-wise comparisons of various criteria generated at step 2 are organized into a square matrix. The diagonal elements of the matrix are 1. The criterion in the ith row is better than criterion in the jth column if the value of element (i, j) is more than 1; otherwise the criterion in the jth column is better than that in the ith row. The (j, i) element of the matrix is the reciprocal of the (i, j) element.

Step 4: The principal eigenvalue and the corresponding normalized right eigenvector of the comparison matrix give the relative importance of the various criteria being compared. The elements of the normalized eigenvector are termed weights with respect to the criteria or sub-criteria and ratings with respect to the alternatives.

Step 5: The consistency of the matrix of order n is evaluated. Comparisons made by this method are subjective and the AHP tolerates inconsistency through the amount of redundancy in the approach. If this consistency index fails to reach a required level then answers to comparisons may be re-examined. The consistency index, CI, is calculated as: $CI = (\lambda max - n)/(n - 1)$, where λmax is the maximum eigenvalue of the judgment matrix. This CI can be compared with that of a random matrix, RI. The ratio derived, CI/RI, is termed the consistency ratio, CR. Saaty suggests the value of CR should be less than 0.1.

Step 6: The rating of each alternative is multiplied by the weights of the sub-criteria and aggregated to get local ratings with respect to each criterion. The local ratings are then multiplied by the weights of the criteria and aggregated to get global ratings.

The AHP produces weight values for each alternative based on the judged importance of one alternative over another with respect to a common criterion.

Outranking

It is pointed out that any method which purports to determine (in some objective manner) the best of a set of decision options might well be claiming too much. While it may be possible to determine unequivocally that, say, option A is better than option B, there may be an option C such that we cannot say whether A is better, worse, or on a level with C. This approach therefore seeks a method, which results in a clear ordering of options when that is appropriate, but does not produce an ordering if it is inappropriate. It assumes that a well-defined set of options exists, and that each option can be unambiguously defined in terms of a set of attributes with respect to which each option can be measured. In mathematical terms, we suppose that there are m attributes and n options, and that a score Xij for the jth option with respect to the ith attribute, is available. We further suppose that a weight W; can be attached to the ith attribute. The writings on outranking pay little attention to where these weights come from or what they mean. There are important and subtle difficulties in weight elicitation. Inputs to the method of outranking clearly involve at least as much elicitation as is needed to construct a multi-attribute value function.

The Weighted-Sum Method (WSM), or the decision matrix approach, is perhaps the earliest method employed. This evaluates each alternative with respect to each criterion and then multiples that evaluation by the importance of the criterion. This product is summed over all the criteria for the particular alternative to generate the rank of the alternative. Mathematically:

$$R_i = \sum_{i,j} a_{ij} \cdot w_j$$

where R_i is the rank of the ith alternative, a_{ij} is the actual value of the ith alternative in terms of the jth criterion, and w_j is the weight or importance of the jth criterion. Subjectivity, bias, and prejudice in giving these ratings and weights cannot be eliminated or evaluated in this method. The additive utility assumption on which this method is based creates problems when the units of the multiple criteria differ from one other. Such methods are often defined as discrete decision-making.

The Multi-Criterion Decision-Making Movement

Decision-making can be considered as the choice, on some basis or criteria, of one alternative among a set of alternatives. A decision may need to be taken on the basis of multiple criteria rather than a single criterion. This requires the assessment of various criteria and the evaluation of alternatives on the basis of each criterion and then the aggregation of these evaluations to achieve the relative ranking of the alternatives with respect to the problem. The problem is further compounded when there are several or more experts whose opinions need to be incorporated in the decision-making. It is lack of adequate quantitative information, which leads to dependence on the intuition, experience, and judgment of knowledgeable persons called experts. We can define a generic decision-making problem as consisting of the following activities:

- Studying the situation.
- Organizing multiple criteria.
- Assessing multiple criteria.
- Evaluating alternatives on the basis of the assessed criteria.
- Ranking the alternatives.
- Incorporating the judgments of multiple experts.

The problem can be abstracted as how to derive weights, rankings, or importance for a set of activities according to their impact on the situation and the objective of decisions to be made. This is the process of Multiple-Criteria Decision-

Making (MCDM). The MCDM problems have been studied under the general classification of Operations Research (OR) problems, which deal with decision-making in the presence of a number of often conflicting criteria. The intellectual roots of this approach lie in the study of Linear Programming (LP). The invention of the simplex algorithm for discovering the optimum of a linear function when there are linear constraints on the variables was a most significant step in the development of management science. For the first time it was recognized that many logistic problems could be considered in this way and considerable savings were achieved by applying linear programming in a wide range of contexts. Its methods are still principally centered on the application of concepts from mathematical programming to the multiple criterion problem. The methods are generally suited to problems where there is a continuum of options (e.g. control variables that may take any values in a given range) rather than a set of discrete alternatives.

The field of MCDM is divided into Multi-Objective Decision-Making (MODM) and Multi-Attribute Decision-Making (MADM). When the decision space is continuous, MODM techniques such as mathematical programming problems with multiple objective functions are used. Sometimes the final solution in these concepts is accepted as reaching the Pareto set, which does not provide a unique solution. On the other hand, MADM deals with discrete decision spaces where the decision alternatives are predetermined. Many of the MADM methods have a common notion of alternatives and attributes. Alternatives represent different choices of action available to the decision-maker, the choice of alternatives usually being assumed to be finite. Alternatives need to be studied, analyzed, and prioritized with respect to the multiple attributes with which the MADM problems are associated. Attributes are also referred to as goals or decision criteria. Different attributes represent different dimensions of looking at the alternatives, and may be in conflict with each other, may not be easily represented on a quantitative scale. One possible solution here could be a multiattribute utility function representing the DM's preferences as is in the utility theory.

3. DECISION SUPPORT BASIS: DECISION-MAKING PROCESS

The domain structure of decision-making is presented on Figure 1. It is not only the area of learning or knowledge about this topic, but the field of special cognitive activity known as decision-making activity. In correspondence with the underlying principles of the Activity Theory, an activity is described through "hierarchical structure of activity" that concerns the activity ontology (Leontiev, 1978). The activity existence is presented on three levels. At the highest level, it is oriented to a motive that determines the necessary potential for activity performance. Each motive is an object (material or ideal) that satisfies a need or a purpose (objective, mission) that requires the existence of an activity. On the second level, the activity exists as action(s). An action is determined by a goal (objective), which presumes a conscious problem solving form of activity realization (goal-directed realization of activity). An activity can be presented by several actions and the course of an action can consists of several actions. On the lowest level, the actions are carried out through operations that are determined by the actual conditions, in which an activity is performed. The operations present the process of acting (doing) and convert the actions into facts.

The hierarchical organization of a decision-making activity necessitates a framework to align different strategic initiative in this domain with decision-making vision. This framework has to satisfy the following requirements:

- It is consistent, robust and stable; it helps the management of strategic initiatives to

Figure 4. A framework to align decision support-driven initiatives with decision-making vision

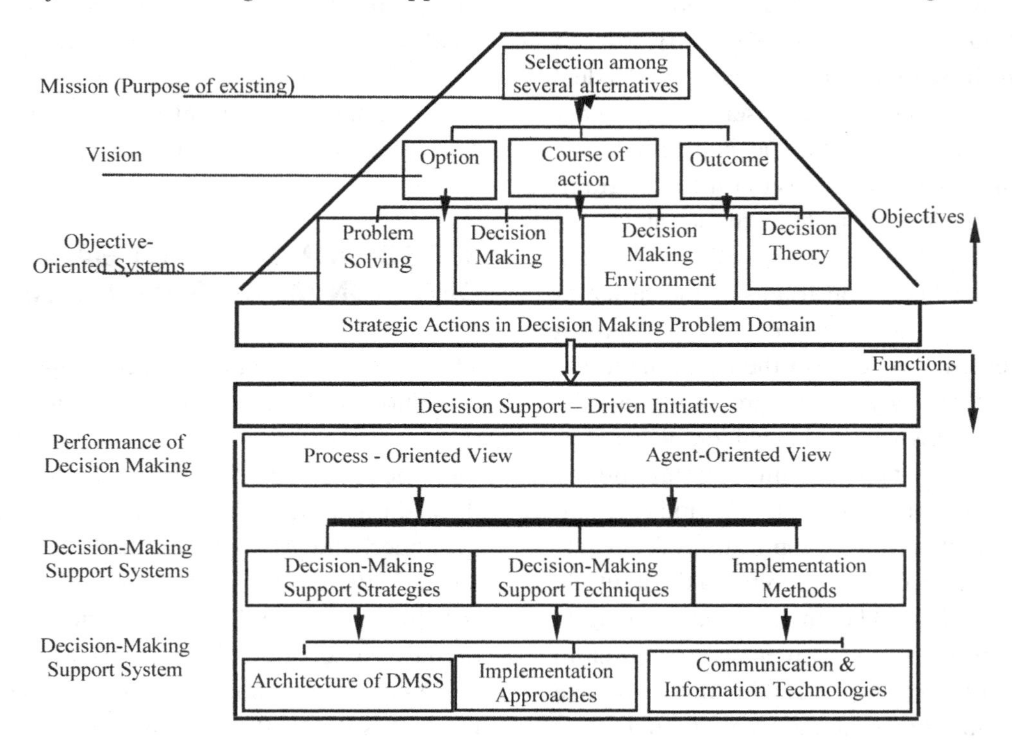

remain focused on their implementation without getting deflected by internal or external chaos that could be caused by changes have come in them;

- It determines strategic actions over their life cycle (course of action);
- It acts as a shared communication vehicle across the organization that concerns the realization of strategic initiatives;
- It is flexible enough to operate in a constantly changing environment;
- It enables accurate delivery of key information while aligning different layers of activity organization.

The framework for aligning decision support-driven initiatives with decision-making vision is presented on Figure 4. Its construction passes through the following steps:

Step 1: Mission and vision of decision-making activity and its organization are articulated to ensure the constant purpose of existence and progress towards excellence in research. Mission describes the purpose of existence of decision-making activity and its organization and vision defines the results to be achieved in carrying out this activity.

Step 2: Define four systems that have relation to the objective of decision-making activity and give system-oriented view of its domain. The systems that are the main component of decision-making domain guarantee to reveal all aspects of the activity. They are presented in the previous parts of this chapter—problem solving, decision-making, decision environment, and rational decision-making.

Step 3: Determine main goals (strategic objectives) and corresponding actions (strategic actions) that describe the decision-making activity. One of main goals during decision-making performance is the support of decision-makers. That is why "decision support" and development of decision support systems are strategic objectives that require some strategic action. The way of its achievement is result of initiatives that are presented as decision support-driven initiatives. They determine the course of the strategic action necessary for development of Decision-Making Support Systems (DMSS).

Step 4: The first stage of DMSS development is construction of the conceptual framework of DMSS. It depends on the performance of decision-making activity. There are two approaches to its representation: process-oriented and agents-oriented. We use process-oriented description of decision-making for conceptual framework of DMSS.

Step 5: Determination of main features of decision support systems—strategies to DMSS construction, main techniques used in support of decision maker's tasks, and possible implementation methods in development of DMSS. The strategy of DMSS construction depends on objectives that have to be achieved through the system development. The objectives fix which task(s) or stage(s) of decision-making process should be supported by the DMSS.

Step 6: The main aspects of DMSS realization are *system architecture, implementation approaches,* and used *information and communication technologies.*

Decision-making process is considered as "the sensing, exploration and definition of problems and opportunities as well as the generation, evaluation, and selection of solutions" and the implementation and control of the actions required to support the course of action selected (Huber & McDaniel, 1986). Herbert Simon is considered a pioneer in the development of models of human decision-making activity. He conceptualizes the performance of this activity as linear process, moving through the following three stages (Simon, 1997). This Simon's definition of the process is considered as the canonical model of decision-making process:

- **Intelligence:** The identification of a problem (or situation) that requires a decision and the collection of information relevant to the decision;
- **Design:** Creating, developing, and analyzing alternative courses of action through the construction of a model of an existing real-world system;
- **Choice:** Selecting a course of action from those available.

During the "intelligence" phase it is performed three tasks: problem recognition, data gathering and objectives determination. The "design" phase bases on the performance of the following tasks—design of criteria, design of choice, and design of decision model. The "choice" phase consists of evaluation (solution modeling) that bases on evaluation procedures and prioritizing, sensitivity analysis and selection. Decision-making is thus cast as problem solving, the process-oriented model provides a representation of "the problem" which can be "solved by" implementing a prescribed course of action identified as "preferred."

As a full problem-solving process, subsequent authors extend the original Simon's model, to include "Implementation" and "Learning" stages (Forgionne & Kohli, 2000). In the new fourth stage, decision(s) must be communicated ("Result Presentation"), and implemented and controlled through "Task Planning" and "Task Monitoring" activities. Outputs of this stage are implemented decisions. The purpose of the last fifth stage ("Learning") is to assess the value contributions to the decision-making process and the value of the outcomes and to learn from mistakes and successful events. It is achieved by the performance of activities of "Outcome-Process Analysis" and "Outcome-Process Synthesis." This stage is very often named "Review Stage." A similar process-oriented model of decision-making that passes through five phases is suggested by Huber (1980). The phases of this decision-making process and the task that guarantee their fulfillment are the following:

- **Analysis:** Problem identification; problem definition and problem diagnosis;
- **Generation:** Generation of alternatives, criterions and scenarios; design of behaviour model;
- **Election:** Evaluation and selection;
- **Implementation:** Decision planning and decision execution;
- **Control:** Monitoring and control, and verification.

Another famous model of decision-making process is given by Sage (1981). It has three stages—formulation, analysis, interpretation. The interpretation phase covers the third and forth stages of the other models. The formulation stage includes problem definition, value system design, and systems synthesis. Analysis is fulfilled by systems modeling and optimization. The main tasks of interpretation are evaluation, selection, and planning of action.

On the base of existing models of decision-making process a generic model is suggested (Mora, et al., 2008). It has four phases—intelligence, design, choice, implementation and learning. As a result of performance of the tasks of the first phase (problem detection, data gathering, problem formulation) a problem schema is constructed. The output of the "design" phase (model classification, model building and model validation) a model scheme is produced. The "choice" phase (evaluation, sensitivity analysis, and selection) produces a decision scheme. Decision implemented is a result of the "implementation" phase (result presentation, tasks planning, tasks monitoring). The "learning" phase (outcome process analysis, outcome process synthesis) ends in an organizational memory of knowledge. As suggested by Simon the proposed generic DMP model treats the decision steps within each phase as an iterative process rather than a sequential one.

In process-oriented modeling of decision-making activity, it is described by actions that are carried out on certain stage of the decision-making process. The achievement of the goal of each action is a result of performance of one or several tasks. Task analysis concerns presentation of a task as a hierarchy of sub-tasks. However, the performance of an action very often could be presented as a network of tasks, where each task is in (not hierarchical) relationships with other tasks that use its result or provide it with necessary resources. Task execution can be considered as responsibility of certain agent (actor). An agent is determined by its roles - task that he has to carry out. The representation of decision-making activity through an organization of agents ensures an agent-based modeling of decision-making. This view on decision-making is very useful when we consider an organizational decision-making activity. Decision-making in organization is becoming more and more multi-actors and complex. An attempt to agents-based modeling of decision-making is made by Csaki (2008). He tries to present all agents that have some role in performance of decision-making activity. The list of roles (agents) includes client, decision analyst, decision maker, decision taker, decision owner, expert, facilitator, problem owner, problem solver, proposal, stakeholder, and user.

In order to capture and understand the organizational context and organizational decision-making activity, it is recommended usage of network analysis (Adam & Pomerol, 2008). Network analysis views actors (agents) as participants in complex systems involving many other actors whose behaviour and actions may affect an individual's behaviour. It may be the key to our understanding of the internal and external fabric of relationships that exist in and around organizations. This approach to agent-based presentation of organizations suggests a radically new focus on the qualitative aspects of the relationships existing between executives at the level of the whole organization, that is, looking at the web of all relationships existing in this organization as opposed to the more traditional emphasis on the characteristics of a specific decision-making process tackled by a single manager or a small group of managers. Thus, the network approach to organizations can constitute a useful analytical basis to guide the actions of DSS developers and enable a more insightful study of the contexts in which DSS systems are being implemented.

4. DECISION-MAKING SUPPORT SYSTEMS: CHARACTERISTICS AND CAPABILITIES

The purpose of a Decision-Making Support System (DMSS) is to provide partial or full support for the decision-making phases (intelligence, design, choice, and monitoring) of decision-making process and the tasks that are performed during each phase. Usually, Decision Support Systems (DSS) are regarded as DMSS. The former is defined as a broad class of information systems that utilizes database or model-base resources to provide assistance to decision makers through performance of their activities—decision analysis, modeling, and output. Many terms are used to describe decision support systems—business intelligence, collaborative systems, computationally oriented DSS, data warehousing, model-based DSS. They are used to describe decision support capabilities and are important in making sense about what technologies have been deployed or are needed. A technology-oriented categorization of DSSs is given by Power (2008). It is determined by the dominant technology component that drives or provides the decision support functionality. Five generic categories based on the dominant component are discussed: communications-driven, data-driven, document-driven, knowledge-driven, and model-driven decision support systems. The author pretends that these categories can classify DSSs currently in use.

Communications-driven DSSs present systems that are built using communication, collaboration, and decision support technologies. Data-driven DSSs include file-drawer and management reporting systems, data warehousing and analysis systems, Executive Information Systems (EISs). They emphasize access to and manipulation of large databases of structured data and especially a time series of internal company data and sometimes external data. Simple file systems accessed by query and retrieval tools provide the most elementary level of functionality. Data warehouse systems that allow the manipulation of data by computerized tools tailored to a specific task and setting or by more general tools and operators provide additional functionality. Document-driven DSSs integrate a variety of storage and processing technologies to provide complete document retrieval, summarization, and analysis. Knowledge-driven DSSs can suggest or recommend actions to managers. These DSSs contain specialized problem-solving expertise based upon artificial intelligence and statistics technologies. The expertise consists of knowledge about a particular domain, understanding of problems within that domain, and skill at solving some of these problems. Model-driven DSSs include systems that use accounting and financial models, representational simulation models, and optimization models. They emphasize access to and manipulation of a model. A simple algebraic model with "what-if" analysis provides the most elementary level of model-driven DSS functionality.

The weakness of the presented categorization of DSSs is that it takes into consideration the bottom up approach to their development.

According to it, the realization of these systems is result of application of existing information and communication technologies in the field of decision-making, especially for supporting of decision-making actions. The purpose and use of DSSs serve only for their determining features (parameters). In conformity with our viewpoint, a great number of decision support systems are produced with the objective to provide mechanisms to help decision makers get through a sequence of stages during decision-making process in order to reduce uncertainly and processing of ambiguous decisions. It is clear that the basis for defining DSS is determined by two factors:

- The perceptions of what a DSS does (supports decision-making in unstructured problems) and the objective(s) of this support-task(s), in which a decision maker has to be assisted during decision-making process;
- Ideas about how the DSS's objectives can be accomplished (e.g., the needed computer, information and communication technologies).

Our approach to definition and classification of DSSs is oriented to the first factor. That is why, in definition of these supporting systems we use the term Decision-Making Support System (DMSS) instead DSS.

In a top-down approach to development of DMSS, the determination of decision-making support strategy is very important. This strategy depends on the objective that has to be achieved in supporting of decision maker. In correspondence with the generic decision-making process, we can determine four major objectives of decision maker that he has to reach with the help of appropriate DMSS—construction of problem scheme, construction of decision model scheme, construction of decision scheme, and decision implementation. Since these objectives drive development of DMSSs, the corresponding

Figure 5. Categorization of decision-making support

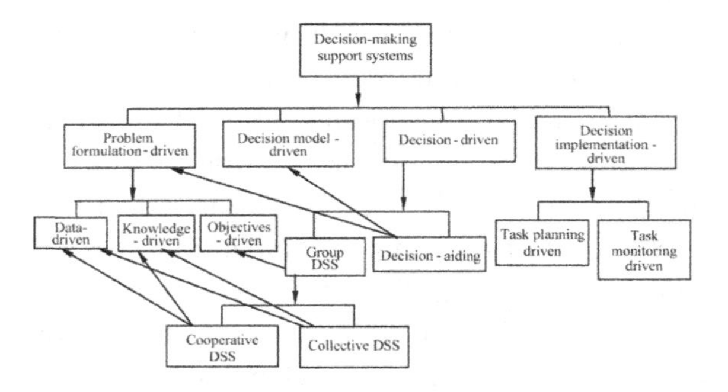

DMSS come respectively under the following categories: problem formulation-driven, decision model-driven, decision-driven, and decision implementation-driven (Figure 5). We can divide implementation-driven DMSS in two classes, since it is possible to develop separately DMSSs for supporting task planning and task monitoring. Model-driven DMSS coincides with determined model-driven DSS. It is, by definition, functionally based on one or more quantitative models and is designed such that a user can manipulate model parameters in order to analyze decision alternatives A DMSS may be classified as model-driven, if the model can be accessible to a non-technical user via an easy-to-use user interface, the model should provide a simplified, easily understandable representation of the decision situation, and the intention should be that the DMSS itself is to be used repeatedly in the same or similar decision situation. Its main component is a *model base* that is a repository for the formal models of the decision problem and the methodology for developing results (simulations and solutions) using these formal models.

Decision-driven DMSS are divided into two classes—decision aiding DSS and group DSS. The development of decision aiding DSS is considered only in the context of Personal Decision Analysis (PDA), quantifies judgment, and processes it logically. The means used to quantify judgment are known as decision tools that realize certain decision technology. There is little "useful" research, in the sense of leading toward decision technology that advances the interests of the decider, through decisions that are either more sound or more effectively communicated. A decision technology has to (1) use all the knowledge a decider normally uses; (2) call for inputs that people can readily and accurately provide; (3) produce output that the decider can use (Brown, 2008). It is necessary to notice that, though recent research has made important scientific advances, little has influenced decision aiding practice, and little current research attacks the problems that are still holding back successful decision aiding.

The development of successful decision aiding tools requires more research in the following aspects of decision technology: decision strategy, principles of decision tools design and development of specific tools. The main objective of decision strategy has to be the best integration of informal analysis into informal reasoning without disrupting it. The principles of decision tool design are the following: how judgment-intensive should decision models be, in given circumstances; how well can people make hypothetical value judgments or hypothetical factual judgments. The development of specific tools concerns the next problems:

- Some mathematical algorithms call for inputs that fit people's cognition better than others. Inference by Bayesian updating, requires assessors to make hypothetical likelihood judgments. Are people better at making the resulting posterior assessments directly?
- What should be the form of utility elicitation-holistic, decomposed into additive pieces, or further decomposed into factual and value judgments as in Multiattribute Utility Analysis.

The DMSSs that realize certain decision aiding (decision technology) could be considered in the light of descriptive, normative, and prescriptive approaches. Since they are designed in the context of PDA, they have to take into account the professional human activity analyzed from the viewpoint of descriptiveness. However, descriptiveness without analysis cannot support the decision-making process. Since the aim is to eliminate the uncertainty and the errors in the decision forming process of the subject, a formal semantically and logically sound approach is required. The removal of uncertainty and errors is done not only at the initial information (e.g. in the expert preferences) but also from the point of view of the formal analysis and synthesis of the solution, reflected in the model and procedures implemented in the system. Therefore, this aspect of DMSSs in its essence reflects commonly accepted principles or models (axioms and theories) and is by nature normative.

For DMSS, intended for construction of individual value or utility functions (Keeney, 1988, 1993) the normativity is in:

- Formal models for existence of multicriterial utility functions;
- Von Neumann axiomatic as principles of logically sound behaviour;
- The logic of mathematical actions, done on the basis of mathematical models (Fishburn, 1970).

Here, as an area of application is included the whole arsenal of the Bayesian approach of Prof. Raiffa for decision-making, as well as the individual particularities of the concrete problem (Raiffa, 1968; Keeney, 1993). Therefore, normativity is under reasonable and sound processing of the initial information and under forming the decision in the formal part of the system as logically consistent prescription. This normativity is included also in the concept of the system, because better solution is sought, and not only repetition and description.

There are decisions aiding tools that help to take decisions with prescriptions, recommendations. There exist several reasons not to limit the consideration only to the normativity in these prescriptions:

1. Every theory or model is a limited form of reflection of reality, because such is human perception. The person determining the decision or expert is inconsistent in reality and due to this even the commonly recognized as reasonable axioms in the normative approach of von Neumann are violated in the everyday life. In the decision-making process, the human is more sophisticated and inconsistent in logical sense. Therefore, if one system is only descriptive, it will bear the same contradictions, which are present in what is being described by it.
2. If the decision technology is only normative, it will not account for the uncertainty and the possibility for errors in the expert data and preferences.
3. If the decision technology is prescriptive, then it has to account for the scientific achievements, the probabilistic and subjective uncertainty, and the errors in the expert human preferences, assessments, and decisions. Failing to do that will produce only and autonomous or expert system, for example. Thus, it has to be more powerful and more flexible in its concept with regard to the formal problems and models than the classical normative (axiomatic) approach in the decision-making, which has been widely (often unfoundedly) criticized in the last decades..

In the proposed (suggested) in this book approach the intention is to account the possibility and richness of ideas and solutions of the normative axiomatic approach in the formation of prescriptions for the decision maker. It uses stochastic approximation approach as a possibility of analysis and limitation or complete elimination of the uncertainty and the errors in the expert opinions, expressed as preferences. The aim is to derive description of the human behaviour that is closer to reality, maximal utilization of the mathematical decision theory in its aspects, reflected in the theory of measurement, utility theory, subjective probability theory, and Bayesian analysis approach. The intention is to diminish or eliminate the uncertainty in human preferences, due mainly to the cardinal nature of human concepts and the ways of their expression. For this purpose are used the achievements of the stochastic programming. In such a way one decision-driven DMSS, encompassing all these aspects could achieve better reflection of the prescriptive approach.

In our classification scheme, we present Group Decision Support Systems mainly as decision-driven in order to consider individual decision taking vs. group decision taking. It is necessary to mark that individual and group decision taking techniques are very useful in problem formulation and decision modeling, i.e. at all stages of decision-making process. In correspondence with the consideration of a group as collective or cooperative group, there are respectively cooperative and collective decision support systems (Zarate, 2008). A collective DSS is designed for collective decision-making process, in which several decision makers are involved that could happen in three kinds of situations that are defined as face-to-face, distributed synchronous, and distributed asynchronous situations. They determine the corresponding collective DSS, as well:

1. **Face to Face Decision-Making:** It is a very classical situation, in which different decision makers are implied in the decisional process and meet them around a table. It is supported by GDSS rooms.
2. **In Distributed Synchronous Decision-Making:** Different decision makers implied

in the decision-making process are not located in the same room but work together at the same time. This kind of situation is enough known and common in organizations and it could be supported by Electronic Meeting Systems (EMS), videoconferences, telephone meetings, and so forth.

3. **Asynchronous Decision-Making:** Different decision makers are implied in the decision-making process and they come in a specific room to make decisions but not at the same time. The specific room could play a role of memory for the whole process and also a virtual meeting point.

There is another style of group decision-making that is defined as distributed asynchronous decision-making. In accordance with it, different decision makers do not necessarily work together at the same time and in the same place. Each of them gives a contribution to the whole decisional process. This is a new kind of situation that requires cooperation of decision- makers. For this purpose, cooperative decision support systems must be designed.

Group DSS (GDSS) expends DSS to include tools consistent with the role and function of groups in decision-making processes. These efforts are driven by acknowledgement that groups are often employed to solve problems, especially in situations exhibiting greater complexity. Usually, the focus in development of GDSS is on a shift of emphasis from group decision support to simply group support emerged. Often, the function of these group support systems is to synthesize whatever perspectives exist of a problem into a single perspective so that DSS approaches to problem solving can be used. For example, brainstorming capabilities are used to surface multiple perspectives of a problem, but group support system rating and voting functions effectively elevate the aspect(s) most widely held in the group to the status of being the aspect(s) most important in the decision-making behaviour of the group. These systems are not designed to maintain the integrity of the various perspectives of a problem being considered. A process that loses a perspective of a problem situation probably loses information valuable in constructing an approach to dealing with the problem in a sustainable, long-run fashion (Paradice & Davice, 2008).

There are two techniques for supporting the creation of multiple perspectives: data organization and "virtualization of relationships," and work group discussion and communication. Data organization including data filtering, categorizing, structuring, and idea formation creates relationships among pieces of information. Then virtualization of those relationships helps to generate multiple perspectives. This technique reflects the belief that an individual's creativity comes partially from tacit knowledge. The systems that support work group discussion and communication provide tools for modeling arguments and thus facilitate discussions of different issue perspectives. They also provide tools for analyzing arguments and the positions taken by stakeholders. They ensure valuable decision-making aids and facilitate perspective generation activities through group discussion and communication.

The early days of the Decision Support Systems (DSS) movement implicitly focused most heavily on the choice and design phases of Simon's model. They may be more effective in helping decision makers to make good choices when support for problem formulation is provided. Research validates the notion that support for problem formulation and structuring, and objectives determination leads to better decisions. In objectives-driven DMSS it is necessary to include a module for description and structuring of the main goal, reflecting the intention of the system, into sub-goals and criteria describing them quantitatively. An analysis of the sub-goals of the type of factor analysis is required to eliminate the overlapping of sub-goals and diminishing to a minimum the over-determination in the formal problems, describing the problem. For multifactor problems, it is necessary to acquire the information about the way the sub-factors influence each other with respect to the main goal. Such information determines the structure and the form of the objective function (value, utility). When extracting expert preferences iteratively, capability for accumulation of the initial information must be allowed, as well as the ability to correct it when needed. This iteration concerns also the description and construction of the individual value or utility functions, which according to the decision maker or the expert describe structurally and quantitatively the goal and subgoals, adequately characterizing the main goal.

5. IMPLEMENTATION OF DECISION-MAKING SUPPORT SYSTEMS

There are two methods for implementation of decision-making support system. The objective of the first method is the development of centralized DMSSs. The second method guarantees distributed DMSSs. This method is more used in the present, since it ensures the following advantages for distributed DMSS:

- A distributed DMSS takes advantage of decentralized architecture;
- This kind of DMSS can survive on an unreliable network;
- They are suitable for mobile applications;
- They are appropriate for Web-based implementation;
- Distributed DMSS can guarantee necessary resources for the tasks of decision-making process that common for use.

The centralized DMSS supports usually one user (decision maker) and provides three sets of capabilities in the areas of dialog, data, and modeling. The system should have access to a wide variety of data and it should provide analysis and modeling in a variety of ways. Its main component is the user interface. The other two components are *database* and *model base*. The models in the model base are linked with the data in the database. Models can draw coefficients, parameters, and variables from the database and enter results of the model's computation in the database. These results can then be used by other models later in the decision-making process. The management of these components and the user interface of the system are designed as a component.

A distributed DMSS can be realized as service-oriented or agent-based system. The service-oriented system bases on the idea, that organizations, decision-makers, and other participants in decision-making process share resources, which can support their activities. This approach to system implementation has been launched under different concepts—"Web services," "network services," "open network environment" (Carlson, 2008). A result of this approach is that previous proprietary architectures, which maintain unique, internal for an organization DMSS, are substituted by an open architecture giving possibility for supporting the participants in a DM process with data, information, knowledge and decision aiding tools

Figure 6. Architecture of distributed DMSS

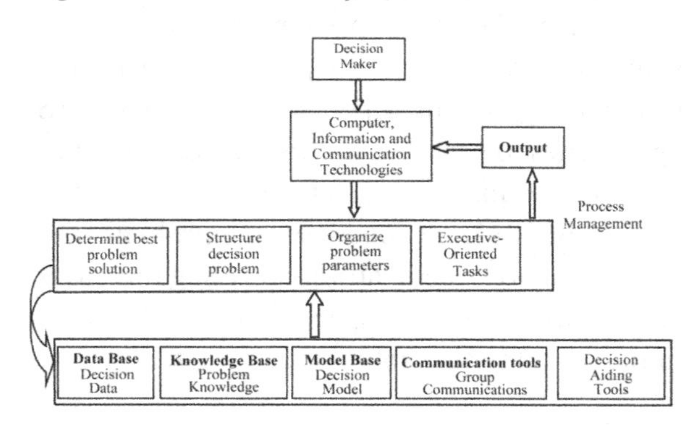

from different types of external service providers. Internet-based, distributed systems have become essential in modern organizations. When combined with artificial intelligence techniques such as intelligent agents, such systems can become powerful aids to decision makers. These newer intelligent systems have extended the scope of traditional decision-making support systems to assist users with real-time decision-making, multiple information flows, dynamic data, information overload, time-pressured decisions, difficult-to-access data, distributed decision-making. As a class, they are called intelligent decision support systems (Philips-Wren, 2008).

A general architecture of distributed DMSS is given on Figure 6. It is influenced by a general architecture of decision support systems that is presented by Fogionne and Russel (2008).

This architecture includes the main components that a system realizing all decision support strategies has to involve: data base, knowledge base, model base, tools for realization of group communications and personal decision aiding tools. All resources that help decision maker in performance of decision-making process are provided to him through the components that are responsible for carrying out of the appropriated tasks.

6. THE BOOK VIEWPOINT AND ORGANIZATION

The purpose of this book is to present an application and demonstration of a new mathematical technique and possibilities for mathematical Value driven descriptions of complex systems, in which human speculative knowledge is decisive for the final solution. The human thinking and preferences have quantitative nature that makes the problems in the domain of the complex systems to be considered as quantitative, i.e. difficult for analytical description. Value driven design can be defined as a development paradigm, in which required human value considerations are engineered into best practices, activities and management (Collopy, 2009). Value-driven design enables design optimization by providing designers with an objective function. The value-based presentation of objective function includes all the important attributes of a system being designed, and outputs a score (Hall & Davis, 2007; Castagne, et al., 2009). At the whole system level, the objective function, which performs this assessment of value, is called a value model. The objective is strict logical mathematical approach for human-adapted management and control design in complex processes with definitive human participation in the final solutions.

The problem of preference rules and their implementation in rule-based information systems is certainly one of the important topics in Decision-Making and Artificial Intelligence, as witnessed by some activity in this area, from the standpoint of utility theory (Keeney, 1988; Klein, 1994), or cognitive psychology (Pinson, 1987; Aranda, 2005). According social-cognitive theories people's strategies are guided both by internal expectations about their own capabilities of getting results, and by external feedback (Bandura, 1986). Internal human expectations and assessments are generally expressed by qualitative preferences. Probability theory and expected utility theory address decision-making under these conditions (Fishburn, 1970; Keeney, 1993). This uncertainty is of subjective and probability nature. A possible approach for solution of these problems is stochastic programming. The uncertainty of the subjective preferences could be taken as an additive noise that could be eliminated, as is typical in the machine learning based on the stochastic programming (Pavlov, 2011; Aizerman, 1970).

The presentation of human preferences analytically with utility functions is a good possible approach for their mathematical description. It is the first step in realization of a human-adapted value driven design process and decision-making, whose objective is to avoid the contradictions in human decisions and to permit mathematical calculations in these fields. Validate mathematical utility evaluation is difficult, since the evaluation is based on quantitatively expression of the DM's preferences. If the subjective and probability uncertainty of DM preferences is interpreted as some stochastic noise, stochastic programming can be used for recurrent evaluation of the utility function, with noise (uncertainty) elimination. In fact, this is pattern recognition of the positive and negative DM's answers in regarding to his preferences. The utility evaluation is human-computer dialog between a decision-maker and computer-based evaluation tool. It concerns mathematically machine learning, since its basis is the axiomatic approach to decision-making theory and stochastic approximation. The latter presents unique stochastic recurrent procedures easy for computer programming. The final results are full mathematically grounded descriptions and modeling of complex systems like "technologist-dynamical model." Following from this the orientation of the book is toward the branch of the Model-Driven Decision-Making and Decision Support in the taxonomy created by Daniel Power (Power, 2002). A Model-Driven Decision Support emphasizes access to and manipulation of a statistical, financial, optimization, or simulation model. Model-Driven Decision Support use data and parameters provided by users to assist decision makers in analyzing a situation.

There is a gap, in some aspects socio-technical gap, between existing theory and actual practice in real-life decision-making (Levy, 2010). In recent years, bridging this gap has been a challenge in many areas of research. It is a great divide between the social aspects aimed to be supported and those that are actually supported, due in part to technical limitations and in part to the complexity of the contexts where decision-making and decision support must be provided. In Decision Support Systems, this challenge has raised several important questions concerned with the account and encapsulation of social aspects of managerial decision-making as well as with the representation of certain human cognitive aspects, such as intuition or insights within computational systems. In this manner the problem of constructing analytically with flexible mathematical techniques value, utility functions in the aspects of the decision-making theory is a now a day problem and a challenge (Bertoni, 2011; Hall, 2007). Value-based management concepts are prevalent in theory and practice since value creation is commonly considered the paramount business goal. Value-based management mainly applies data-driven concepts to support decision-making disregarding model-driven approaches (Courtney, 2001; Hahn, 2012). Development of value based models permits mathematical description and even Value-based Management as optimal mathematical solutions of the complex systems „Manager-mathematical description of the process" (Hall, 2007; Castagne, et al., 2009; Pavlov, 2011).

In our approach, we do not give priority of the existence of the utility function, but of its constructive evaluation. Utility is mathematical abstraction presented within the limits of normative approach. We reveal the existence of a mathematical expectation measured in the interval scale on the basis of empirical preferences. This solution is new in the field of decision-making and permits iterative control design in the domain of the complex problems and systems by iterative analytical stochastic approximation of the utility function. In the book are given stochastic methods for description, representation, evaluation and analytical description of expert information as part of complex systems, in which human participation is crucial for the final decision. These methods allow adaptability in the mathematical representation of empirical knowledge and special skills expressed at the level of expert preferences.

Chapter two has an introductory nature for the principles and some concepts from probability theory. We consider the concept probability in two aspects: objective probability and human expectation. The most widespread interpretation of objective probability is the frequency of occurrence of certain event with respect to another. The human expectation is defined as probability of occurrence of given event from the point of view of subjective perception, based on empirical experience, intuition and professional skills and knowledge. Special attention is devoted to the concept "conditional mathematical expectation" because it is fundamental in the proof of important convergence theorems for the stochastic procedures for evaluation of the subjective value and utility on the basis of the expressed preferences. The concept "generalized gradient" is discussed and recurrent procedures based on it, theorems and proofs regarding the convergence of the recurrent procedures are provided. The choice of the theorems is determined by the fact that they have a broader nature for a class of problems of the stochastic programming used for the evaluation of the subjective probabilities. Mathematical description of the notion of preference is considered and the problem for evaluating subjective preferences under conditions of risk and subjective uncertainty is defined. So defined, this problem is multiplied in problems for evaluation of subjective value and utility and the subjective probability based on stochastic programming is guaranteed. Presenting strict mathematical proofs, we give the mathematically prepared reader an opportunity to enter deeper in the mechanisms of the proposed approaches and methods for evaluating subjective knowledge, empirical and verbally expressed as preferences.

Chapter three has an introductory nature with respect to the decision-making theory. It begins with mathematical description of the notions measurement and scale. The scales have fundamental significance for the decision-making theory. The main achievements of the so-called discrete or group decision-making, which have methodological significance, are presented. The concepts of "value" and "utility" are interpreted from the point of view of the theory of measurement (scaling). Special attention is paid to the concept "subjective probability," both from more application-oriented position of comparing uncertain events by their participation in lotteries (discrete probability distributions) and from strictly formal and mathematically correct position of the theory of measurement and scaling. The chapter ends with example for evaluation of subjective probabilities and examples of utility functions, which have methodological generality.

Chapter four takes into consideration the main results and theorems for the existence of value and utility function. The objective is the consideration results and theorems, which in our opinion are methodological with respect to the utility theory's use and application in practical problems. This book's section concludes with three examples of application of utility theory in practice. The first makes a comparison between utility theory and fuzzy sets theory as outlined in an example from the help menu of MATLAB (MathWorks Inc.), one of the most used information systems. The other examples demonstrate the capabilities of the utility theory for mathematical description of hard to formalize problems from various areas of human activity. The chapter requires certain persistence and mathematical inclination, although the clear definitions and description of the used mathematical concepts.

Chapter five may be regarded as continuation of chapter two. In detail are considered direction of the stochastic programming, used in the stochastic procedure for analytical construction of value and utility functions and subjective probabilities. Proofs of all used theorems are provided. A question of special interest are important topics from the so-called "Potential function method." One of its strongest results is given, suitable for machine (computer) learning from mathematical point of view. This theory is relatively unknown in the scientific literature, in particular in the field of decision-making. Because of this we have provided the used theoretical results with detailed proofs. Regarding this chapter mathematical affinity and skills are required to absorb the material contained. Two types of algorithms for analytical description of value and utility functions under risk and uncertainty and under exact expert preferences are given.

Chapter six presents the main theorems and methods for analytical representation of utility functions and a theorem for the convergence of the proposed recurrent procedures. The use of the proposed stochastic procedures for utility function construction is described from the viewpoint of Kahneman and Tversky and their Prospect Theory. A formula for evaluation of the approximation of the utility function as "empirical risk" according to Vapnik's theory is given. Example of bio-process application is demonstrated through the evaluation of the specific growth rate. This is an important parameter that is difficult for utility qualitative assessment by the biotechnologist.

Chapter seven takes into consideration stochastic procedures for evaluation of subjective probability. They are based on theoretical results, outlined in chapter five and more specifically the Robins-Monro method—a classical result in the stochastic programming. The described procedures are used also for the decomposition of multiattribute utility functions to functions of fewer variables. This decomposition is fundamental for the use of utility theory in practice and in modeling of more complex and difficult for formalization processes.

Chapter eight has a direction different than the preceding. It is devoted to the theoretical description of the of Multilinear Extrapolation (MLE) approach of Rastrigin. This method is effective and suitable for making prognosis and evaluations in complex systems, including as analytical mathematical relationships. It allows adaptation of the prognoses with easy to implement iterative procedures and easy generation of training samples with applied significance. Such adaptation is also allowed by stochastic recurrent procedures. This makes the methods convenient as basis for development of adaptive, complex systems with human participation in the final decision. All the basic facts from the linear algebra needed for strict mathematical description of MLE are given. The chapter concludes with several examples of prognoses from the real practice, which reveal the particularities, flexibility, and the strengths of Rastrigin's MLE.

Chapters nine, ten, and eleven are devoted to examples of application of the approaches and methods considered previously. Each of them may be read independently of others according to the interests and orientation of the readers. Chapter nine shows the applicability of the proposed approaches and methods for complete description of the complex system "human-biotechnologist, bio-process." This is done by analytic description of the optimal control criterion of a known dynamic model as utility function of the biotechnologist. A powerful analytical technique from the optimal control theory including the Pontryagin maximum principle and slide mode control is used. Methods from differential geometry for system reduction are used as well. It is clarified how after reaching analytical description of a complex system, optimal solutions may be sought in mathematical sense. The final solution is obtained with synergetic use of several methods from different areas of mathematics. This, in general sense is true for all similar complex problems from practice. This chapter requires mathematical knowledge and skills for its understanding. However, here is given a complete solution of a practically significant problem, which has not been formally described.

Chapter ten regards the teaching/learning process as decision-making process. The teacher plays the role of decision maker. He has to decide how to ensure effective interactions with individual learners. The optimization of this interaction is possible if all learning resources are offered to a learner in appropriate for him/her form, i.e. they have to be usable. This adaptation of the curriculum and exam material to the cognitive abilities (competences) of individual learners ensures personalized learning and must be a teacher's objective. Since the technology-enhanced learning provides capabilities for easier adaptation of learning resources to the current competence of student, this problem is considered in context of e-learning usability. The e-learning environment and its resources have to be maximally suitable for use both for the teacher and the student. Determining the competence of a learner to use learning resources in one way or another is done on the basis of measuring his/her preferences. This chapter demonstrates the capabilities of the new approach to describe complex, socially determined concepts and processes with the formal methods of the decision-making theory. For its understanding are required only the techniques given in chapters four, six, and seven at the level of applicability of the proposed recurrent stochastic procedures.

Chapter eleven is demonstration of managing complex process from stock-breeding. The process is non-described completely and formally to the moment. For the proposed model description with a complex of formal methods and mathematical models all the techniques presented in the preceding chapters are used. They allow the construction of mathematically empowered decision support system in the considered area. Here are included evaluation of the scale by which the initial information is obtained, factor analysis for eliminating the overlapping, formal model based on MLE and the utility theory for description of the processes of the considered field of stock-breeding. The so-constructed decision support system in poultry farming allows for making prognoses about the processes, including by accounting for the judgments and goals of the particular farmer, as prescription. The chapter uses techniques described in the preceding chapters at the level of applicability of the proposed mathematical procedures.

Chapter twelve together with the stated in chapter one describes the structure and methods for the development of DSS (on the base of data mining and knowledge discovery) using decision aiding tools in data mining process. The role of information and expert systems in problem determination and solution modeling is replaced by the use of experts, from which may be extracted data for evaluating a given situation by measuring their preferences with respect to certain parameters, based on their experience and observations. Chapter twelve gives a brief description of a prototype of decision support system, which allows the construction of value and utility functions of the individual user. This system allows de facto training of the computer in the same preferences as these of the individual user, without the participation of other mediators. Mathematically it is backed up by the methods from the preceding chapters.

Chapters first, tenth, and twelfth are written by Rumen Andreev with the partial participation of Yuri Pavlov in chapters one and twelve. The remaining chapters, from the second to the ninth and eleventh are written by Yuri Pavlov.

The book includes the use of decision-making and decisions as control/management part of adaptive and complex systems. The development of adaptive management systems needs mathematical description of human empirical knowledge as criteria for monitoring and management of various problems and for design of their control. In the book is demonstrated that a possible mathematical tool is the normative approach in decision-making theory, following the prescriptive view of application of the theoretical results.

The book has two important aspects. In the first aspect, it suggests a common mathematical conception and appropriate numeric methods for development of model driven Decision-making based on value models. In that aspect the book provides with general mathematical paradigm and lots of numerical methods the practitioners for development of Value driven models in new applications in sociology, medicine, agriculture, control design in economics. The approach to value or utility evaluation proposed in the book is machine learning and permits a new point of view for development of information systems for individual utility assessment and decision support. The proposed stochastic procedures and discussed applications could help practitioners for development of new applications in different areas of human activities (Hall, 2007; Castagne, et al., 2009; Courtney, 2001; Manos, 2004; Eom, 2006; Hahn, 2012). In that position, the book will be of interest to advanced students and professionals working in the subject of decision theory, control, and management of complex system, developers of information systems, as well as to economists and other social scientists. For this group of readers we could suggest the following sequence of chapters and paragraphs. The readers could start with chapters one and the definitions of chapter two. Special attention merits the last paragraphs of chapter two were was proposed a mathematical description of the methodology for analytical representation of the human preferences. Chapters three and four introduce the readers in the problematic of the normative approach in the decision-making. The proofs of the theorem could be omitted if needed. In chapters six, seven, and eight are described the fundamental numeric methods of interest in the book. The examples of value model design in these chapters together with the chapters ten, eleven, and twelve reveal the mathematical paradigm proposed in the book for value driven design and value-based decision-making and adaptive decision support. Chapter nine could be difficult for this group of readers, but in this chapter is given a full mathematical description of the complex system "technologist-dynamical model." In this chapter is shown the power of the proposed mathematical approach and the possibilities to pass in complex problems of human social, ecological, or economical activity from monitoring and management to optimal control and exact solutions in mathematical aspect (Castagne, 2009; Bertoni, 2011). The proposed methods are unique and permit incorporation of machine learning with mathematical exactness in the decision-making and decision support practices.

Following the ideas of Professor Ralph Keeney, the main assumption in each management or control decision is that the values of the subject making the decision are the locomotive force, and as such, they are the main moment in supporting the decisions (Keeney, 1988, 1993). The values are the guiding force in supporting the decisions and due to this are determining for the formation of the decisions. In information systems in general and especially in the expert systems, the values are implicitly and heuristically included. It is meant that there is no explicit objective function to allow for flexible behaviour of the decision maker when forming the decisions. Such objective value function allows for quantitative analysis and removal of logical inconsistencies and errors (Castagne, 2009; Collopy, 2009; Bertoni, 2011). It is meant that such objective function will nuances the viewpoints with accounting the values with mathematical precision. In chapter twelve, we describe a prototype of a decision support system for individual's utility evaluation as is proposed by Professor Keeney in his paper (Keeney, 1988).

The second aspect of the book reflects the authors' opinion that there are cases when it is necessary to provide mathematical proofs in details, so that readers/scholars with mathematical experience and mathematical background could study in depth the principles and methods outlined in this book and eventually develop further some of them. The reason is that the used stochastic programming approach, the theory of Potential function method, is not well known to the wide scientific community. The authors believe that the book will be of interest for this group of professionals. In this manner, the book is not a simple enumeration of value driven models in different areas of human activities, but guidance how to be designed with mathematical precision value driven models and value driven decision-making in complex processes. Chapters two, three, four, five, six, seven, and eight together with the proofs of the theorems reveal the mathematical side of the proposed in the book value based approach in decision-making and decision support.

The analytical description of DM as utility function together with the model description of the investigated process could give a complete mathematical representation of the system "technologist (DM) – dynamic process." Such models ensure exact mathematical descriptions of problems in various areas which quantitative modeling is difficult: economics, biotechnology, ecology, and so on. These models guarantee that all powerful optimal control theory could be applied in such complex subject areas and exact mathematical solutions.

REFERENCES

Ackoff, R. L. (1981). The art and science of mess management. *Interfaces*, *11*(1), 20–26. doi:10.1287/inte.11.1.20

Ackoff, R. L. (1998). *Ackoff's best: His classic writings on management*. New York, NY: John Wiley & Sons.

Adam, F., & Pomerol, J.-C. (2008). Understanding the influence context on organizational decision-making. In Adam, F., & Humphreys, P. (Eds.), *Encyclopedia of Decision-Making and Decision Support Technologies* (pp. 922–929). Hershey, PA: IGI Global. doi:10.4018/978-1-59904-843-7.ch104

Aranda, G., Vizcaíno, A., Cechich, A., & Piattini, M. (2005). A cognitive-based approach to improve distributed requirement elicitation processes. In *Proceedings of the 4th IEEE Conference on Cognitive Informatics (ICCI 2005)*, (pp. 322-330). Irvine, CA: IEEE Press.

Bandura, A. (1986). *Social foundations of thought and action: A social cognitive theory*. Englewood Cliffs, NJ: Prentice-Hall.

Baron, J. (2008). *Thinking and deciding* (4th ed.). Cambridge, UK: Cambridge University Press.

Bell, D. E., Raiffa, H., & Tversky, A. (1988). *Decision-making: Descriptive, normative and prescriptive interactions*. Cambridge, UK: Cambridge University Press. doi:10.1017/CBO9780511598951

Bennet, A., & Bennet, D. (2008). The decision-making process for complex situations in a complex environment. In Burstein, F., & Holsapple, C. W. (Eds.), *Handbook on Decision Support Systems* (pp. 1–14). New York, NY: Springer-Verlag. doi:10.1007/978-3-540-48713-5_1

Bertoni, M., Eres, H., & Isaksson, O. (2011). Criteria for assessing the value of product service system design alternatives: An aerospace investigation. In *Proceedings of Functional Thinking for Value Creation* (pp. 141–146). Springer. doi:10.1007/978-3-642-19689-8_26

Beynon, M., Rasmequan, R., & Russ, S. (2002). A new paradigm for computer-based decision support. *Decision Support Systems*, *33*, 127–142. doi:10.1016/S0167-9236(01)00140-3

Brown, R. (2008). Decision aiding research needs. In Adam, F., & Humphreys, P. (Eds.), *Encyclopedia of Decision-Making and Decision Support Technologies* (pp. 141–147). Hershey, PA: IGI Global.

Carlson, S. A. (2008). An attention based view on DSS. In Adam, F., & Humphreys, P. (Eds.), *Encyclopedia of Decision-Making and Decision Support Technologies* (pp. 38–45). Hershey, PA: IGI Global. doi:10.4018/978-1-59904-843-7.ch004

Castagne, S., Curran, R., & Collopy, P. (2009). Implementation of value-driven optimisation for the design of aircraft fuselage panels. *International Journal of Production Economics*, *117*(2), 381–388. doi:10.1016/j.ijpe.2008.12.005

Checkland, I. J. (1999). *Svstetns thinking, systems practice: Includes a 30-year retrospective*. Chichester, UK: John Wiley & Sons.

Checkland, P. (1981). *Systems thinking, systems practice*. New York, NY: Wiley.

Collopy, P., & Hollingsworth, P. (2009). *Value-driven design. AIAA Paper 2009-7099*. Reston, VA: American Institute of Aeronautics and Astronautics.

Courtney, J. (2001). Decision-making and knowledge management in inquiring organizations: Toward a new decision-making paradigm for DSS. *Decision Support Systems*, *31*(1), 17–38. doi:10.1016/S0167-9236(00)00117-2

Courtney, J., & Paradice, D. B. (1993). Studies in managerial problem formulation systems. *Decision Support Systems*, *9*(4), 413–423. doi:10.1016/0167-9236(93)90050-D

Csaki, C. (2008). The mythical decision maker: Model of roles in decision-making. In Adam, F., & Humphreys, P. (Eds.), *Encyclopedia of Decision-Making and Decision Support Technologies* (pp. 653–660). Hershey, PA: IGI Global. doi:10.4018/978-1-59904-843-7.ch073

Eden, C., & Ackermann, F. J. (2001). SODA - The principles. In I. Rosenhead & Mingers (Eds.), *Rational Analysis for a Problematic World Revisited*. Chichester, UK: John Wiley & Sons, Ltd.

Eom, S., & Kim, E. (2006). A survey of decision support system applications (1995–2001). *The Journal of the Operational Research Society*, *57*, 1264–1278. doi:10.1057/palgrave.jors.2602140

Fishburn, P. (1970). *Utility theory for decision-making*. New York, NY: Wiley.

Forgionne, G., & Russel, S. (2008). The evaluation of decision-making support systems' functionality. In Adam, F., & Humphreys, P. (Eds.), *Encyclopedia of Decision-Making and Decision Support Technologies* (pp. 329–338). Hershey, PA: IGI Global. doi:10.4018/978-1-59904-843-7.ch038

Forgionne, G. A., & Kohli, R. (2000). Management support system effectiveness: Further empirical evidence. *Journal of the Association for Information Systems*, *1*(3), 1–37.

Fransella, F., Bell, R., & Bannister, D. (2003). *A manual for the repertory grid technique* (2nd ed.). Chichester, UK: John Wiley & Sons, Ltd.

Greeno, J. G. (1978). Nature of problem-solving abilities. In Estes, W. K. (Ed.), *Handbook of Learning and Cognitive Processes*. Hillsdate, NJ: Erlbaum.

Groner, R., Groner, M., & Bischof, W. F. (1983). *Methods of heuristics*. Hillsdale, NJ: Erlbaum.

Hahn, G. J., & Kuhn, H. (2012). Designing decision support systems for value-based management: A survey and an architecture. *Decision Support Systems*, *53*(3), 591–598. doi:10.1016/j.dss.2012.02.016

Hall, D., & Davis, R. (2007). Engaging multiple perspectives: A value-based decision-making model. *Decision Support Systems*, *43*(4), 1588–1604. doi:10.1016/j.dss.2006.03.004

Hastie, R., & Dawes, R. M. (2001). *Rational choice in an uncertain world*. Thousand Oaks, CA: Sage.

Huber, G. P. (1980). *Managerial decision-making*. New York, NY: Scott, Foresman, and Company.

Huber, G. P., & McDaniel, R. (1986). The decision-making paradigm of organizational design. *Management Science*, *32*(5), 572–589. doi:10.1287/mnsc.32.5.572

Keeney, R. (1988). Value-driven expert systems for decision support. *Decision Support Systems*, *4*(4), 405–412. doi:10.1016/0167-9236(88)90003-6

Keeney, R., & Raiffa, H. (1993). *Decision with multiple objectives: Preferences and value trade-offs* (2nd ed.). Cambridge, UK: Cambridge University Press. doi:10.1109/TSMC.1979.4310245

Kelly, G. A. (1955). *The psychology of personal constructs*. New York, NY: Norton.

Kepner, C. H., & Tregoe, B. B. (1981). *The new rational manager*. Princeton, NJ: Princeton Research Press.

Klein, D., & Shortliffe, E. (1994). A framework for explaining decision-theoretic advice. *Artificial Intelligence*, *67*(2), 201–243. doi:10.1016/0004-3702(94)90053-1

Klein, M. R., & Methlie, L. (1990). *Expert systems: A DSS approach*. Reading, MA: Addison Wesley.

Knight, F. H. (1921). *Risk, uncertainty and profit*. New York, NY: Houghton Mifflin.

Leontiev, A. N. (1978). *Activity, consciousness, personality*. Englewood Cliffs, NJ: Prentice Hall.

Levy, M., Pliskin, N., & Ravid, G. (2010). Studying decision processes via a knowledge management lens: The Columbia space shuttle case. *Decision Support Systems*, *48*(4), 559–567. doi:10.1016/j.dss.2009.11.006

Linstone, H. A. (1999). *Decision-making for technology executives: Using multiple perspectives to improve performance*. Boston, MA: Artech House. doi:10.1109/TEM.2000.865908

Manos, B., Ciani, A., Bournaris, T., Vassiliadou, I., & Papathanasiou, J. (2004). A taxonomy survey of decision support systems in agriculture. *Agricultural Economics Research*, *5*(2), 80–94.

Mora, M., Cervantes, F., Forgionne, G., & Gelman, O. (2008). On frameworks and architectures of intelligent decision-making support systems. In Adam, F., & Humphreys, P. (Eds.), *Encyclopedia of Decision-Making and Decision Support Technologies* (pp. 680–690). Hershey, PA: IGI Global. doi:10.4018/978-1-59904-843-7.ch076

Morgan, G. (2006). *Images of organization*. Thousand Oaks, CA: Sage Publications, Inc.

Paradice, D., & Davice, R. A. (2008). DSS and multiple perspectives of complex problems. In Adam, F., & Humphreys, P. (Eds.), *Encyclopedia of Decision-Making and Decision Support Technologies* (pp. 286–295). Hershey, PA: IGI Global. doi:10.4018/978-1-59904-843-7.ch033

Pavlov, Y. (2011). Preferences based stochastic value and utility function evaluation. In *Proceeding of Conferences InSITE 2011,* (pp. 403-411). Novi Sad, Serbia: InSITE.

Perry, M. (2003). Distributed cognition. In Carroll, J. M. (Ed.), *HCI Models, Theories, and Frameworks: Toward an Interdisciplinary Science* (pp. 193–223). San Francisco, CA: Morgan Kaufmann. doi:10.1016/B978-155860808-5/50008-3

Philips-Wren, G. E. (2008). Inteligent agents in decision support systems. In Adam, F., & Humphreys, P. (Eds.), *Encyclopedia of Decision-Making and Decision Support Technologies* (pp. 505–512). Hershey, PA: IGI Global. doi:10.4018/978-1-59904-843-7.ch058

Pinson, S. (1987). A multi-attribute approach to knowledge representation for loan granting. In *Proceedings of the 9th Joint Conference on Artificial Intelligence (IJCAI 1987)*, (pp. 588-591). IJCAI.

Power, D. J. (2002). *Decision support systems: Concepts and resources for managers*. Westport, CT: Quorum Books.

Power, D. J. (2008). Decision support system concept. In Adam, F., & Humphreys, P. (Eds.), *Encyclopedia of Decision-Making and Decision Support Technologies* (pp. 232–235). Hershey, PA: IGI Global. doi:10.4018/978-1-59904-843-7.ch027

Raiffa, H. (1968). *Decision analysis: Introductory lectures on choices under uncertainty*. Reading, MA: Addison-Wesley. doi:10.2307/2987280

Rogers, P., & Blenko, M. (2006). Who has the D? How clear decision roles enhance organizational performance. *Harvard Business Review*, *84*(1), 53–61.

Saaty, T. L., & Vargas, L. G. (1984). The legitimacy of rank reversal. *Omega*, *12*(5), 513–516. doi:10.1016/0305-0483(84)90052-5

Sage, A. (1981). Behavioural and organizational considerations in the design of information systems and processes for planning and decision support. *IEEE Transactions on Systems, Man, and Cybernetics*, *11*(9), 640–678. doi:10.1109/TSMC.1981.4308761

Sauter, V. L. (1999). Intuitive decision-making. *Communications of the ACM*, *42*(6), 109–115. doi:10.1145/303849.303869

Simon, H. (1986). *Research briefings 1986: Report of the research briefing panel on decision-making and problem solving*. Washington, DC: National Academy of Sciences.

Simon, H. A. (1997). *Administrative behaviour: A study of decision-making process in administrative organizations* (4th ed.). New York, NY: Free Press.

Smith, G. (1988). Towards a heuristic theory of problem structuring. *Management Science*, *35*(12), 1489–1506. doi:10.1287/mnsc.34.12.1489

Tversky, A., & Kahneman, D. (1992). Advances in prospect theory: Cumulative representation of uncertainty. *Journal of Risk and Uncertainty*, *5*(4), 297–323. doi:10.1007/BF00122574

Vapnik, V. (2006). *Estimation of dependences based on empirical data, information science and statistics* (2nd ed.). Berlin, Germany: Springer.

Vári, A., & Vecsenyi, J. (1984). Selecting decision support methods in organizations. *Journal of Applied Systems Analysis*, *11*, 23–36.

Watt, D. (2004). Value based decision models of management for complex systems. In *Proceedings-2004 IEEE International, Engineering Management Conference*, (Vol. 3, pp. 1278 – 1283). IEEE Press.

Westcott, M. (1968). *Toward a contemporary psychology of intuition: A historical and empirical inquiry*. New York, NY: Holt, Rinehart & Winston, Inc.

Zarate, P. (2008). Cooperative decision support systems. In Adam, F., & Humphreys, P. (Eds.), *Encyclopedia of Decision-Making and Decision Support Technologies* (pp. 109–115). Hershey, PA: IGI Global. doi:10.4018/978-1-59904-843-7.ch013

Chapter 2
Mathematical Preliminaries

ABSTRACT

The elaboration and the utilization of models of human behavior and the incorporation of human preferences in complex systems aims to develop decision-making with a merger of empirical knowledge (subjective preferences) with the mathematical exactness. People's preferences contain a characteristic of uncertainty due to the cardinal type of the empirical expert information. The appearance of this uncertainty has subjective and probability nature. The necessity of a merger of empirical knowledge with mathematical exactness causes difficulties due to this uncertainty and the cardinal type of expression of the human preferences. Decision-making under uncertainty is addressed in mathematics by probability theory and expected utility theory. These two together are known as decision theory.

The authors suppose that these uncertainty or errors have a random nature and may be represented as random variables with mathematical expectation zero and bounded variance, different in the different practical problems. They define two sets $\mathbf{A}_{u*}$ *and* $\mathbf{B}_{u*}$ *over the set of alternative* $\mathbf{X}$*, of the positive and negative preferences. In the conditions of exact preferences, it is fulfilled* $(\mathbf{A}_{u*} \cap \mathbf{B}_{u*}) = \varnothing$*), which is in accordance with the normative theory. However, in practice, due to errors, the uncertainty in the expert references, and the threshold of indistinguishability, the two sets intersect* $(\mathbf{A}_{u*} \cap \mathbf{B}_{u*}) \neq \varnothing$*). Then arises the problem, can the noise (errors and the uncertainty in the preferences) be eliminated as this is done in the stochastic programming?*

The correct assessment of the level of informativity and usability of these types of knowledge requires careful analysis of the terms measurement, formalization, uncertainty, probability, and admissible mathematical operations, under the respective scale, which does not distort the initial empirical information. This chapter is of an introductory nature in one of the vastest fields of mathematics, namely "probability

DOI: 10.4018/978-1-4666-2967-7.ch002

theory and mathematical statistics" and their application in decision-making. After the discussion of fundamental notions and theorems in Probability theory, the authors reveal some of the fundamental techniques establishing the convergence of the recurrent stochastic procedures in the Stochastic programming, which will permit analytical description of the value and utility functions. The analytical description of the expert's preferences as value or utility function will allow mathematically the inclusion of the decision maker in the description of the complex system "Technologist-process".

1. SOME PROBABILISTIC CONCEPTS

This chapter is of introductory nature in one of the vastest fields of mathematics, namely "probability theory and mathematical statistics" and their application in decision-making. Since this book is predominantly devoted to methods and adaptive approaches for decision-making under risk and uncertainty conditions, we will begin by introducing the concepts uncertainty and probability with one revelation of the famous Italian scientists de Finetti. De Finetti's treatise on the theory of probability begins with the provocative statement "probability does not exist", meaning that probability does not exist in an objective sense. Rather, probability exists only subjectively within the minds of individuals. This opinion of the Italian scientists leads us to philosophical and conceptual thoughts regarding the place of the mathematics in the decision-making process, regarding our capabilities for perceiving the surrounding reality and our ability to reflect it as knowledge and skills for altering our living environment. This revelation leads us far in new horizons before the human thinking and the creative capacity.

There are different approaches with respect to the uncertainty in the human preferences in the decision-making process. The process itself is classified in different ways, individual or group decisions making, decision-making under determinacy or risk, descriptive or normative approach for decision-making, etc.

When the decision-making process is described mathematically, it is interpreted as a mathematical problem within the framework of a particular mathematical theory. At this stage, a conceptual correspondence between the real evaluated processes and the axiomatic foundation of the respective theory is sought. Here the leading role is played by the professional knowledge, intuition, as well as ability to construct adequate theory or models, reflecting the investigated phenomenon. At the next stage, depending on the available information or the ability to encompass the different aspects of the described or investigated processes we introduce the classification "in conditions of determinacy or risk." We say that we make decisions in condition of deterministic situation if each action leads to unambiguous outcome. Here as a fundamental mathematical apparatus may be used linear and non-linear programming in their different aspects, as part of the vast mathematical discipline called "Operations research" (Bazara, 1979; Zaichenko, 1988). The decision-making in deterministic situation, in mathematical sense, consists of finding an extremum of certain index in an adequate mathematical model of the real process. The right choice of this index in accordance with the available information and measurement scales is crucial. If this stage is missed and the types of measurement scales are not taken into account during the modeling and the decision-making, then the whole process of mathematical modeling and decision may be faulty (Pfanzagl, 1971).

In lesser nominal scales, for instance, scales of ordinal order combinatorial models and algorithms are used. Often time the only optimization algorithms here are of general nature, of the "branch and bound" type. At this stage of modeling and decisions there are difficulties not only of computational kind, but also of conceptual nature. For example, here is the place of the famous Arrow's theorem, also known as Arrow's Paradox (Litvak, 1982; Ekeland, 1979).

In greater scales, as the interval scale and ratio scale, until recently there were no strictly proven analytical methods for practical application of the famous theorems for existence in the theory of measurement and the utility theory. Here, to a greater extent are known algorithms described in the books of the famous scientists Keeny and Raiffa, which contain elements of empiricism and are more likely associatively transferred from the representation theorems in utility theory (Keeney, 1993; Raiffa, 1968). With respect to their application aspect, there are often debates and objections reflected in the well known "Allais Paradox" and in the research of many scientists like Khanneman, Tversky, Machina, etc (Kahneman, 1979; Machina, 1982; Mengov, 2010). Obviously a mathematical theory that will establish strictly numerical methods for practical application and the use of the theoretical advancements in the field of mathematical decision-making theory, represented most vividly in the theory of measurement and utility theory, is needed.

When making decisions under risk and uncertainty, the situation is quite different. For each action generated by the decision, there is a set of potential outcomes (Fishburn, 1970). In mathematical aspect, this is a discrete or continuous probability distribution over a set of alternatives. This uncertainty is evaluated by objective probability distributions, obtained through real empirical data and measurements or through subjective estimates of the decision maker for the uncertainty in the final outcome. Usually here is assumed a description by functions of type $f(x,\theta)$, where $x \in X$, the set of possible finite outcomes or results. The second variable θ reflects the uncertainty in the final result. This variable θ is an element of the probability space $(\boldsymbol{\Theta}, \Im, \boldsymbol{P})$, where $\boldsymbol{\Theta}$ is the set of events $\theta \in \boldsymbol{\Theta}$, $\Im$ is σ–algebra, and $\boldsymbol{P}$ is probability measure (Parthasarathy, 1978; Shiryaev, 1980, 1989). The space $\boldsymbol{\Theta}$ will be called space of nature states (Fishburn, 1970).

Let us go into more details for the needed notions from the Probability Theory (Gikhman, 1988; Shiryaev, 1989). In the probability theory, the possible outcomes of statistical experiment are called elementary events θ, which belong to the set $\boldsymbol{\Theta}$ of the elementary events $\theta \in \boldsymbol{\Theta}$. Let $\boldsymbol{A} \subseteq \Theta$, where $\boldsymbol{A}$ is a subset of the space of elementary events. If during the conduction of an experiment an elementary event $\theta \in \boldsymbol{A}$ is observed, we say that the event $\boldsymbol{A}$ has been realized. If $\theta \notin \boldsymbol{A}$, we say that the event $\boldsymbol{A}$ has not been realized. From practical point of view not every set is of interest. But for example belongingness to a measured value to the closed interval $[a, b]$ is sensible. There exists a set $\Im$ of subsets of $\boldsymbol{\Theta}$, whose elements are sets, having practical meaning. Under events we will understand the elements $\boldsymbol{A}$ of the set $\Im$, $\boldsymbol{A} \in \Im$. A natural condition for the elements of $\Im$ is that the union of two sets $\boldsymbol{A}$ and $\boldsymbol{B}$ to belong to $\Im$, $(\boldsymbol{A} \cup \boldsymbol{B}) \in \Im$. This is interpreted as the event, realization of either of the two sets $\boldsymbol{A}$ and $\boldsymbol{B}$. In the same way the intersection of two sets $\boldsymbol{A}$ and $\boldsymbol{B}$, which is the simultaneous realization of the two events belongs to $\Im$, $\boldsymbol{A} \cap \boldsymbol{B} \in \Im$. The occurrence of event $\boldsymbol{A}$ and the non-occurrence of the event $\boldsymbol{B}$ is interpreted as the complement of the event $\boldsymbol{B}$ in the event $\boldsymbol{A}$, $(\boldsymbol{A} \setminus \boldsymbol{B}) \in \Im$. We introduce the concept Boolean algebra (Parthasarathy, 1978; Shiryaev, 1989). This is a set $\Im$ of subsets of $\boldsymbol{\Theta}$, for which it is fulfilled:

- $\boldsymbol{A}, \boldsymbol{B} \in \Im \Rightarrow (\boldsymbol{A} \cup \boldsymbol{B}) \in \Im$ and $(\boldsymbol{A} \cap \boldsymbol{B}) \in \Im$;
- $\boldsymbol{A}, \boldsymbol{B} \in \Im \Rightarrow (\boldsymbol{A} \setminus \boldsymbol{B}) \in \Im$;
- The empty set $\varnothing$ and the entire space $\boldsymbol{\Theta}$ of elementary events belong to $\Im$.

Now we will introduce measure $\boldsymbol{P}$ over the set $\Im$, through which we estimate the possibility for realization of every element of $\Im$. We consider the family $\{\boldsymbol{A}_i, i \in I\}$ subsets of the set $\boldsymbol{\Theta}$, where I is a set of indices. The elements of this family are pair wise non-intersecting, if $(\boldsymbol{A}_i \cap \boldsymbol{A}_j) = \varnothing$, i≠j, i,j$\in I$.

1.1. Definition

Let $\Im$ is a Boolean algebra of subsets of the set $\boldsymbol{\Theta}$. The mapping $\boldsymbol{P}$: $\boldsymbol{\Theta} \rightarrow [0, \infty]$ is called finitely additive, if it is fulfilled:

$\boldsymbol{P}(\boldsymbol{A} \cup \boldsymbol{B}) = \boldsymbol{P}(\boldsymbol{A}) + \boldsymbol{P}(\boldsymbol{B})$ и $\boldsymbol{A}, \boldsymbol{B} \in \Im$, $(\boldsymbol{A} \cap \boldsymbol{B}) = \varnothing$.

This mapping is called countably additive, if for the sequence $\{\boldsymbol{A}_i\}$ of pair wise non-intersecting sets, belonging to $\Im$ is valid the equality:

$$\boldsymbol{P}(\bigcup_{i=1}^{\infty} \boldsymbol{A}_i) = \sum_{i=1}^{\infty} \boldsymbol{P}(\boldsymbol{A}_i), \quad \bigcup_{i=1}^{\infty} \boldsymbol{A}_i \in \Im.$$

The mapping $\boldsymbol{P}$: $\boldsymbol{\Theta} \rightarrow [0, \infty]$ is called probability measure over the set $\Im$, if $\boldsymbol{P}(\boldsymbol{\Theta}) = 1$.

For the so defined finite probability measure obviously the following is fulfilled:

- If $\boldsymbol{A} \subseteq \boldsymbol{B}$, then $P(\boldsymbol{A}) \leq P(\boldsymbol{B})$;
- If $\boldsymbol{A}_1, \boldsymbol{A}_2, \ldots, \boldsymbol{A}_K \in \Im$, then
  $\boldsymbol{P}(\bigcup_{i=1}^{k} \boldsymbol{A}_i) \leq \sum_{i=1}^{k} \boldsymbol{P}(\boldsymbol{A}_i)$.

The above property is called finite semiadditivity. If the sum is infinite, it is called countable semiadditivity. In the case of equality for non-intersecting sets, it is called countable additivity.

1.2. Definition

Under Boolean probability space we understand $(\boldsymbol{\Theta}, \Im)$, where $\boldsymbol{\Theta}$ is a set of elementary events, and $\Im$ is Boolean algebra over $\boldsymbol{\Theta}$. If in the Boolean algebra $\Im$ the union and intersection of countable set of elements of $\Im$ belong to $\Im$, then we call it σ–algebra. Boolean probability space we call $(\boldsymbol{\Theta}, \Im, \boldsymbol{P})$, where $\boldsymbol{P}$ is finitely additive probability measure over $\Im$. If $\Im$ is σ–algebra, then $(\boldsymbol{\Theta}, \Im)$ is called measurable space. In this case $(\boldsymbol{\Theta}, \Im, \boldsymbol{P})$ is called probability space and $\boldsymbol{P}$ is countably-additive measure.

The function $\varphi(.)$: $\boldsymbol{\Theta} \rightarrow]-\infty, \infty[$ is called real random variable ($\Im$-measurable), if it taker values in the space of the real numbers and for it, it is fulfilled $\{w/\varphi(w) \leq z\} \in \Im$ for every real number z. If we consider the n-dimensional Euclidean space $\boldsymbol{R}^n$, then $\varphi(w)$ is n-dimensional random vector. Every random vector function $\varphi(w)$ defines over $\boldsymbol{R}^n$ a probability space. It is seen that the class $\mathfrak{R}$ of subsets of $\boldsymbol{R}^n$, for which $\varphi^{-1}(\boldsymbol{B}) \in \Im$, $\boldsymbol{B} \subseteq \boldsymbol{R}^n$ is σ–algebra, if $\Im$ is σ–algebra. This σ–algebra $\mathfrak{R}$ is called generated by the random variable $\varphi(.)$ σ–algebra. The following formula $\boldsymbol{\mu}(\boldsymbol{B}) = \boldsymbol{P}(\varphi^{-1}(\boldsymbol{B}))$ defines probability measure over $(\boldsymbol{R}^n, \mathfrak{R})$ and thus is a probability space $(\boldsymbol{R}^n, \mathfrak{R}, \boldsymbol{\mu})$.

Let Ω is an arbitrary class of subsets of $\boldsymbol{\Theta}$. The smallest σ–algebra, containing this class of subsets is called σ–algebra generated by the class Ω. For every family $\{\varphi_i(.), i \in I\}$ of n-dimensional random variables, by $\aleph(\varphi_i(.), i \in I)$ we denote the σ–algebra generated by the class of subsets $\{\varphi_i^{-1}(\boldsymbol{B}) \in \Im, \boldsymbol{B} \subseteq \boldsymbol{R}^n, i \in I\}$.

We call mathematical expectation of the random variable $\varphi(.)$ by definition:

$$\mathrm{M}(\varphi) = \int \varphi(w) \boldsymbol{P}(\mathrm{d}w) = \int \varphi \mathrm{d}\boldsymbol{P}.$$

In the above formula, the integral is understood as Lebegues' integral, and if the random variable is partially continuous, it may be perceived as Riemman's integral also. If we use the distribution function $\mathrm{T}(z) = \boldsymbol{P}(\{w/\varphi(w) \leq z\})$ of the random variable $\varphi(w)$, then the mathematical expectation can be written as positive Radon measure or Stieltjes integral over the real space:

$$\mathrm{M}(\varphi) = \int z \mathrm{dT}(z).$$

If $\varphi(w)$ is a vector function $\varphi(w) = (\varphi_1(w), \varphi_2(w), \ldots, \varphi_n(w))$ then the distribution function has the form $\mathrm{T}(z) = \mathrm{T}(z_1, z_2, \ldots, z_n) = \boldsymbol{P}(\{w/\varphi_1(w) \leq z_1, \varphi_2(w) \leq z_2, \ldots, \varphi_n(w) \leq z_n\})$ and is called joint distribution function.

We will give some definitions for the convergence of sequences of random variables (Gikhman, 1988; Shiryaev, 1989).

1.3. Definition

The sequence of random variables $\varphi_i(w)$, i=1, 2, 3,... converges almost everywhere (e.w.) or with probability 1 to the random variable $\varphi(w)$, if the following condition is fulfilled $P(\{w/ \lim_i (\varphi_i(w))=\varphi(w))\})=1$.

Another definition for the almost everywhere convergence is given by the following formula:

$$P(\lim \varphi_i = \varphi) = P(\bigcap_{k=1}^{\infty} \bigcup_{N=1}^{\infty} \bigcap_{i=N}^{\infty} \{|\varphi_i - \varphi| \leq \frac{1}{k}\}) = 1.$$

The following statement of almost everywhere convergence is also valid:

$$\lim_{N\to\infty} P(\bigcap_{i=N}^{\infty} \{|\varphi_i - \varphi| \leq \frac{1}{k}\}) = 1, \forall k, k \in N.$$

The meaning of the term almost everywhere convergence is that the sequence will not tend to the limit random variable only over set with measure 0.

1.4. Definition

The sequence of random variables $\varphi_i(w)$, i=1, 2, 3,... converges in probability to the random variable $\varphi(w)$ if for every ε, $(\varepsilon>0)$ the following condition is fulfilled $\mathbf{lim}(P(\{w/\|\varphi_i(w)-\varphi(w)\|>\varepsilon\}))=0$.

If the sequences of random variables $\varphi_{i1}(w)$, $\varphi_{i2}(w)$,..., $\varphi_{in}(w)$, i=1, 2, 3,.... tend to the random variables $\varphi_1(w)$, $\varphi_2(w)$,..., $\varphi_n(w)$ in probability and the function $\Phi(x_1, x_2,..., x_n)$ is continuous, then $\Phi(\varphi_{i1}(w), \varphi_{i2}(w),..., \varphi_{in}(w))$ converges in probability to the random variable $\Phi(\varphi_1(w), \varphi_2(w),..., \varphi_n(w))$.

1.5. Definition

The sequence of random variables $\varphi_i(w)$, i=1, 2, 3,... converges in mean square sense (m.s.) to the random variable $\varphi(w)$, if the following condition is fulfilled $\mathbf{lim}(M(\|\varphi_i(w)-\varphi(w)\|^2))=0$.

We have the following relations between the three types of convergence:

(e.w. Convergence) $\Rightarrow$ (Convergence in probability) $\Leftarrow$ (Convergence in m.s).

One of the fundamental concepts in Probability theory is the concept of conditional mathematical expectation. Let us consider once more the probability space $(\Theta, \Im, P)$ and two events A and B belonging to $\Im$. Then the conditional probability for occurrence of event B under the condition that the event A is realized, is given by the well known formula

$$P(B / A) = \frac{P(A \cap B)}{P(A)}$$

under the condition that $P(A)\neq 0$. Now we will consider the more general case of conditional mathematical expectation with respect to a random variable. The following theorem is valid (Elliott, 1982):

1.6. Theorem

Let $(\Theta, \Im, P)$ be a probability space and $\varphi(.)$ is a measurable function defined over the measurable space $(\Theta,\Im)$ with values in the measurable space $(E, \mathfrak{R})$.

Let us denote by Q the probability measure induced over $(E, \mathfrak{R})$ by the function $\varphi(.)$, i.e. $Q(A)=P(\{\varphi_i^{-1}(A), A\in\mathfrak{R}\})$. Let Y is P-integrable random variable defined over $(\Theta,\Im)$. Then there exist integrable random variable Z over $(E, \mathfrak{R})$, for which for every $A, A\in\mathfrak{R}$ the following is fulfilled:

$$\int_A Z(x)dQ(x) = \int_{\varphi^{-1}(A)} Y(\omega)dP(\omega).$$

If there exists another random variable Z_1 over $(E, \mathfrak{R})$, for which the above formula is valid, then it is fulfilled that almost everywhere $Z= Z_1$ (a.e.).

Proof

The proof is relatively simple and short and thus we will state it, following a source form the literature (Elliott, 1982). If there exists another random variable Z_1, for which it is fulfilled $\int_A Z(x)dQ(x) = \int_{\varphi^{-1}(A)} Y(\omega)dP(\omega)$ for every A, $A\in\mathfrak{R}$, then it is true $\int_A (Z(x) - Z_1(x))dQ(x) = 0$ for every A, $A\in\mathfrak{R}$. From the definitions for σ–algebra and random variable it follows the uniqueness, namely $Z= Z_1$ (a.e.) in $(E, \mathfrak{R})$, with respect to the measure Q.

Now we will prove the existence of such a function. Let us assume that Y e is square-integrable function, i.e. $Y\in L_2(\Theta,\Im, P)$. Let us suppose that S is also a square-integrable function, i.e $S\in L_2(E, \mathfrak{R}, Q)$. Then to every function S, $S\in L_2(E, \mathfrak{R}, Q)$ we can juxtapose the number:

$$\beta(S) = \int_{\Theta} S \circ \varphi(\omega)Y(\omega)dP(\omega).$$

The constructed function $\beta(.)$ is bounded linear functional over $L_2(E, \mathfrak{R}, Q)$. The space $L_2(E, \mathfrak{R}, Q)$ is self-conjugated. Therefore, from Riesz Theorem follows the existence of element Z from this space, namely a square-integrable function Z, $Z\in L_2(E, \mathfrak{R}, Q)$, such that for it is fulfilled:

$$\beta(S) = \int_{\Theta} S \circ \phi(\omega)Y(\omega)dP(\omega) = \int_E Z(x)S(x)dQ(x),$$

for every function S, $S\in L^2(E, \mathfrak{R}, Q)$.

The random variable Z is $\mathfrak{R}$ measurable. If as S we take the characteristic function of any of the sets A, $A\in\mathfrak{R}$, then the equality in the condition of the theorem is fulfilled. From condition of the theorem follows that if Y is a positive function, then its corresponding function Z is also a positive function.

In the general case Y is only integrable, $Y\in L(\Theta,\Im, P)$. We can represent Y through positive functions Y^+ and Y^-, for which $Y=Y^+-Y^-$. For this reason, we suppose in the proof that Y is the integrable positive function. Let $Y_n=Y\wedge n$, i.e. is bounded from above by the number n. Obviously the following is fulfilled $Y_n\in L_2(\Theta,\Im, P)$. To these functions correspond function Z_n, $Z_n\in L_2(E, \mathfrak{R}, Q)$. The functions Z_n are positive and monotonously increasing. From Lebesgues theorem for monotonous convergence follows the equality:

$$\int_A Z(x)dQ(x) = \int_{\varphi^{-1}(A)} Y(\omega)dP(\omega).$$

Each passage to the limit of measurable functions leads to measurable function. In this way we found that in the general case the measurable function $Z= Z^+- Z^-$ with respect to the σ–algebra $\mathfrak{R}$, which satisfies the conditions of the theorem.

We will call the function Z conditional mathematical expectation of the random variable Y, with respect to random variable φ(.). We emphasize again that in the definition the conditional mathematical expectation is defined almost everywhere.

In the above proof, we can make one extension. Let's assume that the σ–algebra $\mathfrak{R}$ contains the σ–algebra $\mathfrak{R}_\varphi$ generated by φ(.), $\mathfrak{R}_\varphi\subseteq\mathfrak{R}$. What remains is the assumption that $(\Theta,\Im, P)$ is probability space and φ(.) is a measurable function defined over the measurable space $(\Theta,\Im)$ with values in the measurable space $(E, \mathfrak{R})$, i.e. $\varphi^{-1}(B)\in\Im$ for every B, $B\in\mathfrak{R}$. Such mappings in the theory of measure are called Borelian random variables (Parthasarathy, 1978). If we carefully go through the proof of the theorem we will notice that the conditional mathematical expectation Z is obtained in those cases as a projection of the square-integrable functions $L_2(E, \mathfrak{R}, Q)$ over $L_2(E, \mathfrak{R}_\varphi, Q)$, where $\mathfrak{R}_\varphi$ is the σ–algebra generated by φ(.). This conclusion may be seen from the following formula shown in the previous proof:

$$\beta(S) = \int_\Theta S \circ \varphi(\omega) Y(\omega) dP(\omega) = \int_E Z(x) S(x) dQ(x),$$

for every function S, $S \in L^2(E, \mathfrak{R}_\varphi, Q)$.

Let χ be characteristic function of the set A, $A \in \mathfrak{R}_\varphi$. Let Z be measurable function, random variable over $(E, \mathfrak{R})$, $Z \in L_2(E, \mathfrak{R}, Q)$. From the definition of projection over Hilbert space for every set A, $A \in \mathfrak{R}_\varphi$ ($A \in \mathfrak{R}_\varphi \Rightarrow A \in \mathfrak{R}$) follows the following:

$$\int_A Z(x) dQ(x) = (Z, \chi(A)) = (Z, P^S(\chi(A))) = (P^S(Z), \chi(A)) = \int_A M(Z / \mathfrak{R}_\phi) dQ(x),$$

where P^s is projection over the subspace $L_2(E, \mathfrak{R}_\varphi, Q)$ of the basic space $L_2(E, \mathfrak{R}, Q)$. The function $M(Z / \mathfrak{R}_\varphi)$ we will call conditional mathematical expectation of the random variable Z with respect to the σ–algebra $\mathfrak{R}_\varphi$.

1.7. Definition

Let there be given the σ–algebra $\mathfrak{R}$ and σ–subalgebra $\mathfrak{R}_\varphi$, $\mathfrak{R}_\varphi \subseteq \mathfrak{R}$. Let Z be measurable function, random variable over $(E, \mathfrak{R})$ and $Z \in L_2(E, \mathfrak{R}, Q)$. Then the projection of Z, $P^s(Z)$ over the Hilbert subspace $L_2(E, \mathfrak{R}_\varphi, Q)$ we will call conditional mathematical expectation of the random variable with respect to the σ–algebra $\mathfrak{R}_\varphi$. We will denote this conditional mathematical expectation as $M(Z / \mathfrak{R}_\varphi) \in L_2(E, \mathfrak{R}_\varphi, Q)$. The conditional mathematical expectation is unique with precision of set with measure zero, or in other words is defined as function almost everywhere.

From the above definitions and proof directly follows the famous formula for conditional mathematical expectation of random variable Z with respect to the probability event B:

$$M(Z / B) = \frac{1}{P(B)} \int_B Z(x) dQ(x) = \int_B Z(x) dQ_B(x),$$

where

$$Q_B(A) = \frac{Q(A \cap B)}{Q(B)}.$$

We will prove a widely used theorem, revealing the properties of the conditional mathematical expectation (Eliott, 1982).

1.8. Theorem

Let S be $(E, \mathfrak{R}_\varphi)$ integrable random variable and Z be $(E, \mathfrak{R})$ random variable, such that their product is integrable. Then it is fulfilled:

$$M(M(SZ / \mathfrak{R}_\phi)) = SM(Z / \mathfrak{R}_\phi) \ a.e.$$

Proof

Let us assume that S is simple function, i.e. it takes only countable number of values, which we denote by $\{a_i, i \in N\}$. Let $A_i = S^{-1}(a_i) \in \mathfrak{R}_\varphi$. Then for every set B, $B \in \mathfrak{R}_\varphi$ is fulfilled:

$$\int_B M(SZ / \mathfrak{R}_\phi) dQ = \int_B SZdQ = \sum_i \int_{B \cap A_i} a_i Z dQ = \sum_i a_i \int_{B \cap A_i} ZdQ = \sum_i a_i \int_{B \cap A_i} M(Z / \mathfrak{R}_\phi) dQ = \int_B SM(Z / \mathfrak{R}_\phi) dQ.$$

Every non-negative integrable variable is a limit of a sequence of simple functions. Using the Lebesgues theorem for monotonous convergence we obtain the proof of the theorem.

The main properties of the mathematical expectation may be seen in the literature (Elliott, 1982; Parthasarathy, 1978; Shiryaev, 1989).

One of the most important concepts when considering sequences of random variables is the concept of martingale, following the ideas of Doob (Eliott, 1982; Ermolev, 1976; Aizerman, 1970). Let $(E, \mathfrak{R}, Q)$ is a probability space and S_i, i=1,2,3.... is a sequence of random variables, for which $M(S_i)<\infty$.

1.9. Definition

The sequence S_i, i=0,1,2,3.... is called supermartingale if

$$M(S_{i+1} / S_1, S_2, .., S_i) \leq S_i,\ i = 0,1,2,3....$$

The sequence S_i, i=0,1,2,3.... is called submartingale if

$$M(S_{i+1} / S_1, S_2, .., S_i) \geq S_i,\ i = 0,1,2,3....$$

The sequence S_i, i=0,1,2,3.... is called martingale if

$$M(S_{i+1} / S_1, S_2, .., S_i) = S_i,\ i = 0,1,2,3....$$

We will extend a bit the above concept. Let $\mathfrak{R}_i$, i=0,1,2,3.... is a non-decreasing sequence of σ-sub-algebra of the σ-algebra $\mathfrak{R}$, $(\mathfrak{R}_i \subseteq \mathfrak{R}_{i+1})$, i=0,1,2,3..... The sequence S_i, i=0,1,2,3.... is called consistent or adapted with the family $\mathfrak{R}_i$, i=0,1,2,3...., if for every i=0,1,2,3.... S_i is $\mathfrak{R}_i$ measurable. Then the above definition may be extended as follows.

1.10. Definition

The sequence S_i, i=0,1,2,3.... is called supermartingale if

$$M(S_{i+1} / \mathfrak{R}_i) \leq S_i,\ i = 0,1,2,3....$$

The sequence S_i, i=0,1,2,3.... is called submartingale if

$$M(S_{i+1} / \mathfrak{R}_i) \geq S_i,\ i = 0,1,2,3....$$

The sequence S_i, i=0,1,2,3.... is called martingale if

$$M(S_{i+1} / \mathfrak{R}_i) = S_i,\ i = 0,1,2,3....$$

1.11. Lemma (Doob) (Ermolev, 1976)

Let $r(i)$, i=0,1,2,3...K are a finite number of numbers. Let the number of the transitions $h(K)$ in the sequence of numbers $r(i)$, i=0,1,2,3...K, lying to the left and to the right of the [a, b]. We can call this transition intersections of the interval [a,b] from below to above or from left to right. Then it is fulfilled:

$$(b-a)h(K) \leq (a - r(K))^{+} + \sum_{i=0}^{K-1} J(i)(r(i+1) - r(i)).$$

In the formula $J(i)$, i = 0,1,2,3...K takes as values 1 or 0 and these values are entirely determined by the numbers $r(i)$, i=0,1,2,3...K.

Proof

Let t_1 is the first moment, for which $r(i)$ is smaller than (a), and t_2 is the first following moment for which $r(i)>b$. The moments t_3 and t_4, t_5 and t_6 follow, etc. Let in the transitions $[t_{2h}, K]$ the sequence $r(i)$ no longer intersect the interval [a, b] from the left to the right. Then it is fulfilled:

$$(b-a)h(K) \leq \sum_{i=1}^{h} (r(2i) - r(2i-1)).$$

In another notation the formula becomes:

$$(b-a)h(K) \leq \sum_{i=0}^{2h} J(i)(r(i+1)-r(i)),$$

where

$J(i)$, i = 0,1,2,3…,K

thus it is determined that in the intervals $[t_{2i-1}, t_{2i}]$ there is value 1, and in the other cases 0. It is taken into account that if $r(t_{2h+1})$ exists, that it is less than (a) and then $J(i)$ is 1 in the interval $[t_{2h+1}, K]$. From the so determined values $J(i)$, i = 0,1,2,3…,K it follows:

$$(b-a)h(K) \leq (r(2h+1)-r(K))^{+} + \sum_{i=0}^{K-1} J(i)(r(i+1)-r(i)).$$

Since $r(t_{2h+1})$ is smaller than (a) if exists, then it is fulfilled:

$$(b-a)h(K) \leq (a-r(K))^{+} + \sum_{i=0}^{K-1} J(i)(r(i+1)-r(i)).$$

Now we will prove a theorem often used in the stochastic programming (Ermolev, 1976).

1.12. Theorem

Let the sequence of random values S_i, i=0,1,2,3…. is consistent or adapted with the family σ–algebras $\mathfrak{R}_i$, i=0,1,2,3…. and $(\inf(M(S^{-}_{i}))>-\infty)$, where $S^{-} = \min(S,0)$.

Let there also be fulfilled

$$M(S_{i+1} / \mathfrak{R}_i) \leq S_i + g_i,\ i = 0,1,2,3....$$

We suppose that g_i, i=0,1,2,3…. are positive, consistent or adapted with the family σ–algebras $\mathfrak{R}_i$, i=0,1,2,3…. and

$$\sum_{i=0}^{\infty} M(g_i) < \infty$$

Then with probability 1 exists a finite limit of the sequence S_i, i=0,1,2,3…. $\left(\left|\lim_{i\to\infty} S_i\right| < \infty\right)$.

Proof

In accordance with Doob's lemma we juxtapose the numbers $S_i(w)$ $(S(i))$, i=0,1,2,3…K, $w\in\Theta$. To this sequence of numbers we juxtapose $J(i,w)$, i = 0,1,2,3…K and $h(K, w)$. Then the inequality is valid:

$$(b-a)h(K,w) \leq (a-S(K))^{+} + \sum_{i=0}^{K-1} J(i,w)(S(i+1)-S(i)).$$

We take the mathematical expectation from both sides of the above inequality, which in accordance with the properties of the conditional mathematical expectation gives us the following formula:

$$(b-a)M(h(K,w)) \leq M(a-S(K))^{+} + \sum_{i=0}^{K-1} MM(J(i,w)(S(i+1)-S(i)) / \mathfrak{R}_i) = M(a-S(K))^{+} + \sum_{i=0}^{K-1} M(J(i,w)M(S(i+1) / \mathfrak{R}_i - S(i))).$$

By taking into account the properties of the sequence S_i, i=0,1,2,3…. in the condition of the theorem we obtain:

$$(b-a)M(h(K,w)) \leq M(a-S(K))^{+} + \sum_{i=0}^{K-1} M(g_i).$$

If we assume that the sequence $S_i(w)$, i=0,1,2,3… does not converge for almost every w, $w\in\Theta$ then there will exist numbers a and b, such that for them it will be fulfilled:

$$A_{ab} = \{w \,/\, \underline{\lim}_{i\to\infty} S_i(w) < a < b < \overline{\lim}_{i\to\infty} S_i(w)\},$$
and $P(A_{ab}) \neq 0$.

The sequence $h(i, w)$ is monotonously increasing and tends to ($\propto$) for $w \in A_{a,b}$.

Let $h(.)$ is the limit of the sequence $h(i,.)$. Then $P(A_{a,b})=0$, if it is fulfilled $P(h=\propto)=0$. The last will be true when $M(h)<\propto$. And so everything boils down to evaluation of $M(h)$.

We apply the lemma of Fatou (Gikhman, 1988; Shiryaev, 1989) to the formula

$$(b-a)M(h(K,w)) \leq M(a - S(K))^{+} + \sum_{i=0}^{K-1} M(g_i).$$

As a result we obtain:

$$M \lim_{i\to\infty} h(i) = M(h) = \lim_{i\to\infty} M(h(i))$$
$$\leq \frac{1}{(b-a)}(\sup_i(M(a - S(i))^{+}) + \sum_{i=0}^{\infty} M(g_i))$$
$$= \frac{1}{(b-a)}(-\inf_i(M(S(i) - a)^{-}) + \sum_{i=0}^{\infty} M(g_i)).$$

From the condition of the theorem we have (**inf**($M(S_i^-)$)>-$\propto$) and in addition $\sum_{i=0}^{\infty} M(g_i) < \infty$. Then from the above inequality it follows that $M(h)<\propto$ and $P(A_{a,b})=0$. The numbers a and b, can be taken as rational and hence countable. Then it is fulfilled:

$$\bigcup_{a,b} A_{ab} = \{w \,/\, \underline{\lim}_{i\to\infty} S_i(w) < \overline{\lim}_{i\to\infty} S_i(w)\}.$$

From the theory of measure, we know that countable union of sets with measure zero is a set with measure zero, which concludes the proof.

From this theorem follow these theorems as corollaries (Ermolev, 1976).

1.13. Theorem

If $\{S_i$, i=0,1,2,3...., $\Re_i$, i=0,1,2,3....$\}$ is supermartingale, then for the existence with probability 1 of final limit of the sequence S_i, i=0,1,2,3.... $\left(\left|\lim_{i\to\infty} S_i\right| < \infty\right)$ is sufficient (inf($M(S_i^-)$)>-$\propto$), where $S^- = \min(S,0)$.

1.14. Theorem

Let S_i, i=0,1,2,3.... be centered random variables $\left(M(S_{i+1} / S_1, S_2, .., S_i) = 0,\ i = 1,2,3....\right)$, for which $\sum_{i=0}^{\infty} M(S_i)^2 < \infty$. Then the series $\sum_{i=0}^{\infty} S_i < \infty$ converges almost everywhere.

Now we are prepared to look at some parts of the stochastic programming.

2. STOCHASTIC PROGRAMMING

Under risk and uncertainty, to every action corresponds a set of potential outcomes. In mathematical aspect, this is discrete or continuous probability distribution over the set of alternatives (Keeney, 1993; Fishburn, 1970). The description is done through functions of type $f(x,\theta)$, where $x \in X$, the set of outcomes, and the second variable θ reflects the uncertainty or risk in the final result. This variable θ is element of the probability space (Θ, $\Im$, P), where Θ is the set of events θ, $\theta \in \Theta$, $\Im$ is σ-algebra, and P is probability measure. The space Θ is the space of nature states (Fishburn, 1970). In the general case the space of nature states Θ can depend on the space of the variables X, but in most cases, this dependence is negligible (Ermolev, 1976). In these problems the function $f_o(x,\theta)$, which reflects the choice criterion and the functions $f(x,\theta)$, which reflect the decision constraints depend on θ, $\theta \in \Theta$. Thus in the stochastic programming problems there is far greater freedom and variety than in the problems of the non-linear programming. For instance if in the process of investigation the variable θ becomes known, then the problem of the stochastic programming turns into problem of the non-linear programming:

$$\max(f \circ (x,\theta)), f_i(x,\theta) \leq 0, i = 1 \div n, \ x \in X. \quad (2.2.1)$$

In the general case each measurement can reveal only part of the uncertainty and then we have iterativity of the kind: "measurement, solution, measurement, solution, measurement ....." Here in this case in a new classification is introduced depending on whether the information accumulation process begins with *solution* or *measurement*. Ermolev (1976) talks about operative or perspective problems, one-stage or multistage problems, etc. In some aspect this classifications are not strict. Such iterativity exists in the Bayesian approach described in the famous book of Professor Hovard Raiffa (1968). It must not be forgotten that for these problems there is also the main question about the measurement scales of the empirical data and for the admissible mathematical operations, corresponding to the measurement scales. For example, for the scales weaker than the ratio scale, the mathematical expectation is inadmissible as an operation over the measured data.

Without going into details in the mentioned classifications, we will lay out some fundamental problems of the stochastic programming. At this stage, the aim is to reveal the basic practices used in the proofs of the theorems in this area of mathematics. If the variable (x) is determined and is chosen before observing the nature state (θ), then the problem (2.2.1) must be considered as a problem of the stochastic programming. Often times (2.2.1) is understood in the following manner. Let $F_o(x)=M(f_o(x,\theta))$ has the sense of mathematical expectation. The same is true for $F_i(x)=M(f_i(x,\theta))$. Then the problem (2.2.1) may turn into the following problem in the area of the stochastic programming:

Minimize the mathematical expectation $F_o(x)=M(f_o(x,\theta))$, *under constraints*

$$F_i(x)=M(f_i(x,\theta)),\ M(f_i(x,\theta))\leq 0,\ i=1\div n,\ x\in \mathbf{X}.$$

In the general case $F_i(x)$, $i=1\div n$, $x\in \mathbf{X}$ are called regressions. In such problems, if we view the presence of the uncertainty θ, as noise or errors in the measurement or as expression of the preferences of the subject making the decision (decision maker), than the regressions $F_i(x)$, $i=0\div n$ are interpreted as the real values or the exact expert preference.

Other interpretations of the uncertainty and risk are also possible in the problems of stochastic programming. As an example the objective function may be defined in the following way $Q_o(x)=\mathbf{P}(f_o(x,\theta) \geq f)$, where $\mathbf{P}$ has the meaning of probability measure. With regard to the constraints the situation is the following: $Q_i(x)=\mathbf{P}(f_i(x,\theta)\leq 0)-p_i$, $Q_i(x)\geq 0$, $i=1\div n$, $x\in X$.

The above description does not cover all the varieties in the set up of the problems in the stochastic programming. For example, moments of higher order may be introduced to the constraints of the above problem, as well as additional constraints for terms from the area of the probability and statistics. It is possible that the iterativity and stages of the problems in stochastic programming not to be considered only as function in time, but as dependent on other features of the investigated process.

Still a major characteristic of these problems and main difference from the problems of the non-linear programming is that the values of the function objectives and constraints are not known exactly. The derivatives of these functions cannot be determined unambiguously and this raises new requirements for the mathematical description and solutions. The derivative itself, or the gradient in the vector case, is perceived as a generalized concept. This makes the stochastic programming problems unique and different from the problems of the non-linear programming (Ermolev, 1976; Aizerman, 1970: Gatev, 1978). If the function $F(x)$ is convex, then a generalized gradient is called any vector $F^*_x(x)$ for which the equality holds:

$(F(z)- F(x)) \geq (F^*_x(x), z-x)$.

In the general case, the generalized gradient is a set of infinitely many vectors and each of them to some degree reflects the concept of supporting plane in the non-linear programming. Then by analogy with the gradient method problems of the non-linear programming, we can consider the sequence of points x_s:

$$x_{s+1} = x_s - \rho_s \gamma_s F^*_x(x_s), \quad s = 0, 1, 2, 3, ... \qquad (2.2.2)$$

In the above formula x_0 is an arbitrary initial approximation, ρ_s is the length of the step, and $F^*_x(x_s)$ is one of the generalized gradients. Due to the indeterminacy of the generalized gradient in the problems of the stochastic programming, the procedure is not monotonously decreasing as is the case in non-linear programming. This requires that the step is chosen by new rules, different from the similar gradient procedures in the non-linear programming. For example, one of the most common requirements in the scientific literature devoted to the topic has the form (Aizerman, 1970; Gatev, 1978):

$\rho_s \geq 0$, $\rho_s \to 0$, $\sum \rho_s = \infty$, $\gamma_s > 0$ and $\| \gamma_s F^*_x(x_s) \| \leq$ cte.

If we optimize the convex function $F(x)$ for $x \in X$, the recurrent procedure may be improved by including $\pi(z)$ projection to a given set X^* or subset of X:

$x_{s+1} = \pi_{X^*} (x_s - \rho_s \gamma_s F^*_x(x_s))$, $s=0,1, 2, 3.....$,
$\pi_{X^*}(z) \in X$, and $\| y - \pi_{X^*}(z) \| \leq \| y - z \|$.

For instance, as $\pi_X(z)$ in the literature source (Ermolev, 1976) is proposed the solution of the following extremal problem:

$\| x- z \|^2 = \mathbf{min}, x \in X$.

In the stochastic programming the generalized gradient is replaced with a random variable. We consider the sequence of random points x_s, $s = 0, 1, 2, 3...$ Then by analogy with the gradient methods of the non-linear programming we consider the quasi-gradient recurrent stochastic procedure:

$x_{s+1} = \pi_X (x_s - \rho_s \gamma_s \xi_s)$, $s = 0, 1, 2, 3$...........

In this case x_0 is an arbitrary initial point, ρ_s is the magnitude of the step, and γ_s is a norming coefficient. The interesting part is related to the random variable (ξ_s). Its conditional mathematical expectation is related to the gradient or the generalized gradient of the function $F(x_1, x_2, ..., x_n)$.

We will lay out two fundamental convergence theorems described in Ermolev (1976), since in these theorems some important specificities and techniques for proof of convergence of recurrent stochastic procedures are revealed. We denote by X^* the set of points in which the continuous function $F(x)$ takes minimal value. It is supposed that there exists $x \in X^*$, such that $\| x \| \leq$ cte.

2.1. Theorem

If for every number L there exits such number K_L, that

$$\| F_x^*(x) \| \leq K_L$$

for

$$\| x \| \leq L;$$

$$\gamma_i > 0, \quad \gamma_i \| F_x^*(x_i) \| \leq cte;$$

$$\rho_i \geq 0, \ \rho_i \to 0, \ \sum_{i=0}^{\infty} \rho_i = \infty,$$

Then there exists subsequence $F(x_{i\kappa})$ of the sequence $F(x_i)$, such that

$$\lim F(x_{ik}) = F(x^*), \ k \to \infty, \ i.e.$$
$$\lim_{i\to\infty} \min_{k<i} F(x_k) = F(x^*).$$

Proof

From the condition of the theorem and the definition of the projection $\pi(.)$ follows the following inequality:

$$\left\|x^* - x_{i+1}\right\|^2 \leq \left\|x^* - x_i + \rho_i\gamma_i F_x^*(x_i)\ \right\|^2$$
$$= \left\|x^* - x_i\right\|^2 + 2\rho_i\gamma_i(F_x^*(x_i)\ , x^* - x_i)$$
$$+(\rho_i\gamma_i)^2\left\|F_x^*(x_i)\ \right\|^2$$
$$\leq \left\|x^* - x_i\right\|^2 + 2\rho_i\gamma_i \mid (F_x^*(x_i)\ , x^* - x_i) + (\rho_i)K \mid .$$

with K is denoted a constant. For every ε, (ε>0) and for every (i) the following cases are possible:

$$2(\mathbf{F}_x^*(\mathbf{x}_i)\ , \mathbf{x}^* - \mathbf{x}_i) + (\rho_i)\mathbf{K} \leq -\varepsilon,$$
$$2(\mathbf{F}_x^*(\mathbf{x}_i)\ , \mathbf{x}^* - \mathbf{x}_i) + (\rho_i)\mathbf{K} > -\varepsilon.$$

We will prove that there does not exist N, such that for every (i>N) the first inequality is fulfilled. Indeed, if this were true then it is fulfilled

$$\left\|x^* - x_{i+1}\right\|^2 \leq \left\|x^* - x_i\right\|^2.$$

From the conditions of the theorem we have $\left\|F_x^*\ (x_i)\right\| \leq cte$ and we can take $\gamma_i > \gamma > 0$. These yields:

$$\left\|\mathbf{x}^* - \mathbf{x}_{i+1}\right\|^2 \leq \left\|\mathbf{x}^* - \mathbf{x}_i\right\|^2 - \varepsilon\rho_i\gamma$$
$$\leq \left\|\mathbf{x}^* - \mathbf{x}_N\right\|^2 - \varepsilon\gamma\sum_{i=N}^{\infty}\rho_i$$

Passing under the limit gives that the norm $\left\|x^* - x_i\right\|^2$ diminishes infinitely which is impossible. Therefore, for every ε, (ε>0) exists, sufficiently large, (i_k), for which we have:

$$2(\mathbf{F}_x^*(\mathbf{x}_{ik})\ , \mathbf{x}^* - \mathbf{x}_{ik}) + (\rho_{ik})\mathbf{K} > -\varepsilon.$$

Since $\rho_i \to 0$, then we can accept that for every ε>0 exists such K_ε, that for k>K_ε the previous inequality can be described as follows:

$$(\mathbf{F}_x^*(\mathbf{x}_{ik})\ , \mathbf{x}^* - \mathbf{x}_{ik}) > -\varepsilon.$$

Therefore the following is fulfilled:

$$\mathbf{F}(\mathbf{x}^*) - \mathbf{F}(\mathbf{x}_{ik}) \geq (\mathbf{F}_x^*(\mathbf{x}_{ik})\ , \mathbf{x}^* - \mathbf{x}_{ik}) > -\varepsilon.$$

Since $F(x^*)$ takes minimal value, the required convergence follows:

$$\lim_{k\to\infty} \mathbf{F}(\mathbf{x}_{ik}) = \mathbf{F}(\mathbf{x}^*).$$

Now we will consider the case when the set X^* of extremal points is bounded.

2.2. Theorem

Let us suppose that in addition to the previous theorem, the set of extremal points is bounded. Then the sequence $F(x_i)$ tends to $F(x^*)$.

Proof

We evaluate $\left\|x^* - x_i\right\|$ for $(i_\kappa \leq i \leq i_{\kappa+1})$, κ=1,2,3,.... From the inequality $(F(z) - F(x)) \geq (F^*_x(x),z-x))$ for $(z=x^*)$ follows $(0 \geq (F^*_x(x), x^*-x))$. From this inequality and the inequality

$$\left\|\mathbf{x}^* - \mathbf{x}_{i+1}\right\|^2 \leq \left\|\mathbf{x}^* - \mathbf{x}_i + \rho_i\gamma_i\mathbf{F}_x^*(\mathbf{x}_i)\ \right\|^2$$
$$= \left\|\mathbf{x}^* - \mathbf{x}_i\right\|^2 + 2\rho_i\gamma_i(\mathbf{F}_x^*(\mathbf{x}_i), \mathbf{x}^* - \mathbf{x}_i$$
$$+(\rho_i\gamma_i)^2\left\|\mathbf{F}_x^*(\mathbf{x}_i)\right\|^2,$$

for i_κ follows:

$$\left\|\mathbf{x}^* - \mathbf{x}_{i_{k+1}}\right\|^2 \leq \left\|\mathbf{x}^* - \mathbf{x}_{i_k}\right\|^2 + C\rho_{i_k^2}.$$

In this formula **C** is a constant, since we can always assume $\gamma_i \leq$const. For $i \neq i_\kappa$, $\kappa=1, 2,...$ it is fulfilled:

$$\left\|x^* - x_{i+1}\right\|^2 \leq \left\|x^* - x_i\right\|^2.$$

For these conditions for i, ($i_\kappa \leq i \leq i_{\kappa+1}$) the inequality is valid:

$$\inf_{x^*}\left\|\mathbf{x}^* - \mathbf{x}_{i_k+1}\right\|^2 \leq \inf_{x^*}\left\|\mathbf{x}^* - \mathbf{x}_{i_k})\right\|^2 + \mathbf{C}\rho_{i_k^2}.$$

Since the set X^* is bounded, and $F(x)$ is continuous function and $\lim_{k\to\infty} F(x_{ik}) = F(x^*)$, then for some subsequence of x_{ik}, $k = 0, 1, ...$, we have:

$$\inf_{x^*}\left\|\mathbf{x}^* - \mathbf{x}_{i_k}\right\|^2 \to 0, \mathbf{k} \to \infty.$$

Then from the previous inequality it follows:

$$\inf_{x^*}\left\|\mathbf{x}^* - \mathbf{x}_i\right\|_2 \to 0, \mathbf{i} \to \infty, \lim_{i\to\infty} \mathbf{F}(\mathbf{x}_i) = F(\mathbf{x}^*).$$

This concludes the proof.

With this practices and initial theorems, we revealed some of the fundamental techniques establishing the convergence of the recurrent stochastic procedures in the stochastic programming.

3. PREFERENCE'S RELATIONS

In investigating the systems and developing models in the multifaceted human activity it is needed to pose and solve not only well structured and formally described problems, but also weakly structured problems described in natural language in which heuristic methods and subjective expert estimates are used. The main achievement of the system analysis is the development of methods for transition of these non-formal and weakly structured problems to more general system paradigms in which formalized procedures may be used. It is normal to use basic mathematical concepts and theories, which allow the broadest semantic meaning and permit evaluation of human knowledge and skills even at the lowest level of manifestation and measurement (preferences). The correct assessment of the level of informativity and usability of these types of knowledge requires careful analysis of the terms measurement, formalization, and admissible mathematical operations under the respective scale, which do not distort the initial empirical information. This process becomes especially complex when in the investigation we have to account for the existence of human subject in the system, making the final or the determining solution. In hard and weakly formed scientific fields like medicine, biology, sociology, etc., often times the only way of expression of the empirical subjective knowledge is their manifestation as qualitative preference when comparing two or more possibilities. Moreover, the explicit expression of this knowledge through preferences is related to admission of errors from the subjects, due to the influence of external or internal psychological factors and the cardinal type of expression. Such are for example the inability to distinguish of close situations under qualitative assessment of certain phenomenon, inconsistency in the preferences, etc. Or put briefly, in these types of problems, we have to include subjects as experts, who are assumed to be able to simplify or even solve the problem or part of it, on more intuitive level, and their participation is reflected in their preferences.

Let us summarize. The elaboration and the utilization of models of human behavior and the incorporation of human preferences in complex systems aims to develop decision-making with a merger of empirical knowledge (subjective preferences) with the mathematical exactness. People preferences contain characteristic of uncertainty due to the cardinal type of the empirical expert information. The appearance of this uncertainty has subjective and probability nature. The necessity of a merger of empirical knowledge with mathematical exactness causes difficulties. decision-making under uncertainty is addressed in mathematics by probability theory and expected utility theory. These two together are known as decision theory. In addition, the utility theory deals with expressed subjective preferences.

Thus, at the lowest level of formalization, when the problem is described verbally, the DM is tasked with choosing from several variants the one that is more preferred by him. Usually the alternatives are compared two by two and is sought the one that is more preferred in all comparison with the others. There are two fundamental comparative value concepts, namely "better" (strict preference) and "equal in value to" (indifference). The relations of preference and indifference between alternatives are usually denoted by the symbols ($\succ$) and ($\approx$). In accordance with a long-standing philosophical tradition, $A \succ B$ is taken to represent "B is worse than A," as well as "A is better than B."

There are several commonly accepted postulates for choice at this level. The first postulate is the capability to realize comparison, to express relation to all possible couples. This means that for every couple A and B it is true $A \succ B$ *or* $B \succ A$, or $A \approx B$ (equivalent or cannot take solution). The second postulate is transitivity of the preference, i.e. if $A \succ B$ and $B \succ C$ then it is true $A \succ C$. The same is true for the indifference relation, if $A \approx B$ and $B \approx C$ then it is true $A \approx C$. In addition, here start the difficulties. In human preferences, in complex multifactor problems, the transitivity is most often violated. Often time the preferences are closed in cycles $A \succ B$, $B \succ C$, $C \succ A$. In other words in practices there are preferences for which transitively, $A \succ C$ and $C \succ A$. The problem at the discrete level for finite number of alternatives leads to definitions of partial relation, for which it is possible to lack comparability between couples of alternatives. Either the cyclic connection is broken in a way that least alters the initially expressed preferences, etc. For example an average ratio in the sense of measure is sought, as the median of Kemeny-Snell or belongingness to a set with particular extremal properties as the Pareto set (Litvak, 1982). At this level, due to the small informativity of the discrete orders, there are unexpected or hard to accept solutions as the Arrow theorem (Ekeland, 1979; Litvak, 1982).

In the indifference relation, additional natural obstacle is the threshold of non-distinguishability between two events, inherent for every person in more complex situation. Even more complex and difficult is the problem of adequate description of the preferences in the stronger axiomatic, mathematical theories, as the theory of measurement or utility theory. The problem does not lie in the commonly accepted postulates as transitivity and acyclicity of their preferences or their entirety (completeness). In theoretical plan there are strict and logically sound mathematical theories, yielding theorems of existence or theorems for expression (representations) of preferences in stronger scales as the interval or the ratio scales or the absolute scale (Litvak, 1982; Fishburn, 1970; Keeney, 1993). The problem arises in practice with the application of the theoretical results. The Allais paradox for the comparison of lotteries in the gambling approach of von Neumann expected utility is widely known (Mengov, *2010*). Another example is when we compare consecutively a large number of non-distinguishable alternatives, but between the first and the last alternative the preferences from indifference changes to refusal or acceptance. For example if we compare consecutively 1 teaspoon of sugar with 1.1 teaspoons of sugar obviously $1 \approx 1.1$. If we continue with the comparisons $1.1 \approx 1.2$; $1.2 \approx 1.3$; $1.3 \approx 1.4$;..... we get to $1 \approx 123$.

Therefore, the associative transaction of theoretical results from the normative approach is difficult and not well founded. There is a need of theory and numerical methods that would combine the beautiful theoretical results of the theory of measurement and utility theory with the qualitative nature of expression of empirical knowledge through preferences and taking into account the inherent uncertainty in the expression of the empirical knowledge as preferences.

Now we will give short formal description of the preferences and then we will describe our concept for solving the problem, which is one of the aims of this book. We suppose that the formal preference relation $(x \succ y)$ and its related indifference relation $((x \approx y) \Leftrightarrow \neg((x \succ y) \vee (x \prec y)))$ fulfill the following postulates (Fishburn, 1970):

- The preference and indifference relation are transitive:$((x\approx y) \wedge (y\approx t)) \Rightarrow (x\approx t))$, $((x\succ y) \wedge (y\succ t)) \Rightarrow (x\succ t))$, $((x\approx y) \wedge (y\succ t)) \Rightarrow (x\succ t))$, $((x\succ y) \wedge (y\approx t)) \Rightarrow (x\succ t))$;
- The indifference relation is equivalence;
- For any two alternatives (x, y) one and only one of the following three possibilities is fulfilled $(x\succ y)$ or $(x\succ y)$ or $(x \approx y)$.

We assume that the subject in the process of expressing preferences makes mistakes or has uncertainties in the choice. We suppose that this uncertainty or errors have random nature and may be represented as random variable with mathematical expectation zero and bounded variance, different in the different practical problems. We assume that this uncertainty is expressed in each comparison independently of the other comparisons (Pavlov, 2005). We define two sets $\mathbf{A}_{u*}$ and $\mathbf{B}_{u*}$ over the set of alternative $\mathbf{X}$, of the positive preferences and negative preferences. Let are the sets:

$$\mathbf{A}_{u*}=\{(x,y)\in\mathbf{X}^2/\ (u^*(x))>u^*(y)\},$$

$$\mathbf{B}_{u*}=\{(x,\ y)\in\mathbf{X}^2/\ (u^*(x))<u^*(y)\}.$$

In the conditions of exact preferences it is fulfilled $(\mathbf{A}_{u*}\cap\mathbf{B}_{u*})=\varnothing)$, which is in accordance with the normative theory. However, in practice, due to errors, the uncertainty in the expert preferences and the threshold of indistinguishability the two sets intersect $(\mathbf{A}_{u*}\cap\mathbf{B}_{u*})\neq\varnothing)$. Then arises the problem, whether there exists function of two variables (x,y), which would divide in the best stochastic way the sets $\mathbf{A}_{u*}$ and $\mathbf{B}_{u*}$, $(\mathbf{A}_{u*}\cap\mathbf{B}_{u*})\neq\varnothing)$ (Aizerman, 1970; Pavlov, 1989, 2005; Herbrich, 1998). Can the noise (errors and the uncertainty in the preferences) be eliminated as this is done in the stochastic programming? Can this function be split in two symmetric parts, each having the properties of value function, preserving the preferences $((x,y)\in X^2, x\succ y) \Leftrightarrow (\mathbf{u}^*(x)>\mathbf{u}^*(y))$ in accordance with the normative theory (Pavlov, 2005). The solution of these problems is the aim of this book.

In this manner, we have posed the decision-making problem as a problem of constructing the value function based on stochastic recurrent procedures, which can later be used in optimization problems. The analytical description of the expert's preferences as value or utility function will allow mathematically the inclusion of the decision maker in the description of the complex system "technologist-process."

REFERENCES

Aizerman, M. A., Braverman, E., & Rozonoer, L. (1970). *Potential function method in the theory of machine learning*. Moscow, Russia: Nauka.

Bazara, M., & Shetty, C. (1979). *Nonlinear programming: Theory and algorithms*. New York, NY: Wiley.

Ekeland, I. (1979). *Elements d'economie mathematique*. Paris, France: Hermann.

Elliott, R. J. (1982). *Stochastic calculus and applications*. Berlin, Germany: Springer-Verlag.

Ermolev, Y. M. (1976). *Methods of stochastic programming*. Moscow, Russia: Nauka.

Fishburn, P. (1970). *Utility theory for decision-making*. New York, NY: Wiley.

Gatev, G. (1978). *Analysis and synthesis of automatic systems*. Sofia, Bulgaria: Tehnika.

Gikhman, A., Skorokhod, V., & Yadrenko, M. (1988). *Probability theory and mathematical statistics*. Kiev, Russia: Vyshcha Shkola.

Herbrich, R., Graepel, T., Bolmann, P., & Obermayer, K. (1998). Supervised learning of preferences relations. In *Proceedings of Fachgruppentreffen Maschinelles Lernen*, (pp. 43-47). Retrieved from http://research.microsoft.com/apps/pubs/default.aspx?id=65615

Kahneman, D., & Tversky, A. (1979). Prospect theory: An analysis of decision under risk. *Econometrica*, *47*, 263–291. doi:10.2307/1914185

Keeney, R., & Raiffa, H. (1993). *Decision with multiple objectives: Preferences and value trade-offs* (2nd ed.). Cambridge, UK: Cambridge University Press. doi:10.1109/TSMC.1979.4310245

Litvak, B. (1982). *Expert information, analyze and methods*. Moscow, Russia: Nauka.

Machina, M. (1982). Expected utility analysis without the independence axiom. *Econometrica*, *50*(2), 277–323. doi:10.2307/1912631

Mengov, G. (2010). *Decision-making under risk and uncertainty*. Sofia, Bulgaria: Publishing house JANET 45.

Parthasarathy, K. R. (1978). *Introduction to probability and measure*. Berlin, Germany: Springer-Verlag.

Pavlov, Y. (1989). Recurrent algorithm for value function construction. *Proceedings of Bulgarian Academy of Sciences*, *7*, 41–42.

Pavlov, Y. (2005). Subjective preferences, values and decisions: Stochastic approximation approach. *Proceedings of Bulgarian Academy of Sciences*, *58*(4), 367–372.

Pfanzagl, J. (1971). *Theory of measurement*. Wurzburg, Germany: Physical-Verlag.

Raiffa, H. (1968). *Decision analysis: Introductory lectures on choices under uncertainty*. Reading, MA: Addison-Wesley. doi:10.2307/2987280

Shiryaev, A. N. (1989). *Probability*. New York, NY: Springer.

Zaichenko, P. (1988). *Operations research*. Kiev, Russia: Vishta Shkola.

Chapter 3
Preferences-Based Performance Measurement Models

ABSTRACT

In the current chapter, we consider the term "measurement" from a general point of view. The mathematical description on such a fundamental level requires basic mathematical terms, sets, relations, and operations over them in their gradual elaboration to more complex and specific terms.

The authors start with a mathematical description of the notion of scale. From the definition of the measurement and scale, they see that there are infinitely many types of scales. to the authors consider several of them that are basic in the theory of decision-making. Depending on the way by which one derives the empirical information for the preferences of the decision maker and the scales, they define the functions describing or representing the preferences as value or utility function. The most popular rank dependent utility model and its derivative cumulative prospect theory are discussed from the stochastic measurement point of view.

It is shown that the human subjective expectations for the uncertainty events can be described mathematically with the terms of the probability theory and can be inserted into the mathematical theory of von Neumann and Morgenstern. Some examples of utility functions are shown.

1. THEORY OF MEASUREMENT AND SOME BASIC MEASUREMENT SCALES

In the current chapter, we consider the term measurement from a general point of view as we consider necessary notions and mathematical formulations in the framework of these general settings. In a broader plan, the scientific progress is a constant development and refinement of the terms and their mutual relations, with regard to known or learned facts from the objective reality. The mathematical description on such a fundamental level requires basic mathematical terms, as sets, relations, and operations over them, as their gradual elaboration to more complex and specific terms as functions, operators on mathematically structured sets as well, as equivalency of these

DOI: 10.4018/978-1-4666-2967-7.ch003

descriptions with respect to a given real object (Pfanzagl, 1971). In the last aspect of equivalency of the mathematical descriptions, we enter the theory of measurements and scaling (Krantz, 1971, 1989, 1990).

Generally speaking, we can divide the measurements in two basic types, fundamental and derivative from them. The fundamental measurements are in the first stage of the scientific investigation and development, in which several real manifestations formed as new terms are subject to estimation. The derivative measurements appear on the next stages after the basic terms have been well investigated and measured (metrified). These new measures are constructed based on the several basic fundamental scales of metrizations in the investigated scientific domain. We can say that in some sense the theory of measurement and scaling considers the mathematical description of the way of forming the mathematical metrization of the fundamental terms. Of course, the investigations in the real life hardly ever follow the described in such a way the consecutive ways of scientific branches formation.

We start with defining the term relation (Pfanzagl, 1971). Let $\boldsymbol{A}=\{x_1, x_2, x_3,....,x_n, ...\}$ be a set of elements (different occurrences of the fundamental terms which are in the base of the investigated phenomenon). This set, in general setting, can be be infinite, countable or of greater power. The set of all pairs (x_1, x_2) forms the Cartesian product $\boldsymbol{A}\times\boldsymbol{A}$, such that $x_1\in\boldsymbol{A}$ and $x_2\in\boldsymbol{A}$. Every subset of this set is called binary relation. It is possible to consider ternary, quaternary or m-ary relations defined over the set $\boldsymbol{A}\times\boldsymbol{A}\times\boldsymbol{A}...\boldsymbol{A}\times\boldsymbol{A}\times\boldsymbol{A}$ or in other terms $\boldsymbol{A}^m$ is the m-ary Cartesian product of the set $\boldsymbol{A}$. Further, we will mostly consider binary relations. Shortly we are going to consider the binary relation as an ordered pair $(\boldsymbol{A}, \boldsymbol{R})$, and the belonging of the ordered pair (x_1, x_2) to $\boldsymbol{R}$ will be denoted by $x_1\boldsymbol{R}x_2$ or $\boldsymbol{R}(x_1, x_2)=1$. For the m-ary relation the notation is $(x_1, x_2, x_3, ...,x_m)\in\boldsymbol{R}$ or $\boldsymbol{R}(x_1, x_2, x_3, ...,x_m)=1$. We are going to mention some of the basic properties of the binary relations, necessary for the further description.

We start with the term reflexivity (Pfanzagl, 1971; Dixmier, 1967). The relation $\boldsymbol{R}$ is *reflexive* if the main diagonal belongs to the subset $\boldsymbol{R}$, $(x,x)\in\boldsymbol{R}, \forall x\in\boldsymbol{A}$. We get *irreflexivity* when the main diagonal does not belong to $\boldsymbol{R}$.

The binary relation $\boldsymbol{R}$ is *symmetric*, if from $(x, y)\in\boldsymbol{R}$ follows $(y, x)\in\boldsymbol{R}$. *Asymmetric is* when from $(x, y)\in\boldsymbol{R}$ follows $(y, x)\notin\boldsymbol{R}$.

The binary relation $\boldsymbol{R}$ is *antisymmetric*, if from $(x, y)\in\boldsymbol{R}$ and $(y, x)\in\boldsymbol{R}$ follows $(x=y)$.

The binary relation $\boldsymbol{R}$ is *transitive*, if from $(x,y)\in\boldsymbol{R}$ and $(y,z)\in\boldsymbol{R}$ follows $(x\boldsymbol{R}z)$.

The binary relation is $\boldsymbol{R}$ is *negatively transitive*, if from $(x,y)\notin\boldsymbol{R}$ and $(y,z)\notin\boldsymbol{R}$ follows $(x,z)\notin\boldsymbol{R}$ or in other notations $\neg(x\boldsymbol{R}z)$.

The binary relation $\boldsymbol{R}$ is *connected*, if for every pair of elements $(x,y)\in\boldsymbol{A}\times\boldsymbol{A}$ either $(x,y)\in\boldsymbol{R}$ or $(y,x)\in\boldsymbol{R}$, or both are simultaneously true. The binary relation $\boldsymbol{R}$ is *weak connected*, if from $x\neq y$ follows $(x,y)\in\boldsymbol{R}$ or $(y,x)\in\boldsymbol{R}$.

The binary relation R is *partially ordered*, when it is reflexive, antisymmetric, transitive relation.

The binary relation R is *equivalency,* when it is reflexive, symmetric, and transitive relation.

The binary relation R is *linear ordering*, when it is *asymmetric* and *negatively transitive*.

System with relations (SR) is called the set $\boldsymbol{A}$ in conjuction with a set of relations $\boldsymbol{R}_i$, $i\in\boldsymbol{I}$, $\boldsymbol{I}=\{1,2,3,...,n\}$ defined over it (Pfanzagl, 1971). We denote it by $(\boldsymbol{A}, (\boldsymbol{R}_i), i\in\boldsymbol{I})$. In this manner, we introduce structure in the set $\boldsymbol{A}$. This structure may be defined not only through the relations but also via operations such as intersections and unions of sets. Two SR $(\boldsymbol{A}, (\boldsymbol{R}_i), i\in\boldsymbol{I})$ and $(\boldsymbol{B}, (\boldsymbol{S}_i), i\in\boldsymbol{I})$ are of the same type, if the relations $(\boldsymbol{R}_i)$, $i\in\boldsymbol{I}$ and $(\boldsymbol{S}_i)$, $i\in\boldsymbol{I}$ have the same dimensions, i.e. if $(x_1, x_2, x_3, ...,x_{hi})\in\boldsymbol{R}_i$ and $(y_1, y_2, y_3, ...,y_{hi})\in\boldsymbol{S}_i$ for $\forall i, i\in\boldsymbol{I}$. If for $\forall i$, $\boldsymbol{R}_i$, $i\in\boldsymbol{I}$ is h_i*–ary* relation of $\boldsymbol{A}$ then $(h_i)_{i\in I}$ is called type of SR $(\boldsymbol{A}, (\boldsymbol{R}_i), i\in\boldsymbol{I})$.

Let SR (A, (R_i), $i \in I$) be given. Relation of *congruency* is called relation of equivalency ($\approx$) defined over the basic set A, if the property of *substitution* is satisfied, i.e. from the fulfillment of relations $(x_1, x_2, x_3, ..., x_{hi}) \in A^{hi}$ and $(x_j \approx y_j)$ for every j=1, 2, 3 4,...,h_i it follows that

$R_i(x_1, x_2, x_3, ..., x_{hi}) = R_i(y_1, y_2, y_3, ..., y_{hi})$ for $\forall$ i, $i \in I$ [ ].

We say that the relation of equivalency ($\approx_2$) is coarser than the equivalency ($\approx_1$), if the inclusion $(\approx_1) \subseteq (\approx_2)$ is satisfied.

The following proposition holds (Pfanzagl, 1971).

1.1. Proposition

Let SR (A, (R_i), $i \in I$) be given. Then there exist a coarsest relation ($\approx_A$) over SR (A, (R_i), $i \in I$).

Proof

Define ($x \approx_A y$) in the following way. For $\forall$ i, $i \in I$, $j=1,...,h_i$ and $(x_1, x_2, ..., x_{j-1}, x_{j+1} ...., x_{hi}) \in A^{hi-1}$ it is fulfilled

$$R_i(x_1, x_2, . x_{j-1}, x, x_{j+1} ...., x_{hi}) = R_i(x_1, x_2, . x_{j-1}, y, x_{j+1} ...., x_{hi}).$$

Obviously, ($\approx_A$) is a relation of equivalency and congruency and it includes all other congruencies into itself.

That means that if two elements are in congruency relation ($x \approx_A y$), then they are undistinguishable with respect to the properties in the set A, described with the set of relations ((R_i), $i \in I$) or, more precisely, by SR (A, (R_i), $i \in I$). If we factorize the set A by the coarsest congruency ($\approx_A$), then in the factor set $A/\approx_A$ the congruency ($\approx_A$) is in fact equality (=). A SR (A, (R_i), $i \in I$), in which the congruency ($\approx_A$) is coarsest is called *irreducible*. In the case SR ($A/\approx_A$,(R_i), $i \in I$) is irreducible.

We are going to define the term homomorphism. This is the image f, f:$A \rightarrow B$ between two SR (A,(R_i), $i \in I$) and (B, (S_i), $i \in I$) from the same type, for which $\forall$ i, $i \in I$ and $(x_1, x_2, x_3, ..., x_{hi}) \in R_i$ is satisfied:

$$R_i(x_1, x_2, x_3, ..., x_{hi}) \Leftrightarrow S_i (f(x_1), f(x_2), f(x_3), ..., f(x_{hi})).$$

Another definition is also possible:

$$R_i(x_1, x_2, x_3, ..., x_{hi}) \Rightarrow S_i (f(x_1), f(x_2), f(x_3), ..., f(x_{hi})).$$

We are going to use the first definition, the algebraic definition.

Isomorphism is a bijective correspondence which is homomorphism and whose inverse correspondence is also homomorphism (Pfanzagl, 1971; MacLane, 1974; Dixmier, 1967). *Partial endomorphism* is called every injective morphism from SR (A, (R_i), $i \in I$) to SR (A, (R_i), $i \in I$). Let A_0 be a subset of A. We denote by $G_A(A_0)$ all injective homomorphisms, partial endomorphism from SR (A_0, (R_i), $i \in I$) in SR (A, (R_i), $i \in I$). Then the following proposition holds (Pfanzagl, 1971):

1.2. Proposition

Let SR (A,(R_i), $i \in I$) is irreducible, and SR (B,(S_i), $i \in I$) is SR from the same type. Let $\mathfrak{R}$ be the set of all injective homomorphisms from SR (A,(R_i),$i \in I$) in SR (B,(S_i), $i \in I$) and let f_0 be arbitrary injective homomorphism from A into B. Then $\mathfrak{R} = \{\gamma_o f_0 /$ where $\gamma \in G_B(f_0(A))\}$. All homomorphisms belonging to $\mathfrak{R}$ are injective.

Proof

Since the congruency in irreducible SR is equivalence, then every homomorphism is injective. In this case f_0:$A \rightarrow f_0(A)$ is isomorphism from SR (A, (R_i), $i \in I$) into SR ($f_0(A)$, (S_i), $i \in I$). For γ, for which $\gamma \in G_B(f_0(A))$ $\gamma_o f_0$ is homomorphism from SR (A, (R_i), $i \in I$) into SR (B, (S_i), $i \in I$). For every $\theta \in \mathfrak{R}$ the correspondence $\theta_o f_0^{-1}$ is isomorphism from SR (f_0 (A), (S_i, $i \in I$) into SR ($\theta(A)$, (S_i, $i \in I$). Thus way $\theta_o f_0^{-1} \in G_B(f_0(A))$ and obviously it is satisfied $\theta = (\theta_o f_0^{-1})_o f_0$.

We call SR $(A, (Q_i), i \in I)$ k-dimensional number SR, if $A=R^\kappa$, where in the concrete case R is the set of real numbers. Now we can give the exact mathematical definition of the term scale and measurement.

1.3. Definition

We call k-dimensional scale every homomorphism from irreducible empirical system into the number system SR $(A, (Q_i), i \in I)$.

We are going to give a brief informal explanation. The empirical system of relations SR is an object from the reality with the properties described by the relations in SR, while the number SR is a mathematical object. The number SR reflects the interesting for us properties in the real investigated object. Converting this real world measurement from the irreducible empirical system into the mathematical set R^κ we strive for preserving, as much as possible, the complete knowledge for the empirical properties. For that reason, we are using the definition for homomorphism in the form:

$$R_i(x_1, x_2, x_3, \ldots, x_{hi}) \Leftrightarrow S_i\,(f(x_1), f(x_2), f(x_3), \ldots, f(x_{hi}))\ \text{за}\ \forall\, i,\ i \in I.$$

In the definition for scale, the correspondence $A \rightarrow R^\kappa$ is not simply defined. In general sense, there exist entire class of scales converting the irreducible empirical system of relations SR $(A, (R_i), i \in I)$ into the number system SR $(R^\kappa, (S_i), i \in I)$. We denote this class of homomorphisms by $\aleph(A, R^\kappa)$ (injective in its essence, because the empirical system is irreducible and surjective in regards to $f(A)$). If a scale $f_0 \in \aleph(A, R^\kappa)$ is given, then from proposition (2) it follows that we can characterize the whole class of scales $\aleph(A, R^\kappa)$ in the following way

$$\aleph(A, R^\kappa) = \{\gamma f_0 \,/\, \text{where } \gamma \in G_B\,(f_0(A))\},$$

where in this case $B = R^\kappa$.

Or in other words two scales are equivalent with precision to partial endomorphism $\gamma \in G_B(f_0(A))$. The elements of $G_B(f_0(A))$ are called admissible manipulations of the scale f_0.

As an example we are going to denote the measurement of the temperature. We can accept that $f_0(.)$ is the *temperature in Celsius*. Then every partial endomorphism is an affine correspondence of the type

$$\gamma(x)=ax+b,\ a \in R,\ b \in R \text{ and } a>0.$$

Another equivalent temperature scale is the scale in Kelvin. In it the scale is changed by a multiplication with positive number (a) and the zero is shifted by an addition of another number (b).

From the definition of the measurement and scale we can see that there are infinitely many types of scales. In the science literature and in our everyday life several types of scales are constantly discussed. We are going to consider several of them which are basic in the theory of decision-making.

Now we know the mathematical definition of measurement. In informal terms, measurement is an operation in which a given state of the observed object is mapped to a given denotation (Pfanzagl, 1971; Roberts, 1976). We will start with the so-called *nominal scale*, which is an expression of the equivalence of two phenomena only (Litvak, 1982). Let $\mathbf{X}$ be the set of alternatives ($\mathbf{X} \subseteq \mathbf{R}^m$). Let x and y are two alternatives ($(x,y) \in \mathbf{X}^2$). For this weakest scale, the following axioms are valid:

1. $((x \approx y \vee \neg x \approx y) \equiv 1) \wedge ((x \approx y \wedge \neg x \approx y) \equiv 0) \wedge x \approx x$;
2. $(x \approx y \Rightarrow y \approx x)$;
3. $((x \approx y \wedge x \approx z) \Rightarrow y \approx z)$.

Here ($\approx$) denotes equivalence and $\neg(\approx)$ is the opposite (non-equivalence). In essence the above three properties define equivalence relation, which splits the set of alternatives $\mathbf{X}$ into non overlapping subsets, classes of equivalence. In this scale, the only mathematical processing is "pattern recognition." Only the Kronecker symbol may be used here as a measure.

In the case when the observed property allows us not only to distinguish between states but to compare them by preference we use a stronger scale, the *ordering scale*. The preference relation in the ordering scale (x is preferable to y) is denoted by ($x \succ y$). In this scale together with the above three axioms two more are satisfied:

4. $\neg(x \succ x)$ for $\forall x \in \mathbf{X}$, $((x \succ y) \Rightarrow \neg(y \succ x))$;
5. $(x \succ y \wedge y \succ z) \Rightarrow x \succ z$.

If incomparable alternatives exist, then the scale is called a scale of *partial ordering*.

An example for such scale is the Magnitude scale foe earthquake by Richter. This scale is 12 magnitudes and it estimates the energy of the seismic waves according to the destruction that they cause.

Another example is the scaling of the grades in schools. Based on comparison they estimate who of both given students knows better the taught material.

One of the most famous examples of such a scale is the scale for measurement of hardness–scale of Mohs. We say that the mineral M_1 is harder than the mineral M_2, if the first can scratch over the second.

Under these five axioms we can search for an analytical representation through function, the so-called value function u(.). A *value function* is a function u(.) for which it is fulfilled (Keeney, 1976; Fishburn, 1970; Roberts, 1976):

$$((x, y) \in X^2, x \succ y) \Leftrightarrow (u(x) > u(y)).$$

In this definition in addition to axioms (4, 5) weak connectedness is also assumed:

$$\neg(x \approx y) \Rightarrow ((y \succ x) \vee (x \succ y)).$$

The definition could be restricted to the following formula in the case of partial ordering:

$$((x, y) \in X^2, x \succ y) \Rightarrow (u(x) > u(y)).$$

Depending on the type of the function—continuous, partially continuous, or discrete—there exist different types of scale, measuring the above relations. It is well known that a transformation with an arbitrary monotonous function of ordinal scale leads to an ordinal scale. When using those scales, apart from comparison by magnitude, we can search the minimum and maximum of the function as feasible mathematical operations. It may be used as criterion for optimal control in Control theory. However, it has to be emphasized that in this case operations like finding mathematical expectation are unfeasible. Under this scale, we cannot talk about distance between the different alternatives. Here only ordinal evaluations within different mathematical processing of the information may be used.

If with the ordering of the alternatives we can evaluate the distance between them we can talk about *interval scale* (Pfanzagl, 1971; Litvak, 1982). For these scales, the distances between the alternatives have the meaning of real numbers.

For these scales, the central moments and the variance are sensible evaluations and have physical meaning; whereas the mathematical expectation depends on the origin of the scale and thus is unfeasible. The transition from one interval scale to another is achieved with affine transformation:

$$\mathbf{x} = \mathbf{a}\mathbf{y} + \mathbf{b}, (\mathbf{x}, \mathbf{y}) \in \mathbf{X}^2, \mathbf{a} > 0, \mathbf{b} \in \mathbf{R}.$$

Among these type of scales is also the *measurement of the utility function* through the so called "gambling approach." Once more we emphasize that here the calculations are done with numbers related to the distances between the alternatives, and not with the numbers relating to the alternatives themselves. For instance, if we say that a body is twice as warm as another in Celsius, this will not be true if the measurements were in Kelvin.

The next, stronger scale is the *ratio scale*. This is an interval scale with fixed origin. Such is the scale for measuring weight, since zero is always the absence of weight. For these scales in addition to the previous 5 axioms the following additivity axioms are satisfied:

1. $(x=y \wedge z>0) \Rightarrow ((x+z)>y)$;
2. $x+y=y+x$;
3. $(x=y \wedge z=q) \Rightarrow (x+z=y+q)$;
4. $q+(x+y)=(q+y)+x$.

The *absolute scale* is the most powerful. For it the zero and one are absolute and it is a one of a kind and unique scale. There exists no transformation to another type of scale.

2. DISCRETE DECISION MODELS, GROUP DECISION-MAKING, ARROW'S IMPOSSIBILITY THEOREM

The first more serious attempts for the formal description of the process of decision-making have been done during the seventeen century by the French scientist Jean-Sharl de Borda and the marquis de Condorcet. During this period, formal procedures were proposed for determining the will of the group of individuals, when the desire of the separate individuals is known (Hazewinkel, 2002; Litvak, 1982; Roberts, 1976).

We can accept that this is one naive period in the initial development of the theory of decision-making. The reason is that the analytical methods in mathematics had not reached its full development and had not taken its place in mathematics. The methods of Borda and Condorcet and their similar are defined over a discrete set of possibilities or alternatives over this set and a partial or linear ordering (*arrangement*) is given over the scale of the orderings. From here originates one of their names, discrete methods, or rank methods for decision-making. Because they are applied for taking the preferences of group of individuals, they are considered as *group methods of decision-making*.

We are going to give a short description of the method of Borda (Litvak, 1982). Let $\boldsymbol{A}=\{x_1, x_2, x_3, \ldots, x_n\}$ be a discrete set of alternatives over which m arrangements P_i, $i=1,\ldots,m$ are given. Usually the orderings are being interpreted in the terms of preferences. We say x is more preferable to y and denote $x \succ y$. We accept that arrangements are *asymmetric* and *negative transitive*. This is the so called *weak ordering* (Fishburn, 1970; Litvak, 1982). Later we are going to see that in this case the transitivity of the relation of preferences and transitivity of the relation indifference between two alternatives follows from the definition. The set of $\{P_1, P_2, P_m\}$ arrangements defines table or profile of m orderings. The rule of choice of the best alternative according to Borda is the following. Let for the alternative (a) $\boldsymbol{B}_i$(a) be the number of less preferred in P_i alternatives. For every alternative from the set $\boldsymbol{A}$ we determine the number of Borda $\boldsymbol{B}$(a) which has the form:

$$B(a) = \sum_{i=1}^{m} B_i(a).$$

If $\boldsymbol{B}(a)>\boldsymbol{B}(b)$, then we accept that $(a \succ b)$. As a best alternative, we choose the alternative with the greatest number of Borda.

The rest of discrete or rank methods for individual or group decision-making are based on similar methods for computing ranks and estimates on them. The procedures usually are based on estimates similar to those of Borda. It is possible to assign weights on the different arrangements:

$$B(a)=\sum_{i=1}^{m}\lambda_i B_i(a),\ \lambda_i \in \boldsymbol{R}, \lambda_i > 0\ .$$

These procedures are easy from application point of view but they are methodologically weak, because in the general case existence of an additive value function is necessary for their application. From Utility theory, it is known that the requirement for existence of additive utility function is the heaviest condition concerning the mutual dependencies of the arrangements. We are going to give an example where the rank method of Borda type is not admissible.

2.1. Example

Let $\boldsymbol{A}=\{1, 2, 3\}\times\{1, 3, 5\}$. We denote $(z,t) \succ (x,y)$ if and only if $(xy + x^y) < (zt + z^t)$ holds. It is proven that additive utility function does not exist in this case. Let as try to apply rank method of the type of Borda in this case. By definition we have $(1,5) \sim (3,1)$ and $(2, 1) \sim (1, 3)$. If a rank method of Borda type is applicable then we must have:

$$\lambda_1 \boldsymbol{r}_1(1) + \lambda_2 \boldsymbol{r}_2(5) = \lambda_1 \boldsymbol{r}_1(3) + \lambda_2 \boldsymbol{r}_2(1),$$
$$\lambda_1 \boldsymbol{r}_1(2) + \lambda_2 \boldsymbol{r}_2(1) = \lambda_1 \boldsymbol{r}_1(1) + \lambda_2 \boldsymbol{r}_2(1).$$

In this case by $r_i(.)$ we denote the corresponding rank according to first and correspondingly second ordering (arrangement) and by λ_i the corresponding weight coefficient. After summation and cancellation we receive the following relation:

$$\lambda_1 \boldsymbol{r}_1(2) + \lambda_2 \boldsymbol{r}_2(5) = \lambda_1 \boldsymbol{r}_1(3) + \lambda_2 \boldsymbol{r}_2(3).$$

Therefore (2.5)~(3.3), which is not true by the direct application of the formula $(xy + x^y) < (zt + z^t)$. That means that using rank estimates of the type of Borda we cannot obtain the actual ordering. In this example, we used an idea from (Fishburn, 1970).

The demonstrated difficulties and discrepancies have let to investigations and development of better founded approaches of forming the decisions based on linear orderings and some conditions over the discrete set of alternatives (Litvak, 1982). We denote by F $(P_1, P_2,..., P_m)$ the resulting relation, derived using some rule from the set of relations $\{P_1, P_2,..., P_m\}$.

The first condition is:

Let $\{P_1, P_2,..., P_m\}$ and $\{R_1, R_2,..., R_m\}$ are two sets of relations, such that over the subset ***A'*** of the set ***A*** ($\boldsymbol{A}'\subseteq\boldsymbol{A}$) holds $P_i=R_i$, i=1,..,*m*. Then over ***A'*** holds $F(P_1, P_2,..., P_m)= F(R_1, R_2,..., R_m)$. This condition is called condition for *independence.*

The second condition is:

We suppose that the set of linear orderings can be sufficiently varied. Namely for every three alternatives $x_1, x_2, x_3 \in \boldsymbol{A}$, we can find linear orderings P_1, P_2 и P_3, such that:

$(x_1 \succ x_2) \in P_1$, $(x_2 \succ x_3) \in P_1$, and $(x_1 \succ x_3) \in P_1$,

$(x_1 \succ x_2) \in P_2$, $(x_2 \succ x_3) \notin P_2$, and $(x_1 \succ x_3) \notin P_2$,

$(x_1 \succ x_2) \in P_3$, $(x_3 \succ x_2) \notin P_3$, and $(x_3 \succ x_1) \notin P_3$.

The above condition together with the transitivity of the linear orderings defines the second condition and is called *universality.*

The third condition is:

If in some linear order P_ν the preference between alternatives has changed on favor of the resulting relation $F(P_1,P_2,...,P_m)$, then the resulting relation does not change. This condition is called *monotonicity*. Let two of the relations P_ν and P'_ν from the set $\{P_1, P_2,..., P_m\}$ differ only by the preferences between the pair of alternatives $(x_i \succ x_j)$. Then the monotonicity is described in the following way:

$$(((x_i \succ x_j) \in F(P_1,P_2,...,P_m) \wedge (x_i \succ x_j) \in (P'_\nu \backslash P_\nu)) \vee$$
$$((x_i \succ x_j) \notin F(P_1,P_2,...,P_m) \wedge (x_i \succ x_j) \notin (P_\nu \backslash P'_\nu))) \Rightarrow$$
$$F(P_1,P_2,..,P'_\nu,..., P_m) = F(P_1,P_2,..,P_\nu,..., P_m).$$

The fourth condition is:

For every pair of alternatives (x_i, x_j) there exist set of linear orderings $(P_1, P_2,..., P_m)$ and $(R_1, R_2,..., R_m)$, for which the following holds:

$(x_i \succ x_j) \in F(P_1, P_2,..., P_m)$ and $(x_j \succ x_i) \in F(R_1, R_2,..., R_m)$.

The fifth condition is:

There is a lack of dictator. Mathematically this is described in the following way. There does not exist a relation P_ν, for which $P_\nu = F(P_1, P_2,.., P_\nu,.., P_m)$ independently of the set of relations $\{P_1, P_2,.., P_{\nu-1}, P_{\nu+1},.., P_m\}$.

These five conditions are reasonable enough for the choice of a resulting ordering $F(P_1, P_2, \ldots, P_m)$. There are some discussions concerning the first condition only. But here the well known Arrow theorem, also known as Arrow paradox is satisfied (Arrow, 1950; Ekeland, 1979).

2.2. Theorem

There does not exist a weak ordering, which satisfies the five conditions.

The result is striking for the standard human thinking and it has led to a lot of discussions. Some of them are looking for the philosophical explanations for the existence of dictators in the human society based on this mathematical result. For the moment we are going to note only that it was obtained only for the linear ordering scale. Here we can derive one weaker result related to Pareto principle. The resulting relation satisfies the Pareto principle if:

$$\bigcap_{i=1}^{m} P_i \subseteq F(P_1, P_2, \ldots P_m) \subseteq \bigcup_{i=1}^{m} P_i.$$

The union and the intersection in the above formula is over sets, which the relations are. Then the following theorem is satisfied (Litvak, 1982).

2.3. Theorem

If the resulting relation satisfies the conditions one three and four then the resulting relation satisfies the Pareto principle.

One of the strongest results in the *ordering scale* is the median of Kemeny-Snell. For this purpose between the individual weak orders in the set of relations $\{P_1, P_2,..., P_m\}$ a measure of nearness (distance) is introduced. Then it is natural to require the resulting relation $F(P_1, P_2,..., P_m)$ to be as close as possible, according to this measure to the set of relations $\{P_1, P_2,..., P_m\}$. This resulting relation is called the median of Kemeny-Snell:

$$F(P_1, P_2,..., P_m) = \arg\min_{P} \sum_{i=1}^{m} \boldsymbol{d}(P, P_i).$$

The median of Kemeny-Snell can be defined not only over the set of weak orders, but also over the set of the equivalent relations or over the set of partial orderings. The median is the only resulting relation, which satisfies the conditions 2 to 5. It does not satisfy only the first condition, over which there is some discussion.

Measure of nearness (distance) is possible to be introduced for the following relations: equivalency, partial ordering, weak or linear ordering.

We are going to consider the way Kemeny has introduced *measure of nearness* (Kemeny distance) between the relations of linear orders. The first condition is that the measure of nearness $\boldsymbol{d}(P, R)$ satisfies the standard axioms of metric between the weak orders P and R:

Axiom 1:
$$\boldsymbol{d}(P, R) \geq 0 \wedge (\boldsymbol{d}(P, R) = 0 \Leftrightarrow P = R).$$

Axiom 2: $\boldsymbol{d}(P, R) = \boldsymbol{d}(R, P)$.

Axiom 3: The triangle inequality holds:
$$\boldsymbol{d}(P, R) \leq (\boldsymbol{d}(P, Q) + \boldsymbol{d}(Q, R).$$

Box 1.

$$p_{ij} = \begin{cases} 1, \text{ if } (x_i, x_j) \in P, (x_j, x_i) \notin P, \\ 0, \text{ if } (x_i, x_j) \in P, (x_j, x_i) \in P, \\ -1, \text{ if } (x_i, x_j) \notin P, (x_j, x_i) \in P, \\ \theta, \text{ if } (x_i, x_j) \notin P, (x_j, x_i) \notin P, \text{ where } \theta \text{ is a number, defined according to the case.} \end{cases}$$

The case when in triangle inequality there is equality requires different definition. For every relation P we define a matrix of the relation of the type *mxm,* whose elements p_{ij} are defined in Box 1.

We say that the weak order P is between the weak orders R and Q, if for their matrices of the relation the following holds:

$$\min(r_{ij}, q_{ij}) \leq p_{ij} \leq \max(r_{ij}, q_{ij}).$$

Axiom 4: If the above condition is satisfied then the triangle inequality becomes equality.

Axiom 5: If we swap the notations of the alternatives in the basic set ***A***, then the distance between the weak orders does not change.

Axiom 6: We consider two partial orders R and Q, which are distinguishable only by the ordering of the alternatives taking place between (r+1) and (r+k) where (r+k) $\leq m$. We denote by T(R) and T (Q) the weak orders obtained from R and Q via rejection of the rest of alternatives. Then it is true: $d(Q, R) = d(T(Q), T(R))$.

Axiom 7: The minimal positive distance between two weak orders is 1.

The following theorem holds (Litvak, 1982):

2.4. Theorem

The axioms 1-7 define measure of nearness between the weak orders in a unique way. This measure of nearness has the form:

$$d(Q, R) = \frac{1}{2} \sum_{i,j=1}^{m} \left| q_{ij} - r_{ij} \right|.$$

When we work with partial orderings, it is necessary to consider the possible case when two alternatives are not comparable.

The median of Kemeny-Snell is one of the most well founded rules for forming of a resulting relation when considering partial ordering and weak ordering. However, the finding of this median is connected with great computational difficulties. Combinatorial algorithms based on the method of "branches and borders" are known to find the median of Kemeny-Snell (Litvak, 1982).

We have seen that the forming of rules for decisions in the scale of discret orderings is connected with a lot of difficulties. Except for the Arrow paradox, which demonstrates methodological difficulties, we are faced with difficulties from computational character. In the next chapter, we are going to consider ways for the represent the relations of preferences in the stronger scales. In these scales, the Arrow paradox is surmountable via analytical representation of the preferences as value or utility functions (Keneey, 1993).

3. VALUE FUNCTION AND MEASUREMENT SCALE

Depending on the way by which we derive the empirical information for the preferences of the decision maker, we define the functions describing or representing the preferences as value or utility.

We will start with some mathematical definitions and the description of the simplest function, the value function (Fishburn, 1970; Keeney, 1993). Let $\mathbf{X}$ be the set of alternatives ($\mathbf{X}\subseteq\mathbf{R}^m$). The DM's preferences over $\mathbf{X}$ are expressed by ($\succ$). The real expert value function is denoted by u*(.). A "value" function is a function u*(.) (u*: $\mathbf{X}\rightarrow\mathbf{R}$) for which it is fulfilled (Keeney & Raiffa, 1976):

$$((x,y)\in\mathbf{X}^2, x\succ y)\Leftrightarrow(u^*(x)>u^*(y)). \qquad (3.3.1)$$

A weaker case is when the following condition is fulfilled instead:

$$((x, y)\in\mathbf{X}^2, x\succ y) \Rightarrow (u^*(x)>u^*(y)). \qquad (3.3.2)$$

The American scientist Fishburn has proved that for a finite set of alternatives and strict partial ordering (axioms 4, and 5, paragraph 3.1) there always exist such a function (Fishburn, 1970). Obviously, it always exists with precision up to monotonous transformation. In this manner we can move from the language of binary relations and preferences to the language of control criteria. We will use the following definitions. Let $\mathbf{A}_{u*}$ and $\mathbf{B}_{u*}$ are the sets of positive preferences and negative preferences:

$$\mathbf{A}_{u*}=\{(x,y)\in\mathbf{R}^{2m}/\ (u^*(x))>u^*(y)\}, \qquad (3.3.3)$$

$$\mathbf{B}_{u*}=\{(x, y)\in\mathbf{R}^{2m}/\ (u^*(x))<u^*(y)\}. \qquad (3.3.4)$$

If there is a function $\mathbf{F}(x,y)$ of the form $\mathbf{F}(x,y)=\mathbf{f}(x)-\mathbf{f}(y)$, positive over $\mathbf{A}_{u*}$ and negative over $\mathbf{B}_{u*}$, then the function $\mathbf{f}(x)$ is a value function equivalent to the empirical DM's value function u*(.).

The construction of such functions of two variables is one of the ways for evaluating the value functions in ordinal aspect (Pavlov, 1989, 2005, 2010, 2011). Such approach also permits the use of stochastic recurrent techniques for "pattern recognition" for solving the problem. In the deterministic case it is true that $\mathbf{A}_{u*}\cap\mathbf{B}_{u*}=\varnothing$. In the probabilistic case it is true that $\mathbf{A}_{u*}\cap\mathbf{B}_{u*}\neq\varnothing$ (Aizerman, 1970). Evaluation of the value functions in the deterministic case is analyzed in Pavlov (2005, 2011). Evaluation of the value functions in the probabilistic case is also analyzed in Pavlov (1989).

This definition of value function is determined with the help of the set $\mathfrak{R}=\{\gamma_\sigma f_0\ /\ \text{where}\ \gamma\in G_B(f_0(A))\}$ and statement 3.1.2. We are going to estimate in details the properties of the empirical system of relations ($\boldsymbol{X}$,($\approx$),($\succ$)), where ($\approx$) can be considered as the relation "indifference or equivalent," indifference of the decision maker, and ($\succ$) is the relation "prefer" of the decision maker. We will look for equivalency of the empirical system with the system of relations (***R***-real numbers, (=), (>)). We are going to look for the conditions under which the relation of the indifference ($\approx$) is transitive.

Let us look at how things stand from practical point of view. The transitivity of the relations ($\succ$) and ($\approx$) is the most breached in the practice. The violation of the transitivity of the relation ($\succ$) could be interpreted as a lack of information or as a DM's subjective mistake. The violation of the transitivity of the relation ($\approx$) is due to the natural "uncertainty" of the human's preference and the qualitative nature of expressions of the subjective notions and evaluations (Cohen, 1988; Keeney, 1993; Raiffa, 1968).

The *"indifference"* relation ($\approx$) based on ($\succ$) is defined by Fishburn (1970):

$$((x \approx y) \Leftrightarrow \neg((x\succ y) \vee (x\succ y))). \qquad (3.3.5)$$

The assumption of existence of a value function $u(.)$ leads to the *"negatively transitive"* and *"asymmetric"* relation ($\succ$). The *"negative transitivity"* $(\neg(p\succ t) \wedge \neg(t\succ q)) \Rightarrow \neg(p\succ q))$ is equivalent to the following expression (Fishburn, 1970):

$$((p\succ q) \Rightarrow ((p\succ r) \text{ or } (r\succ q) \text{ for } \forall\ (p, q, r) \in\boldsymbol{P}^3)). \qquad (3.3.6)$$

The following proposition is almost obvious.

3.1. Proposition

The DM "*preference*" relation ($\succ$) defined by the utility function $u(.)$ is "*negatively transitive*" $(\neg(p\succ t) \wedge \neg (t\succ q)) \Rightarrow \neg(p\succ q))$.

Proof

We shall use the following equivalence of the definition of the "negative transitivity" $((\neg(p \succ t) \wedge \neg(t\succ q)) \Rightarrow \neg(p \succ q)) \Leftrightarrow ((p \succ q) \Rightarrow ((p \succ r) \vee (r\succ q)$ for $\forall$ $(p, q, r)\in \boldsymbol{P}^3$)) (Fishburn, 1970). Let $(p, q, r)\in \boldsymbol{P}^3$, and $(u(p)>u(q))$. If $(u(p) \le u(r))$, then $(u(q) < u(p) \le u(r)) \Rightarrow (u(q) < u(r)) \Rightarrow (r\succ q))$. If $(u(r) \le u(q))$, then $(u(r) \le u(q) < u(p)) \Rightarrow (u(r) < u(p)) \Rightarrow (p\succ r))$. If $((u(r)< u(p)) \wedge (u(r) >u(q)) \Rightarrow ((p\succ r) \wedge (r\succ q))$.

The following proposition discusses the transitivity of the equivalence relation ($\approx$). This property is violated in practice most often times.

3.2. Proposition

In the case of a "*negative transitivity*" of ($\succ$) the "*indifference*" relation ($\approx$) is transitive $(((x\approx y) \wedge (y\approx t)) \Leftrightarrow (x\approx t)))$.

Proof

(By contradiction) Let the "preference" relation ($\succ$) is "negatively transitive," $(p, q, t)\in \boldsymbol{P}^3$, $(u(p)\approx u(t)$, $u(t)\approx u(q)$ and $\neg(u(p)\approx u(q))$. The expression $(\neg(u(p)\approx u(q)))$ brings to $((u(p)\succ u(q)) \vee (u(p)\prec u(q)))$. The relations $((u(p)\succ u(t)) \vee (u(t)\succ u(q)))$ or $((u(p)\prec u(t)) \vee (u(t)\prec u(q)))$ results from the "negative transitivity." The relations $\neg(u(p)\succ \upsilon(t))$ and $\neg(u(p)\prec u(t))$ results from $(u(p)\approx u(t))$. The relations $\neg(u(t)\succ u(q))$ and $\neg(u(t)\prec u(q))$ results from $(u(q)\approx u(t))$. Since the above cases are impossible the only possibility is $(u(p)\approx u(q))$.

3.3. Corollary

Let the preference relation ($\succ$) is "*irreflexive*" $(\neg(p\succ p))$ and "*negatively transitive*," then the "*indifference*" relation ($\approx$) is an "*equivalence*" (reflexive, symmetric and transitive).

Proof

It is obvious that the "reflexivity" of ($\approx$) follows from the definition of the relation ($\approx$) $(\neg(u(p)\prec u(p))\wedge\neg(u(p)\succ u(p)) \Rightarrow u(p)\approx u(p))$. The transitivity is proved in the Proposition 2. The symmetry of the relation ($\approx$) is obvious.

The irreflexivity of the preferences and the negative transitivity of the preference relation split the set $\boldsymbol{X}$ into non-crossing equivalence classes. The factorized set of these classes is marked by $\boldsymbol{X}/\approx$. The factorized set $\boldsymbol{X}/\approx$, constructed with the relations ($\approx$) and ($\succ$) is irreducible.

We need two definitions. We remember that a "*weak order*" is an asymmetric and "negatively transitive" relation. The transitivity of the "*weak order*" ($\succ$) follows from the "asymmetry" and the "negative transitivity." A "*strong order*" is a "*weak order*" for which is fulfilled $(\neg(x\approx y)\Rightarrow((x\succ y) \vee (x\prec y))$. It is proven in Fishburn (1970) that the existence of a "*weak order*" ($\succ$) over $\boldsymbol{X}$ leads to the existence of a "*strong order*" preference relation ($\succ$) over $\boldsymbol{X}/\approx$.

Consequently the assumption of existence of a value function u(.) leads to the existence of: asymmetry $((x\prec y)\Rightarrow (\neg(x\succ y))$, axiom 4 in paragraph 3.1.), transitivity $((x\succ y) \wedge (y\succ z) \Rightarrow (x\succ z)$, axiom 5 in paragraph 3.1.) and transitivity of the "indifference" relation ($\approx$) (axiom 3 in paragraph 3.1.).

In the following exposition we accept that the relations ($\approx$) and ($\succ$) are transitive. As we mentioned before the scale of orderings was defined by homomorphic monotone functions. But if we are looking for the equivalency between SR ($\boldsymbol{X}$, ($\approx$), ($\succ$)) and SR ($\boldsymbol{R}$-real numbers, (=), (>)) the homomorphisms have to be not only monotone, but continuous as well (Pfanzagl, 1971).

We need several definitions (Pfanzagl, 1971). Open ray is called a set of the type $(\leftarrow,a)=\{x\in X/(a>x)\}$ or $(a,\rightarrow)=\{x\in X/(x>a)\}$. Closed ray is a set of the type $(\leftarrow,a]=\{x\in X/(a\geq x)\}$ or $[a,\rightarrow)=\{x\in X/(x\geq a)\}$. In this case $(>)$ is interpreted as $(\succ)$ in X. Open or closed interval in X is called a set of the type $(a, b)=\{x\in X/(a>x>b)\}$ or correspondingly $[a, b]=\{x\in X/(a \geq x\geq b)\}$, where $a\in X$, $b\in X$.

3.4. Definition

Topology $\mathcal{T}$ is called a family of subsets of the main set X, which contain the empty set, the intersection of every pair of sets from $\mathcal{T}$ and the union of every subfamily of sets from $\mathcal{T}$. The elements of $\mathcal{T}$ are called open sets. A base in the topology $\mathcal{T}$ we called a subfamily of sets from $\mathcal{T}$, such that every set in $\mathcal{T}$ can be represented as union of elements of the base (Engelking, 1977; Kelly, 1955).

3.5. Definition

An interval topology is called the smallest topology generated by the intervals $(a, b)=\{x\in X/(a>x>b)\}$ (Pfanzagl, 1971).

3.6. Definition

The correspondence f from the topological space $(X, \mathcal{T})$ in the topological space (Y, Σ) is continuous if and only if when $f^{-1}(\Sigma)\subseteq\mathcal{T}$. Or in other words every inverse image of open set in (Y, Σ) is open set in $(X, \mathcal{T})$.

As was mentioned above the scale of orderings was defined via homomorphisms, monotone functions. But if we are looking for the equivalency between SR $(X, (\approx), (\Sigma))$ and SR (R-real numbers, $(=), (>)$) the homomorphisms have to be not only monotonic but continuous as well and then the ordering in the real numbers will be reflected in the empirical set X with the properties of the interval topology generated by the relation $(>)$ in R. The relation $(\succ)$ in the empirical set X generates the interval topology $\mathcal{T}_X$. Every monotone correspondence f converts the set X in $f(X)$ and, moreover, it is continuous for the interval topology $\mathcal{T}_{f(X)}$ generated from the relation $(\succ)$ in $f(X)$ defined by the monotonic correspondence. But the topology $\mathcal{T}_{f(X)}$ is coarser than the topology $f(X)\cap\mathcal{T}_R$ generated by the interval topology of the real numbers. Because of this reason we are going to use a stronger scale, called the *continuous scale of orders* in which the homomorphisms are monotone and continuous functions with respect to the interval topology in the set of the real numbers R. We need the following definition:

3.7. Definition

The subset Ω, $(\Omega\subseteq X)$ is subset dense in the topological space $(X, \mathcal{T})$, if every nonempty open set contains at least one element from the set Ω. The topological space $(X,\mathcal{T})$ is separable (separable topological space) if there is countable set dense in $(X,\mathcal{T})$.

The following theorem holds (Pfanzagl, 1971):

3.8. Theorem

The ordered system of relations $(X, (\approx), (\succ))$ can be represented via a continuous scale of orders (monotone continuous functions) if and only if it is separable with respect to the interval topology $(X,\mathcal{T})$, generated by the relation $(\succ)$ and this topology has countable base.

In the above theorem it is sufficient to say that the interval topology $(X,\mathcal{T})$ generated by the relation $(\succ)$ has countable base. If we use the continuous scale of orders, then the convergence in $(X,\mathcal{T})$ has the same properties as in the set of real numbers in their SR (R-real numbers, $(=), (>)$).

In a connected ordered set countable base exists if and only if it is separable. In these sets the interval topology coincides with the interval topology generated by the set of real numbers via monotonic continuous mapping. Then the term for the convergence in measurements coincides with the standard generally accepted term for convergence.

4. UTILITY FUNCTION AND MEASUREMENT SCALE

Now we will consider the more complex case of the utility functions. Let X be the set of alternatives and P is a set of probability distributions over X and $X \subseteq P$. A utility function $u(.)$ will be any function for which the following is fulfilled:

$$(p \succ q,\ (p,q) \in P^2) \Leftrightarrow (\int u(.)dp > \int u(.)dq). \quad (3.4.1)$$

In keeping with von Neumann and Morgenstern (Fishburn, 1970; Keeney, 1993) the interpretation of the above formula is that the integral of the utility function $u(.)$ is a measure concerning the comparison of the probability distributions p and q defined over X. The notation ($\succ$) expresses the preferences of DM over P including those over X ($X \subseteq P$). There are different systems of mathematical axioms that give sufficient conditions of a utility function existence. The most famous of them is the system of *von Neumann and Morgenstern's axioms*:

- **(A1)** The *preferences* relation ($\succ$) is transitive, i.e. if $(p \succ q)$ and $(q \succ r)$ then $(p \succ r)$ for all $p,q,r \in P$;
- **(A2)** *Archimedean Axiom*: for all $p,q,r \in P$ such that $(p \succ q \succ r)$, there exist $\alpha, \beta \in (0,1)$ such that $((\alpha p+(1-\alpha)r) \succ q)$ and $(q \succ (\beta p+(1-\beta)r))$;
- **(A3)** *Independence Axiom*: for all $p,q,r \in P$ and let $\alpha \in [0, 1)$, then $(p \succ q)$ if and only if $((\alpha p+(1-\alpha)r) \succ (\alpha q+(1-\alpha)r))$.

Axioms (A1) and (A3) cannot give solution. Axioms (A1), (A2) and (A3) give solution in the interval scale (precision up to an affine transformation):

$$((p \succ q) \Leftrightarrow (\int v(x)dp \succ \int v(x)dq) \Leftrightarrow (v(x)= au(x)+b,\ a,b \in R,\ a>0)). \quad (3.4.2)$$

It is known that the assumption of existence of a utility (value) function $u(.)$ leads to the "*negatively transitive*" and "*asymmetric*" relation ($\succ$) and to transitivity of the relation ($\approx$). The irreflexivity of the preferences and the negative transitivity of the preference relation split the set X into non-crossing equivalence classes. The factorized set of these classes is marked by $X/\approx$. It is proven in Fishburn (1970) that the existence of a "*weak order*" ($\succ$) over X leads to the existence of a "*strong order*" preference relation ($\succ$) over $X/\approx$.

Consequently the assumption of existence of a utility function u(.) leads to the existence of: asymmetry $((x \prec y) \Rightarrow (\neg(x \succ y))$, relation axiom 4), transitivity $((x \succ y) \wedge (y \succ z) \Rightarrow (x \succ z)$, relation axiom 5) and transitivity of the "indifference" relation ($\approx$) (relation axiom 3). So far we are in the preference scale, the *ordering scale*. The assumption of equivalence with precision up to affine transformation has not been included. In other words, we have only a value function. For value, however, the mathematical expectation is unfeasible, but we underline that the mathematical expectation is included in the definition of the utility function.

In practice, the utility function is measured by the gambling approach (Farquhar, 1984; Keeney, 1993; Raiffa, 1968). There are many different *utility evaluation methods* that are based prevailingly on the "lottery" approach (gambling approach). A "lottery" is called every discrete probability distribution over X. We denote as $<x,y,\alpha>$ the lottery: α is the probability of the appearance of the alternative x and $(1-\alpha)$—the probability of the alternative y. The most used evaluation approach is the following assessment: $z \approx <x,y,\alpha>$, where $(x,y,z) \in X^3$, $(x \succ z \succ y)$ and $\alpha \in [0,1]$. Weak points of this approach are violations of the transitivity of the relations and the so called "certainty effect" and "probability distortion" identified by Kahneman and Tversky (Kahneman, 1979; Cohen, 1988).

Additionally, the determination of alternatives x (*the best*) and y (*the worst*) on condition that $(x \succ z \succ y)$ where z is the analyzed alternative is not easy in the complex situations in the practice. Therefore, the problem of utility function evaluation on the grounds of expert preferences is a topical one. The violations of the transitivity of the relation ($\approx$) also leads to declinations in the utility assessment. All these difficulties explain the DM behavior observed in the Allais Paradox that arises from the "*independence*" axiom (A3) (Fishburn, 1970).

The determination of a measurement scale of the utility function u (.) originates from the previous mathematical formulation of the relations ($\succ$) and ($\approx$). It is accepted that ($X \subseteq \boldsymbol{P}$) and that $\boldsymbol{P}$ is a convex set $((q,p) \in \boldsymbol{P}^2 \Rightarrow (\alpha q+(1-\alpha)p) \in \boldsymbol{P}$, for $\forall \alpha \in [0,1])$. The utility $u(.)$ over X is determined with the accuracy of an affine transformation (interval scale) (Fishburn, 1970):

4.1. Proposition

If $((x \in X \wedge p(x)=1) \Rightarrow p \in \boldsymbol{P})$ and $(((q, p) \in \boldsymbol{P}^2) \Rightarrow ((\alpha p+(1-\alpha)q) \in \boldsymbol{P}, \alpha \in [0,1]))$ are realized, then the utility function $u(.)$ is defined with precision up to an affine transformation (in the case of utility function existence):

$$(u_1(.) \approx u_2(.)) \Leftrightarrow (u_1(.)=au_2(.)+b,\ a>0) \qquad (3.4.3)$$

Now we are in the *interval scale* and here the mathematical expectation is feasible taking in mind the initial point (b) of the interval scale. That is to say, this is a utility function (Fishburn, 1970; Keeney, 1993). Now it becomes obvious why in practice the gambling approach is used for construction of the utility function in the sense of von Neumann. The reason is that to be in the interval scale the set of the alternatives for comparison must be convex. Apart from the definitional aspect, the above properties related to Proposition 3.4.1 have also practical significance. This property is essential for the application of the utility theory, since it allows a decomposition of the multiattribute utility functions into simple functions (Keeney, 1993).

The first condition in the Proposition 4.1 can be interpreted as capability of the DM to imagine one alternative independently of the others. The second condition is capability of the DM to report on the uncertainty of the results. This proposition reveals that the utility measurement scale of the utility function is equivalent to the temperature scale (interval scale). Starting from the gambling approach for the definitions and the presentation of the expert's preferences we use the following sets of positive preferences and negative preferences., motivated by Proposition 4.1:

$$\boldsymbol{A}_{u^*}=\{(\alpha,x,y,z)/(\alpha u^*(x)+(1-\alpha)u^*(y))>u^*(z)\}, \qquad (3.4.4)$$

$$\boldsymbol{B}_{u^*}=\{(\alpha,x,y,z)/(\alpha u^*(x)+(1-\alpha)u^*(y)<u^*(z)\}. \qquad (3.4.5)$$

The approach we are using in our investigation for the evaluation of the utility functions in its essence is the recognition of these sets. Through stochastic recurrent algorithms, we approximate functions recognizing the above two sets (Aizerman, 1970; Pavlov, 2005, 2010).

The new stage in utility functions is that we can take into account not only the value of a given phenomena via the value function $u(.)$, but also take into account the probabilities, the uncertainty of the final solution in the nature and human activities. In this way, we have a measure for the estimation of more complex phenomena in which the expectations and the notions of the decision maker and the indetermination, the uncertainty of the final result of the human activity are reflected.

When developing the Expected utility theory, the intention of von Neumann and of Morgenstern was to define axioms and to prove statements based on them in order to well found logically satisfactory behavior in the economical surroundings. Despite the fact that the three cited axioms above are logically satisfactory, the real human behavior often shows deviations from them. It is easy to violate the condition for transitivity in the relation of equivalency. The example given by Maurice Allais (Nobel prizewinner in the area of decision-making) is well known; i.e. the Expected utility theory cannot describe accurately the behavior of the economical subject in the real life.

A long and not very fruitful discussion for the role of mathematics and for the role of the Bayesian theory of decision-making in the human reality was started. It was talked about normative axiomatic approach, for descriptive approach and for prescriptive approach in the decision-making. The psychological behavior of the economical individual has been investigated. In this aspect, in our days, the scientific results of the Nobel laureates Kaneman and Tversky are well known (Kahneman, 1979). New extensions and axiomatic bases of the developed mathematical theories in decision-making are looked for Birnbaum (2008) and Weber (1994).

Following the research of the Nobel prizewinners Kahneman and Tversky and after the debates about the so-called Allais paradox, extensions and further developments of von Neumann's theory were sought (Allais, 1953). Several non-additive utility theories have been developed in response to the occurring transitivity violations (Kahneman, 1979; Shmeidler, 1989; Weber, 1994; Birnbaum, 2008; Mengov, 2010; Machina, 1987). Among these theories, the rank dependent utility model and its derivative cumulative prospect theory are currently the most popular. In the Rank Dependence Utility (RDU) the decision weight of an outcome is not just probability associated with this outcome. It is a function of both probability and the rank (the alternative) x. For example, the RDU of the lottery (p_1, x_1; p_2, x_2; ...; p_n, x_n) is:

$$RDU = \sum_{i=1}^{n} W(p_i)u(x_i). \qquad (3.4.6)$$

Based on empirical researches several authors have argued that the probability weighting function W(.) has the inverse S-shaped form, which starts on concave and then becomes convex (Kahneman, 1979; Mengov, 2010). It is supposed that $W(p_i)=p_i+\Delta W(p_i)$. The declination of the probability assessment, the probability distortion has a S-shaped form. Such a function is shown in Figure 1.

The developed in this book approach for estimation and analytical description of the utility function allows this analytical construction and gives us strict methodology for evaluating the utility function in the theory of von Neumann as well as in the theory of Kaneman and Tversky.

5. SUBJECTIVE PROBABILITY

In the theory of the expected utility of von Neumann the used probabilities in the definition are objective probabilities, independent of the opinion of the subject making the decisions. The following question arises: How to approach our expectations with respect to the indeterminacy in the real life. Can we describe our subjective expectations for the uncertainty events in the terminology of the probability theory and to insert them in the mathematical theory of von Neumann and Morgenstern?

There are two types of uncertainty. From one side is the uncertainty of the type of tossing coin and choosing head or tail as is in the probability theory (Shiryaev, 1989). On the other side is the uncertainty in the economical state of Bulgaria and Europe during the next ten years.

There exist different opinions regarding the question "can we use our subjective expectations about the indeterminate events in our life" (Raiffa, 1968; Weber, 1994). In this book, we claim that we can make mathematically and logically based decisions as we use our subjective expectation about the outcomes of the uncertainty in the events. In the practical procedures of decision-making and for the human uncertainty in these procedures, the subjective probabilities are formed in the terminology and in the axiomatic of the probabil-

Figure 1. RDU weighting function W(.)

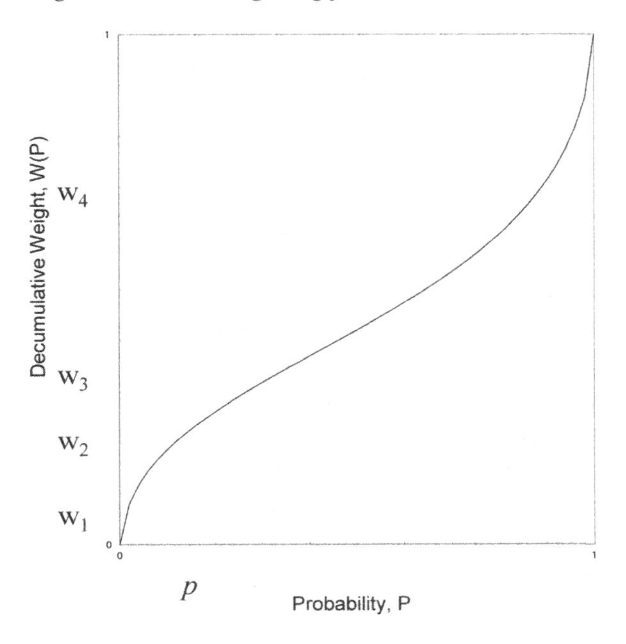

ity theory. The subjective probabilities are used logically sound and practically reasonably for the analysis and the solution of the uncertainty in the problems in the area of decision-making. In the procedures for estimation of the subjective probabilities discussed in the book, we consider the uncertainty and discrepancy of the human thinking when describing the subjective expectations for the uncertainty in the analyzed events.

In the well-known book of professor Raiffa, a foundation for the using of the subjective probabilities in the terms of lotteries is given (Raiffa, 1968). This is intuitive and not exactly mathematically rigid explanation of how we describe using probability our subjective attitudes for the uncertainty in decision-making. We are going to state the definition of the subjective probability of professor Raiffa:

5.1. Definition

Let E be an event from the reality and let L_E be a lottery in which we win with probability p_1 (event – W) and we lose with probability p_2 (event – L). If this lottery is undistinguishable in our notion with respect to the indeterminacy of the occurrence of the event (indifference with respect to taking part in the lottery or in the event with the indeterminate outcome) then the probability equivalent to this lottery is subjective probability with which we describe our attitude toward this uncertainty.

We are going to give a short explanation of the above definition. We win with probability p_1 or lose with probability p_2. On the right hand side of the figure is the occurrence or non-occurrence of the event θ. If our participation in the lottery or occurrence or non-occurrence of the event θ for us is with equal indeterminacy with respect to the final outcome then we accept that our attitude towards the indeterminacy θ can be described with the subjective probability equal to p_1:

$$P_{\text{subjectve probability}}(\theta) = p_1 \times 1 + p_2 \times 0 = p_1.$$

Graphically the lottery L_E and our indifference towards the participation in the lottery or the participation in the event with the indeterminate outcome are described in Figure 2.

Following the idea of professor Raiffa we can treat the subjective probabilities as objective probabilities and we can freely use them in the definitions and the considerations of von Neumann. If the decision maker agrees with the principle of non-contradictory behavior: transitivity in the preferences between lotteries (discrete probability distributions over the set of alterna-

Figure 2. L_E and our indifference towards the participation in the lottery or the participation in the event with the indeterminate outcome

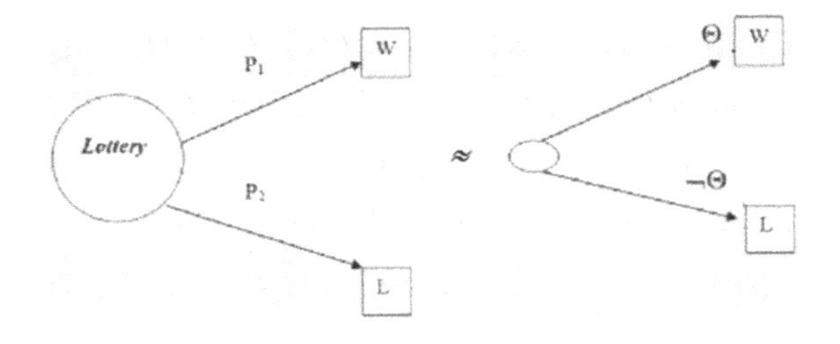

tives), transitivity in the relation of indeterminacy or indifference ($\approx$) and mutual replaceability of the indistinguishable lotteries, then we can treat our subjective uncertainty and the subjective probabilities as objective when we use them in the process of decision-making and in mathematical formulations and proofs.

In the theory of decision-making the approach of using the mathematical definitions is constructive and descriptive-prescriptive. We are not looking for the ideal rational human in his estimates and preferences. We are looking for the expressions of the internal confidence of the subject making decisions when faced with problems for choice and the uncertainty of this choice. We are looking for coordinated and logically sound decision based on internal conviction and knowledge. The decision maker consciously determines the agreement of his subjective reasoning and describes them in strict mathematical definitions and conclusions. In this manner, the decision-making process is freed from internal inconsistency and incoherence. The final result is mathematically based and founded on the knowledge and confidence of the decision maker and is expressed as the most correct (mathematically optimal) choice of activity in the problem under resolvation (Raiffa, 1968; Keeney, 1994, 1988).

Now we are going to approach mathematically exact the description of the term subjective probability following the approach of Pfanzagl (1971). We are going to consider the SR $(\boldsymbol{X},(\approx),(\succ))$ with defined in it order of preferences and mathematical operations describing lotteries and operations on them. As a final solution, we reach to a description in which the subjective probabilities are described with the same set of axioms and rules as the objective probabilities.

Let in SR $(\boldsymbol{X},(\approx),(\succ))$ a weak order is given, i.e. connectivity and transitivity of the relations ($\approx$) and ($\succ$). The relation of indifference ($\approx$) is congruency with respect to all mathematical operations, defined over the set $\boldsymbol{X}$. Every algebraic operation and its final result in $\boldsymbol{X}$ can be described as a relation. We are going to consider the binary operation (*) defined over $\boldsymbol{X}$. For example this can be multiplication, addition or division in the set of real numbers if $\boldsymbol{X}=\boldsymbol{R}$. Then the operation (*) to every pair $(a, b)\in\boldsymbol{X}^2$ corresponds to element (c), $c\in\boldsymbol{X}$. This can be described as 3-ary relation in $\boldsymbol{X}$, $(a, b, c)\in R_*$ or $(a*b=c)$.Then we have SR $(\boldsymbol{X}, (\approx), (\succ), *)$ or in other words $(\boldsymbol{X}, (\approx), (\succ), R_*)$.The relation of indifference ($\approx$) is congruency with respect to every operation defined in $\boldsymbol{X}$. This for the operation (*) will mean the following:

$$(a \approx a' \wedge b \approx b') \Rightarrow (a * b = a' * b').$$

Let Ω be a system of "events," which can be interpreted as a system or the set of subsets of $\boldsymbol{X}$. We say that Ω is a *Boolean algebra*, if here are defined the operations $\cap$, $\cup$ and $\bar{\ }$, which have the properties in Box 2.

We are going to define "ideal," as nonempty subset $\Im\subset\Omega$, for which the following holds:

Box 2.

$$\begin{aligned}
&\mathit{Commutativity}: P\cap Q = Q\cap P,\ P\cup Q = Q\cup P,\\
&\mathit{Associativity}: P\cup(Q\cup R) = (P\cup Q)\cup R,\ P\cap(Q\cap R) = (P\cap Q)\cap R,\\
&\mathit{Distributivity}: P\cap(Q\cup R) = (P\cap Q)\cup(P\cap R),\ \ P\cup(Q\cap R) = (P\cup Q)\cap(P\cup R),\\
&\mathit{Absorption}: (P\cup Q)\cap Q = Q,\ (P\cap Q)\cup Q = Q,\\
&\qquad\qquad (P\cap\overline{P})\cup Q = Q,\ (P\cup\overline{P})\cap Q = Q.
\end{aligned}$$

1. $P \cup Q \in \Im$, for $\forall$ P, $Q \in \Im$,
2. $(P \in \Im \wedge Q \subseteq P) \Rightarrow Q \in \Im$.

To the defined this way ideal corresponds the relation of equivalency ($\approx_{\Im}$), defined by the conditions ($P \approx_{\Im} Q$) if and only if $P \cap \bar{Q} \in \Im$ and $\bar{P} \cap Q \in \Im$. We can construct the ideal $\Im$ using arbitrary event (subset) R and all of its sub events. If instead the subset R we take its complement in *X,* then the ideal $\Im_R$ we are going to define as the set of subsets of *X,* for which the following holds:

$$\Im_R = \{Q \in \Omega / P \cap Q = \varnothing\}.$$

In the literature source (Pfanzagl, 1971), it is proven that ($\approx_{\Im}$) is a relation of congruency with respect to the operations $\cap$, $\cup$ and $\overline{(.)}$.Then we can factorize the set Ω and define the factor algebra $\Omega/\Im$. Every element of this factor algebra is class of equivalency, set of equivalent subsets in *X* with respect to the congruency ($\approx_{\Im}$). The natural homomorphism m(.): $\Omega \rightarrow \Omega/\Im$ preserves the operations $\cap$, $\cup$ and $\overline{(.)}$.The following holds:

(i) $\mathbf{m(P) \cup m(Q) = m(P \cup Q)}$,
(ii) $\mathbf{m(P) \cap m(Q) = m(P \cap Q)}$,
(iii) $\mathbf{\overline{m(P)} = m(\bar{P})}$.

Under these conditions the factor algebra $\Omega/\Im$ are again Boolean algebra and the partial ordering between the equivalency classes is defined as follows:

$$\mathbf{(m(P) \subset m(Q)) \Leftrightarrow (m(P) \cap m(Q) = m(P))}.$$

Equivalent condition is the following (Pfanzagl, 1971):

$$\mathbf{(m(P) \subset m(Q)) \Leftrightarrow (P \cap \bar{Q} \in \Im)}.$$

Then every element of the factor algebra can be interpreted as a conditional (uncertainty) event. We have in mind the realization of Q, $Q \in \Omega$ under the condition that R is realized ($\Im = \Im_R$ in the case). This conditional event we are going to denote as an event from the factor algebra $(Q| R) \in \Omega/\Im$. After a second factorization as we use again an ideal in $\Omega/\Im$, we reach the following equality (Pfanzagl, 1971):

$$((Q| R)| (P| R)) = (Q| P \cap R).$$

We define the lottery (aPb), as a lottery in which we choose (a) if P has occurred or (b), if $\neg$P has occurred (non-occurrence of P).

We define the set *XPX* as the set (*XPX*)={aPb/ a,b$\in$*X,* P$\in\Omega$}. Over it we define an order, which preserves the order of comparison of lotteries over *X* (discrete distributions).

We need some more definitions. We start with the *axiom for unicity,* which is stated in Box 3.

We accept the following *postulate*, defined over the lotteries of the type aPb:

Box 3.

$$(\exists(a,b \in \mathbf{X}) \wedge (aPb \begin{Bmatrix} \succ \\ \approx \\ \prec \end{Bmatrix} aQb)) \Leftrightarrow ((xPy \begin{Bmatrix} \succ \\ \approx \\ \prec \end{Bmatrix} xQy), \forall(x,y \in \mathbf{X})).$$

$$\mathbf{aPb \approx b\overline{P}a, \forall\ a, b \in X,}$$
$$\mathbf{P \approx \varnothing \Rightarrow (aPb \approx a'Pb,, \forall\ a, a' \in X),}$$
$$\mathbf{P \neq \varnothing \Rightarrow (aPb \prec a'Pb,\ a \prec a', \forall\ a, a', b \in X).}$$

It is obvious that this is a meaningful condition. The first condition is obvious. The second condition means that the impossible outcome condition from a lottery does not matter. The third condition says that if in a lottery we put more useful for us alternatives then it becomes more preferable.

Now we are going to define the next axiom, the *axiom for continuality*, which says: The correspondence a→aPb defines continuous monotone correspondence. The same holds for the second variable b.

Then it is known (Pfanzagl, 1971) that for every pair of elements (b, a) from ***X***, for which (b≻a) there exist an element (a*b), such that for it the lotteries aPb≈((a$*_p$b)P(a$*_p$b)) are equivalent. This operation we denote as (a$*_p$b) in SR (***X***, (≈), (≻),$*_p$) or in other notations SR (***X***, (≈), (≻), R_{*p}). It is continuous, increases with respect to both variables, and preserves the equivalency with respect to the uncertainty events. The last means that if P≈Q in terms of the *axiom for unicity* as lottery, then (a$*_p$b)= (a$*_Q$ b) for every alternative in the main set ***X***.

Let us consider the more complicated case of the appearing of the event Q, under the condition that R holds. Let us recall the notation (Q| R)∈Ω/ℑ. We can define lottery (aQ| Rb). We choose the first alternative, if Q holds, under the condition that R is true or the second if Q is false under the condition that R holds. We define two new *postulates*.

The first has the form

$$((a\ *_{Q\ |\ R}\ b)\ *_R (a *\ _{Q|\overline{R}}\ b))$$
$$= ((a\ *_{R\ |\ Q}\ b)\ *_Q (a *\ _{R\ |\ \overline{Q}}\ b)).$$

The second states that for every two alternatives from the main set ***X*** and for every two events Q and R holds:

$$((a\ *_{Q\ |\ R}\ b)\ *_R b) = (a\ *_{R \cap Q}\ b).$$

Under the condition that these *axioms and postulates* hold in (Pfanzagl, 1971) is proven the following main theorem for the utility function.

5.2. Theorem

For every two events P∈Ω and Q∈Ω, where P≠∅ and for every two elements from the set ***X***, a,b∈***X*** the following holds:

$$\mathbf{U}(a\ *_{Q\ |\ R}\ b) = \frac{s(Q \cap R)}{s(R)}\mathbf{U}(a) + (1\text{-}\ \frac{s(Q \cap R)}{s(R)})\mathbf{U}(b).$$

The function s(P) has the meaning of a probability over the set of events P∈Ω, including the set of conditional events (Q| R)∈Ω/ℑ. The following holds s(***X***)=1 and s(∅)=0. This function is additive. The only fact that is not derived for it in order to become classical probability measure is the σ-additivity. Something more, in the conditional lottery (Q|R)∈Ω/ℑ we do not accept independence between Q and R in advance, as it is in the theory of von Neumann and in the derivation of professor Raiffa (Fishbern, 1970; Raiffa, 1968). As a final derivation of this mathematical formulation we can say that we can work with the conditional probability, we can work as with finite additive probability measure.

As a conclusion, we present one example of professor Raiffa for estimation of the subjective probability (Raiffa, 1968). Similar examples have being discussed in both aspects, as a criticism of the mathematical approach in decision-making, but also as a grounding of the necessity of exact methods for the clearance of the process of decision-making from the internal discrepancies.

Figure 3. Green and white urns

100 yellow P(yellow)=0.5 P(green box)=0.5
100 red P(red)=0.5 P(white box)=0.5

(event) = YYRYRYYYRYRY 8-yellow; 4-red
Probability(green box)= ?

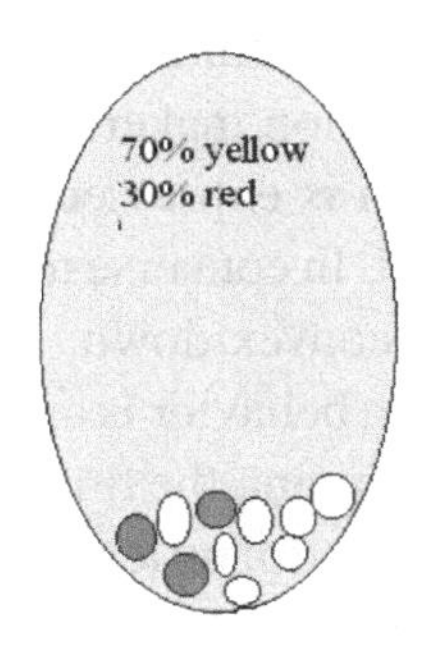

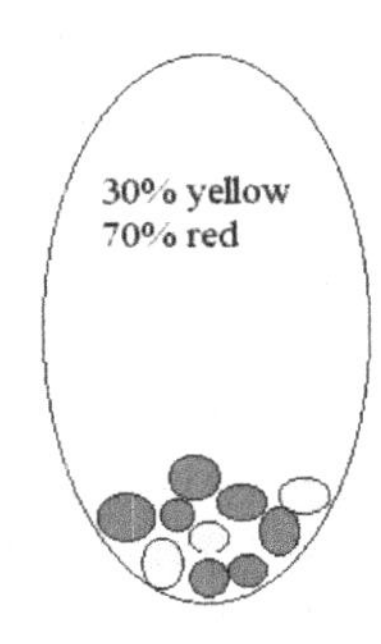

The example is the following. Let us have two urns, white and green, as in the first urn there are 70 yellow and 30 green balls. In the second urn there are 30 yellow and 70 red balls. The urn is in dark room so that we cannot see their colors. From one of them 8 yellow and 4 red balls are taken out. We ask the following question toward the Decision maker. What is the probability that this is the green urn? The process can be followed in Figure 3.

The usual answers for the estimate of the subjective uncertainty event are between 0.6 till 0.8. We are going to use the Bayesian theorem for the exact estimate of the probability. The result is rather unexpected. (see Box 4.)

We can see that similar subjective estimates are relative. They themselves are determined with an indetermination of subjective type. This can be partially accepted as a ground for using of the rank dependent utility model and its derivative cumulative prospect theory. Now we know that in the Rank Dependence Utility (RDU) the decision weight of an outcome is not just probability associated with this outcome. It is a function of both probability and the rank (the alternative) x. For example the RDU of the lottery (p_1, x_1; p_2, x_2; ...; p_n, x_n) is:

$$RDU = \sum_{i=1}^{n} W(p_i)u(x_i). \qquad (3.5.1)$$

The weighting function W(.) has the inverse S-shaped form $W(p_i)= p_i+\Delta W(p_i)$.

An interesting moment is the problem for estimation and approximation of the utility function as a function of two variables. The two variables are the probability in the lottery and the values, following the RDU model.

6. USE CASES

In conclusion of this chapter, we are going to consider the type of one of the most famous in the economical environments utility function. This is the utility function for gains and losses investigated by Kahneman and Tversky in the famous prospect theory (Kahneman, 1979), shown in Figure 4.

It describes how people frame and value a decision involving uncertainty. According to this theory, people look at choices in terms of potential gains or losses in relation to a specific reference point, which is often the purchase price (Kahne-

Box 4.

$$\begin{aligned} &\mathrm{P}(\text{green box}) \\ &\quad = (\mathrm{P}(\text{event / green box}).\mathrm{P}(\text{green box})) / (\mathrm{P}(\text{event / green box}).\mathrm{P}(\text{green box}) \\ &\quad +\mathrm{P}(\text{event / white box}).\mathrm{P}(\text{white box})) \\ &\quad = \left((0,7)^8 \times (0,3)^4 \times 0,5\right) / \left((0,7)^8 \times (0,3)^4 \times 0,5 + (0,7)^4 \times (0,3)^8 \times 0,5\right) = 0.964. \end{aligned}$$

Figure 4. Utility function for gains and losses investigated by Kahneman and Tversky in the famous prospect theory (Kahneman, 1979)

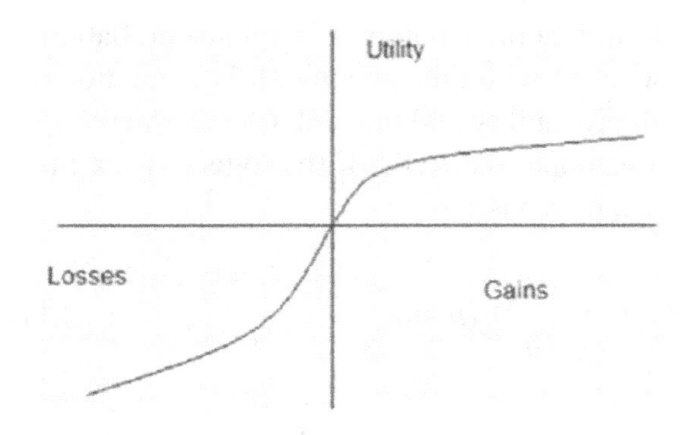

Figure 5. Inflex points in the area of gains and losses

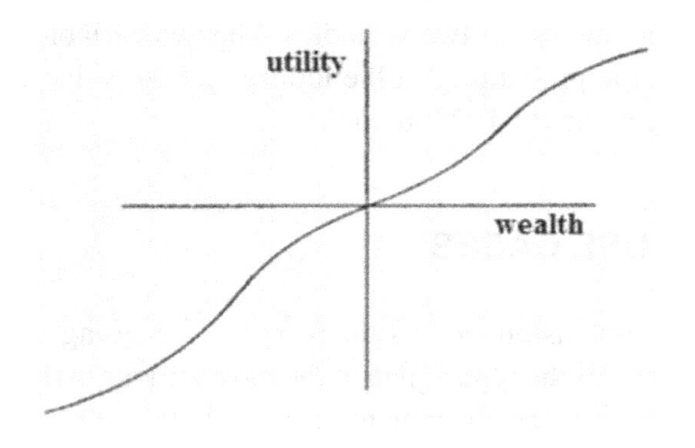

man, 1979; Mengov, 2010). The utility function is concave for gains. This means that people feel good when they gain, but twice the gain does not make them feel twice as good. The utility function is convex for losses. This means that people experience pain when they lose, but twice the loss does not mean twice the pain. The utility function is steeper for losses than for gains. This means that people feel stronger about the pain from a loss than the pleasure from an equal gain-about two and half times as stronger and this phenomenon is referred as loss aversion. This type of utility function was first proposed by Friedman and Savage (Friedman, 1948).

The second moment is that such a function can have inflex points in the area of gains and losses as it is shown in Figure 5.

One of the first scientist investigating functions of the given type is Markowitz (Markowitz, 1952). He has first noted that under the condition of gain the utility function of the decision maker shows inclination toward risk, which is expressed as a convexity in the given interval. In converse under the condition of losses, it is convex down. That means that the standard human behavior is avoiding the risk. Around the initial point, the function is more complex with the presence of inflex points (Kahneman, 1979; Markowitz, 1952; Pratt, 1964).

We have to note that this is averaged utility function. It depictures the average according to the statistics behavior under the condition of gain or loss. In personal plan for separated individuals, such a function can have rather different shape. In concrete problem for decision-making, the shape of the function is important, because it determines the solution. Its change can lead to change of the choice of the optimal alternative in decision problems.

The building of the utility function, adequate of the attitude of the decision maker is a serious mathematical challenge. One of the reasons is that the choices between the lotteries, the preferences are expressed in qualitative aspect. The preferences in the human imaginations are qualitative, with internal uncertainty and are not explicitly expressed easily.

REFERENCES

Aizerman, M., Braverman, E., & Rozonoer, L. (1970). *Potential function method in the theory of machine learning*. Moscow, Russia: Nauka.

Allais, M. (1953). Le comportement de l'homme rationnel devant le risque: critique des postulats et axiomes de l'école américaine. *Econometrica*, *21*, 503–546. doi:10.2307/1907921

Arrow, K. (1950). A difficulty in the concept of social welfare. *The Journal of Political Economy*, *58*(4), 328–346. doi:10.1086/256963

Birnbaum, M. (2008). New paradoxes of risky decision-making. *Psychological Review*, *115*(2), 463–501. doi:10.1037/0033-295X.115.2.463

Cohen, M., & Jaffray, J.-Y. (1988). Certainty effect versus probability distortion: An experimental analysis of decision-making under risk. *Journal of Experimental Psychology. Human Perception and Performance*, *14*(4), 554–560. doi:10.1037/0096-1523.14.4.554

Dixmier, J. (1967). *Cours de mathematiques*. Paris, France: Gautier-Villars.

Ekeland, I. (1979). *Elements d'economie mathematique*. Paris, France: Hermann.

Engelking, R. (1985). *General topology*. Warsaw, Poland: Panstwove Widawnictwo Naukove.

Farquhar, P. (1984). Utility assessment methods. *Management Science*, *30*, 1283–1300. doi:10.1287/mnsc.30.11.1283

Fishburn, P. (1970). *Utility theory for decision-making*. New York, NY: Wiley.

Friedman, M., & Savage, L. (1948). Utility analysis of choices involving risk. *The Journal of Political Economy*, *56*(4), 279–304. doi:10.1086/256692

Kahneman, D., & Tversky, A. (1979). Prospect theory: An analysis of decision under risk. *Econometrica*, *47*, 263–291. doi:10.2307/1914185

Keeney, R. (1988). Value-driven expert systems for decision support. *Decision Support Systems*, *4*(4), 405–412. Retrieved from http://www.sciencedirect.com/science/article/pii/0167923688900036-foot1#foot1 doi:10.1016/0167-9236(88)90003-6

Keeney, R., & Raiffa, H. (1993). *Decision with multiple objectives: Preferences and value trade-offs*. Cambridge, UK: Cambridge University Press. doi:10.1109/TSMC.1979.4310245

Kelly, J. (1955). *General topology*. New York, NY: Ishi Press.

Krantz, D. H., Luce, R. D., Suppes, P., & Tversky, A. (1971). *Foundations of measurement* (*Vol. 1*). New York, NY: Academic Press.

Krantz, D. H., Luce, R. D., Suppes, P., & Tversky, A. (1989). *Foundations of measurement* (*Vol. 2*). New York, NY: Academic Press.

Krantz, D. H., Luce, R. D., Suppes, P., & Tversky, A. (1990). *Foundations of measurement* (*Vol. 3*). New York, NY: Academic Press.

Litvak, B. (1982). *Expert information, analysis and methods*. Moscow, Russia: Nauka.

Machina, M. J. (1987). Choice under uncertainty: Problems solved and unsolved. *The Journal of Economic Perspectives*, *1*, 121–154. doi:10.1257/jep.1.1.121

MacLane, S., & Birkhoff, G. (1974). *Algebra*. New York, NY: The Macmillan Company.

Markowitz, H. M. (1952). Portfolio selection. *The Journal of Finance*, *7*, 77–91.

Mengov, G. (2010). *Decision-making under risk and uncertainty*. Sofia, Bulgaria: Publishing House JANET 45.

Pavlov, Y. (2005). Subjective preferences, values and decisions: Stochastic approximation approach. *Proceedings of the Bulgarian Academy of Sciences*, *58*(4), 367–372.

Pavlov, Y. (2011). Preferences based stochastic value and utility function evaluation. In *Proceeding of Conferences InSITE 2011,* (pp. 403-411). Novi Sad, Serbia: InSITE.

Pfanzagl, J. (1971). *Theory of measurement*. Wurzburg, Germany: Physical-Verlag.

Pratt, J. (1964). Risk aversion in the small and in the large. *Econometrica*, *32*, 122–136. doi:10.2307/1913738

Raiffa, H. (1968). *Decision analysis*. Reading, MA: Addison-Wesley.

Roberts, F. (1976). *Discrete mathematical models with application to social, biological and environmental problems*. Englewood Cliffs, NJ: Prentice-Hall, Inc.

Shiryaev, A. N. (1989). *Probability*. New York, NY: Springer.

Shmeidler, D. (1989). Subjective probability and expected utility without additivity. *Econometrica*, *57*(3), 571–587. doi:10.2307/1911053

Tversky, A., & Kahneman, D. (1991). Loss aversion in riskless choice: A reference dependent model. *The Quarterly Journal of Economics*, *106*, 1039–1061. doi:10.2307/2937956

Weber, E. (1994). From subjective probabilities to decision weights: The effect of asymmetric loss functions on the evaluation of uncertain outcomes and events. *Psychological Bulletin*, *115*(2), 228–242. doi:10.1037/0033-2909.115.2.228

Chapter 4
Elements of Utility Theory

ABSTRACT

The value and utility functions are homomorphisms from the empiric system "Decision maker, preferences" to subset of the set of the real numbers with relation "greater" or "more preferred." The so-called axiomatic (normative) approach considers the conditions for existence of value and utility functions. This chapter considers different variants of theorems for the existence of value and utility functions. These theorems may be assumed as basic in the formulations of problems and the models for development of mathematical approaches and methods for evaluation and analytical polynomial representation of the value and utility functions, since these theorems are conceptually such. In practice, almost all sets used in the construction of the value and utility functions in the practice are discrete or countable. That is why we discuss in detail the case of discrete sets of alternatives.

In most of the cases in practice, multifactor problems have been solved and these investigations need multiattributte utility or value functions. However, the human capabilities for the simultaneous taking a right solution and right estimates and preferences in multifactor problems are not good. In such cases, the satisfaction of several conditions, which are acceptable for most of the real problems, allows the decomposition of the utility function, which is useful for the application's mono factors shape. In this chapter, the authors discuss three fundamental theorems, providing the most general decomposition of multiattributte utility to simpler utility functions.

In the scientific literature, there are numerous examples of investigations and applications of the utility theory in various areas of human activity. The authors discuss in the chapter three areas of applications of theory of measurement and utility theory. As a first area of application and investigation, they make a parallel between the fuzzy sets theory and the utility theory in one example described from the information system MATLAB (MathWorks Inc.).

DOI: 10.4018/978-1-4666-2967-7.ch004

As another nonstandard example of investigation, the authors consider the usage of the utility theory for the estimate of the ecological effect of the decomposition of the bio-substance of the animal breeding farms. In this example, we show the possibility for decomposition of a multiattributte function to simpler functions.

In the science literature there are many examples of investigation and application of the theory of measurements and utility theory in different areas of human activity. The authors use in the following chapters their achievements for applications in optimal control of bioprocesses for a complete mathematical description of complex systems "technologist- dynamical process," and in the agriculture for modeling difficult examples and for formalizing another way.

1. BEGINNINGS: THE ST. PETERSBURG PARADOX

In the previous chapter, we presented the fundamental concepts in the theory of measurement and scaling. There we defined the term value or utility, as a measurement scale of the preferences of the decision maker. Let $(\boldsymbol{X}, (\mathrm{R}))$ is a system with relations (SR) defined over the set $\boldsymbol{X}$ together with preference relation over it $(x \succ y)$ $(\mathrm{R} \equiv (\succ)$, (*strict preference* or *strict order* or *strong order*, weak order and weak connectedness). In this manner, a structure is introduced in the set $\boldsymbol{X}$, $(\boldsymbol{X}, (\succ))$. If there exists homomorphism $\mathrm{u}(.):(\boldsymbol{X}, (\succ)) \rightarrow (\boldsymbol{R}, (>))$, then this homomorphism will be called value function. In the above formulas, we denote by $\boldsymbol{R}$ the set of the real numbers or a subset of it, and with $(>)$ the relation greater, defined over the real numbers. Following (Pfanzagl, 1971) this homomorphism is defined as follows:

$$((\mathrm{x}, \mathrm{y}) \in \mathrm{X}^2, \mathrm{x} \succ \mathrm{y}) \Leftrightarrow (\mathrm{u}(\mathrm{x}) > \mathrm{u}(\mathrm{y})).$$

This value function is also called perfect value function. If the homomorphism is defined by the inclusion $(\boldsymbol{X} \subseteq \mathrm{u}^{-1}(\boldsymbol{R}))$, then the definition of value function is the following:

$$((\mathrm{x}, \mathrm{y}) \in \boldsymbol{X}^2, \mathrm{x} \succ \mathrm{y}) \Rightarrow (\mathrm{u}(\mathrm{x}) > \mathrm{u}(\mathrm{y})).$$

In other words, the value function is a homomorphism from the empiric system (Decision Maker (DM), preferences) to (subset of the set of the real numbers with relation greater). We assume that the set $\boldsymbol{X}$ is countable, i.e. bijective to the set of the natural numbers (the set of integers). Then the following propositions are fulfilled (Fishburn, 1970).

1.1. Proposition

Value function u(.) for the preference relation $(\succ)$ of $\boldsymbol{X}$ exists if and only if the relation is acyclic (i.e. there does not exist $(x_1, x_2, x_3, \ldots, x_h)$, for which $(x_1 \succ x_2, x_2 \succ x_3, x_3 \succ x_4, \ldots, x_{h-1} \succ x_h, x_h \succ x_1))$.

This proposition is obvious and follows directly from the properties of the relation $(>)$ defined over the real numbers and the definition of the homomorphism. We define the relation "indifference, indiscernibility or equivalence" as $((x \approx y) \Leftrightarrow \neg((x \succ y) \vee (x \prec y)))$. In Fishburn (1970), the following theorem is proved.

1.2. Theorem

Let the relation $(\succ)$ is weak order (asymmetric and negatively transitive relation) over the set of alternatives. Then it is fulfilled:

For any two alternatives (x, y) one of the following relations is fulfilled $(x \succ y)$, $(y \succ x)$ or $(x \approx y)$:

- The relation $(\succ)$ is transitive;
- The relation $(\approx)$ is equivalence;
- $((x \succ y) \wedge (y \approx z)) \Rightarrow (x \succ z)$, $((x \approx y) \wedge (y \succ z)) \Rightarrow (x \succ z)$;

- The relation $((x \succ y) \vee (x \approx y))$ "more preferred or equivalent" is transitive and connected;

The relation $(\succ')$ over the factor set $(\boldsymbol{X}/\approx)$ is strong order (weak connected weak order) $(\boldsymbol{a}, \boldsymbol{b} \in \boldsymbol{X}/\approx, (\boldsymbol{a} \succ' \boldsymbol{b}) \Leftrightarrow (\exists\ x, y \in \boldsymbol{X}$, such that $(x \in \boldsymbol{a}) \wedge (y \in \boldsymbol{b}) \wedge (x \succ y))$.

We investigated part of the above propositions in the previous chapter 3. We saw that asymmetricity and the negative transitivity lead to the transitivity of the preference relation $(\succ)$ and transitivity of $(\approx)$. Now we will state a fundamental theorem ensuring the existence of perfect value function—Fishburn's terminology (Fishburn, 1970).

1.3. Theorem

Let the set $(\boldsymbol{X}/\approx)$ is countable. Perfect value function u(.) over $\boldsymbol{X}$ for the preference relation $(\succ)$ exists if and only if, when $(\succ)$ is a weak order.

Proof (Fishburn, 1970)

We number the elements of $(x_1, x_2, x_3, \ldots, x_h, \ldots)$ in the countable set $(\boldsymbol{X}/\approx)$ and let $(r_1, r_2, r_3, \ldots, r_h, \ldots)$ is some enumeration of the rational numbers (some ordering of the rational numbers). We put $u(x_1)=0$. Let us choose on the (h)-th step element (x_h). The following cases are possible:

- For every i, $(i< h)$ is fulfilled $(x_h \succ x_i)$ and then we put $u(x_h)=h$;
- For every i, $(i< h)$ is fulfilled $(x_i \succ x_h)$ and then we put $u(x_h)=-h$;
- There exist (x_j) and (x_i) for which $(x_j \succ x_h \succ x_i)$ and there does not exist (x_k), such that $(x_j \succ x_k \succ x_i)$. In this case we choose the first rational number in the chosen from us initial ordering of the rational numbers, for which it is fulfilled $((u(x_j) \succ r_k \succ u(x_i))$ and we put $u(x_h)=r_k$. The above construction we apply inductively for every (h). If two alternatives $(x,y) \in \boldsymbol{X}^2$ belong to one and the same class of equivalence $(x \approx y)$, then we set $u(x)=u(y)$.

The so-constructed function is a perfect value function, $((x, y) \in X^2, x \succ y) \Leftrightarrow (u(x)>u(y)))$. The converse was proven in the previous chapter 3.

From the above theorems, it follows that the presence of perfect value function leads to grouping of the alternative in indifference classes and each alternative belongs to exactly one such class. This is a corollary of the transitivity of the relation $(\approx)$. If we assume that for the set $\boldsymbol{X}$ is fulfilled $\boldsymbol{X} \subseteq \boldsymbol{R}^2$, then we can split $\boldsymbol{X}$ into indifference classes (classes of indifference or equivalence). Two points from such a class are equally preferred. The economists use in this case the term "indifference curve" (Keeney & Raiffa, 1993) (Figures 1 and 2).

The example for the indifference curves is taken from the investigation of Terzieva et al. (2007). In decision-making practice the relation $(\approx)$ is one of the most violated. In the scientific literature are cited "countless" similar case of violation of the transitivity of $(\approx)$. The reasons may be threshold of indiscernibility, lack of knowledge, errors in the preferences and inconsistencies in the reasoning (inconsistency in the judgments) of the DM and others. It should be expected that by accumulation of knowledge re-

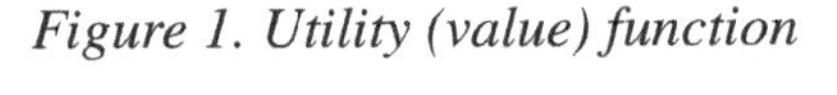

Figure 1. Utility (value) function

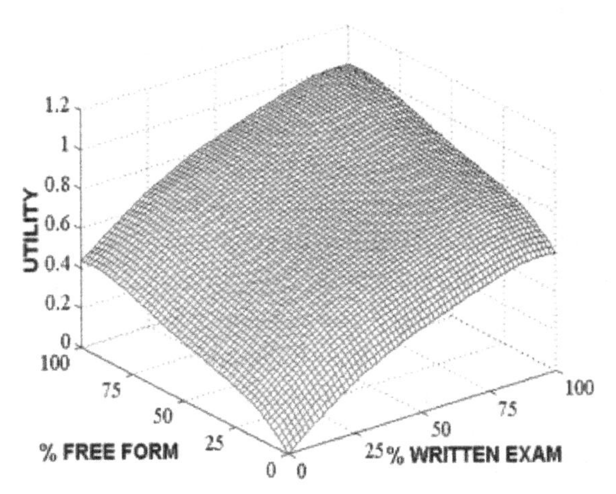

Figure 2. Curve of indifference

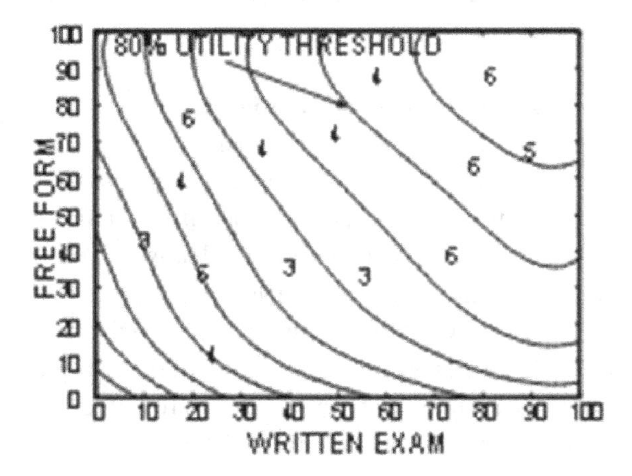

garding the object of investigation, the threshold of indiscernibility of the decision maker will be decreased and the errors in the reasoning will disappear. Of course, for verbally described tasks and qualitative reasoning this process will remain incomplete. If we are in the framework of the normative, axiomatic approach then this is a violation of the axiomatic principles and mathematical constructions and theorems permitting non-transitivity in the relation $(\approx)$ should be sought (Machina, 1987). Similar mathematical constructs may be seen in Machina (1982) and Fishburn (1988). But if the absolute fulfillment of the conditions in the theorems above is not an end in itself, then we are in the prescriptive approach, in which only practically meaningful for the particular problem conditions are verified and for which the theorems for existence of value or utility functions are taken as guiding and not as normative. In this approach we assume that the errors and violations with respect to the transitivity of the relation $(\approx)$ and the relation $(\succ)$ have random nature with mathematical expectation zero and finite variance (dispersion). We also suppose that every comparison (preference) of two alternatives is done independently from the rest of the other similar comparisons. Under this conditions the violation of the transitivity of the relations $(\approx)$ and $(\succ)$ can be eliminated using the techniques of the stochastic programming.

As it was mentioned in the previous chapter, the ordering scale was defined by homomorphisms, monotonous functions. But if we seek equivalence of SR $(\boldsymbol{X}, (\approx), (\succ))$ with SR ($\boldsymbol{R}$-real numbers, $(=), (>)$), the homomorphisms need to be not only monotonous but continuous and then the ordering in the real numbers will be reflected in the empirical set $\boldsymbol{X}$ with the properties of the interval topology generated by the relation $(>)$ in $\boldsymbol{R}$. The relation $(\succ)$ in the empirical set $\boldsymbol{X}$ generates its own interval topology $\mathcal{T}_X$. We know that every monotonous mapping $u(.)$ is continuous for the interval topology $\mathcal{T}_{u(X)}$ generated by the relation $(\succ)$ in $u(\boldsymbol{X})$ and defined through the monotonous transformation. But the topology $\mathcal{T}_{u(X)}$ is coarser than topology $u(X)\cap\mathcal{T}_{R}$, induced by the interval topology of the real numbers. We have equivalence if the interval topology $(\boldsymbol{X},\mathcal{T})$ generated by the relation $(\succ)$ has a countable base. These questions were briefly discussed in chapter 3. If we use the ordering scale defined by continuous function, then in this case the convergence in $(\boldsymbol{X},\mathcal{T})$ will have the same properties as in the set of the real numbers and in their SR ($\boldsymbol{R}$-real numbers, $(=), (>)$). Then the concept of convergence in measurements coincides with the standard and widespread outside the mathematical circles notion of convergence. The theorems for existence of continuous value function may be found in Fishburn (1970), Pfanzagl (1971), and Ekeland (1979).

Fishburn (1970) considers different topological variants of theorems for existence of continuous value functions. These conditions are difficult for practical checking in their topological versions. These topological conditions may be assumed as basic in the formulation of the problems and the models of investigation, since they are conceptually such. In practice, almost all sets used in the description of the choice variants or in the construction of the value functions are discrete or countable. Under these conditions, it can be assumed that they are subset of set with greater cardinality, and over these bigger sets, we search definitions and construction of continuous value functions. For example, it is sufficient to suppose that X is metrizable, locally compact and separable. Over such set we can define positive Radon measures (and in particular Lebegue's integral), as well as the conditions guaranteeing the existence of continuous value or utility function. Then the value function over the discrete or countable set is just a restriction of the more generally defined function and we can freely use more powerful analytical methods (Pavlov, 1989).

Now we will consider the case in which the set X is uncountable and we will give a fundamental theorem for this case. We require the following definition.

1.4. Definition

The subset H of the set X, ($H \subseteq X$), is called dense with respect to the order ($\succ$) (to the preference relation) in X, if for every two alternatives $(x,y) \in X^2$, there exists $z \in H$, for which $((x \succ z \succ y)$.

1.5. Theorem

If in the infinite set X there exists strong order ($\succ$) (weakly connected, weak order), then for the existence of perfect value function it is sufficient and necessary to exist ordered, dense, countable subset H of the set X, ($H \subseteq X$).

Proof (Vilkas, 1990)

In its first part, the proof follows the proof in Vilkas (1990). We enumerate the elements of the countable subset H, dense in X ($H \subseteq X$), (x_1, x_2, x_3, ...,x_h,...). We also suppose that H contains the most and least preferred elements in X, if such exist. We put u(x_1)=0. Let us choose at the (h)-th step the element (x_h). The following cases are possible:

- For every i, (i< h) is fulfilled ($x_h \succ x_i$) and then we put u(x_h)=h;
- For every i, (i< h) is fulfilled ($x_i \succ x_h$) and then we put u(x_h)=-h;

There exist (x_j) and (x_i), (i< h, j< h) for which ($x_j \succ x_h \succ x_i$) and there does not exist (x_k), such that ($x_j \succ x_k \succ x_i$). In this case we choose rational number in our chosen initial order of the rational numbers, for which it is fulfilled ((u(x_j)$\succ$ r_k $\succ$ u(x_i)) and we put u(x_h)= r_k.

Under this construction, we currently have:

$$((x, y) \in H^2, x \succ y) \Leftrightarrow (u(x)>u(y)).$$

The above construction is applied inductively for every (h). Thus we construct a perfect value function over H, $((x, y) \in H^2, x \succ y) \Leftrightarrow (u(x)>u(y))$ and we can extend the domain of the function over tne set ($X \backslash H$). For every $x \in X \backslash H$ we put

$$u(x) = \frac{1}{2}(r_1(x) + r_2(x)),$$
$$r_1(x) = \sup_{x_i \prec x}(u(x_i)),$$
$$r_2(x) = \inf_{x_i \succ x}(u(x_i)).$$

Due to the boundedness (H contains the most and least preferred elements if they exist), $r_1(x) = \sup_{x_i \prec x}(u(x_i))$ and $r_2(x) = \inf_{x_i \succ x}(u(x_i))$ exist. It is clear that the so-defined function u(.) is perfect value function with respect to the strong order over X.

The proof of the necessity follows the proof in Fishburn (1970). Let Q is the set of the rational numbers in R. Let J is the set of segments in the set of the real numbers with rational limit points. For every segment, which contains some value u(x), $x \in X$ we choose an element x, $x \in X$. Let A is a subset in the set X, ($A \subseteq X$), constructed in this manner. Obviously, A is countable.

Now we will consider the following set:

$$K = \{(b,c) \,/\, c \in X, b \prec c, \neg\exists a, a \in A, b \prec a \prec c\}.$$

Therefore (u(b),u(c)) does not contain a segment of the set of the real numbers. This means that any two such intervals, constructed with points belonging to the set K are non-intersecting. Thus, the set K is also countable. Then $K \cup A$ is countable and dense in the set X with respect to the preference order ($\succ$), i.e. for every ((x, y)$\in X^2$, x$\succ$y) there exist $a \in K \cup A$, (u(x)> u(a) >u(y)). This concludes the proof.

These two theorems provide the basis conditions for the existence of value functions. The value functions, as was seen in chapter 3, define by concept the extremal points and therefore, can be used as optimal control criteria or for choosing better variant in operation research problems. However, they cannot provide quantitative measure for how much better one variant is than the other. They cannot be used in problems, where there is mathematical expectation because there are needed measurements in scales more powerful than the interval.

Now, to make the transition to utility functions we will consider a famous example in the scientific literature devoted to decision-making known as "*Saint Petersburg Game*" (Fishburn, 1970). This example has been discussed starting from the first naïve years in the formation of decision-making theory. We participate in a game where a coin is tossed. Let μ_n is the probability of "tails" showing for the first time at the nth-toss of the coin (μ_n=2^{-n}). We can play and bet 100 dollars or quit the game. If we win we win (2^n/n), in the opposite we loss 100 dollars. Now, to use the mathematical expectation we suppose that every measurement or function is in the absolute scale. Then the mathematical expectation for win under infinite participation of the game is:

$$\mathrm{M(win)} \geq \sum_{n=1}^{\infty} \frac{(\frac{2^n}{n} - 100)}{2^n} = \infty.$$

Of course, this is possible if we have initial sum sufficient to play this long and consecutively. Let us assume that we have a value (utility) function in the absolute scale of the type:

$$\mathrm{u(x)} = \mathrm{x} \, / \, (|\mathrm{x}| + 10000), \; -1 < \mathrm{u(x)} < 1.$$

Then following (Fishburn, 1970) it is not recommended to participate in the game, because:

$$M(win) = \sum_{n=1}^{\infty} \frac{(x \, / \, (|x| + 10000))}{2^n} < 0.$$

Therefore, in the decision-making, in modeling complex processes and systems with human participation, the final choice in the Bayesian approach depends on the value system of the decision maker and the precise mathematical measurement of the values (utilities) is not only necessary, but imperative. Now we will make a transition to utility functions.

2. AXIOMATIC APPROACH

Until now, we investigated the existence of function, representing analytically the order of preferences under conditions of determinacy. To every choice of the Decision Maker (DM) corresponds one final outcome x, x$\in X$, where X is the set of the final outcomes from arbitrary feasible action of the DM. Now we will consider a more general scheme of interaction between the DM and reality. Let to each choice of DM correspond (i) possible outcomes, (i=1÷n), each of which occurs with probability p_i, $\left(\sum_{i=1}^{n} p_i = 1\right)$. This means that to every choice corresponds as a final result probability distribution. Following the Bayesian approach, it is reasonable to choose an action that will generate result with maximal mathematical expectation $\sum_{i=1}^{n} p_i u(x_i)$ (Raiffa, 1968). In this case, the function $u(x)$ is function with which are evaluated the final alternative $\boldsymbol{x}$, belonging to the set of final results $\boldsymbol{X}$, $\boldsymbol{x}\in\boldsymbol{X}$. In the expected utility theory (Fishburn, 1970; Pfanzagl, 1971) are considered different mathematical models which guarantee the existence of such function $\boldsymbol{u}(.)$, also called utility function.

We will give a mathematical description of the so-formulated problem. Let $\boldsymbol{X}$ be a set of alternatives and $\boldsymbol{P}$ is a subset of discrete probability distributions over $\boldsymbol{X}$. A utility function is any function u(.) for which it is fulfilled (Fishburn, 1970):

$$(p \succ q,\ (p,q)\in \boldsymbol{P}^2) \Leftrightarrow ((\int \boldsymbol{u}(.)dp > \int \boldsymbol{u}(.)dq),\ p\in \boldsymbol{P},\ q\in \boldsymbol{P}).$$

According to Von Neumann and Morgenstern the above formula means that the mathematical expectation of $\boldsymbol{u}(.)$ is a quantitative measure concerning the expert's preferences for probability distributions $\boldsymbol{P}$ over $\boldsymbol{X}$ (Fishburn, 1970; Roberts, 1976; Tenekedjiev, 2004). The DM's preferences over **P**, including those over $\boldsymbol{X}$, ($\boldsymbol{X}\subseteq\boldsymbol{P}$) are expressed by ($\succ$). The "indifference" relation ($\approx$) is defined by $((x\approx y) \Leftrightarrow \neg((x \succ y)\vee(x \prec y)))$. We will emphasize some important moments in this definition. The first is that the preferences ($\succ$) are defined de facto over the set of probability distributions over $\boldsymbol{X}$. The second moment is that the function $\boldsymbol{u}(.)$ is linear with respect to the probability distributions over $\boldsymbol{X}$. This fact, as we saw in chapter 3 of this book guarantees measurement of our preferences in the interval scale. This means that evaluating the uncertainty in the final result of the preferences, this allows us to make measurements in which to express quantitatively the differences between the various final results. The third moment in this book in particular is that we consider the important for practical needs case of finite discrete probability distributions (card($\boldsymbol{X}$) is finite number). The general case for arbitrary probability distribution may be viewed in details in Fishburn (1970). The fourth moment is that the existence of perfect value not linear function over the set of discrete probability distributions $\boldsymbol{P}$, $((p \succ q),\ (p,q)\in \boldsymbol{P}^2) \Leftrightarrow (\Psi(p)>\Psi(q))$, which preserves the preferences defined over the set $\boldsymbol{P}$ is also possible. Now we will state an important theorem from von Neumann theory, which provides the conditions for the existence of utility functions over the set of finite discrete probability distributions $\boldsymbol{P}$.

2.1. Theorem

Let $\boldsymbol{P}$ is the set of all finite, discrete probability distributions over the set $\boldsymbol{X}$ of final outcomes (card($\boldsymbol{X}$) is finite a number). Let ($\succ$) is a binary preference relation over $\boldsymbol{P}$. In order for the real function $\boldsymbol{u}(.)$, satisfying the condition

$$(p \succ q,\ (p,q)\in \boldsymbol{P}^2) \Leftrightarrow ((\int \boldsymbol{u}(.)dp > \int \boldsymbol{u}(.)dq),\ p\in \boldsymbol{P},\ q\in \boldsymbol{P}),$$

to exist it is necessary and sufficient that the following conditions are fulfilled—von Neumann and Morgenstern's axioms:

- The binary preference relation ($\succ$) is weak order;
- *Independence Axiom*: for all $p,q,r\in \boldsymbol{P}$ and let $\alpha\in(0, 1)$, then $(p \succ q)$ if and only if $((\alpha p + (1-\alpha)r) \succ (\alpha q + (1-\alpha)r))$.
- *Archimedean Axiom*: for all $p,q,r\in \boldsymbol{P}$ such that $(p \succ q \succ r)$, there exist $\alpha,\beta\in(0,1)$ such that $((\alpha p + (1-\alpha)r) \succ q)$ and $(q \succ (\beta p + (1-\beta) r))$;

The function $\boldsymbol{u}(.)$ is unique with precision to positive linear transformation. That is, if there exists a function $\boldsymbol{v}(.)$, satisfying the condition:

$$(p \succ q,\ (p,q)\in \boldsymbol{P}^2) \Leftrightarrow ((\int \boldsymbol{v}(.)dp > \int \boldsymbol{v}(.)dq),\ p\in \boldsymbol{P},\ q\in \boldsymbol{P}),$$ then $\boldsymbol{u}(x)=\boldsymbol{a}\boldsymbol{v}(x)+\boldsymbol{b}$, $(\boldsymbol{a}, \boldsymbol{b})\in \boldsymbol{R}^2$, $\boldsymbol{a}>0$.

Proof (Fishburn, 1970)

We saw from chapter 3 of the present book that the presence of utility function and, in particular case, the presence of perfect value function always defines the binary preference relation ($\succ$), as a weak ordering—transitivity of ($\succ$) and of ($\approx$), $((x \approx y) \Leftrightarrow \neg((x \succ y) \vee (x \prec y)))$.

The complete proof of this theorem may be found in Fishburn (1970). Here we will prove that the utility function is determined with precision up to affine transformation. We assume that the functions $u(.)$ and $v(.)$ are equivalent utility functions, $u(.) \approx v(.)$, which means:

$$(\forall(p,q) \in P^2, (p \succ q)) \Leftrightarrow ((\int u(.)dp > \int u(.)dq) \wedge (\int v(.)dp > \int v(.)dq)).$$

Since the two functions $u(.)$ and $v(.)$ are utility functions, they simultaneously fulfill the following conditions:

$$(*)\ (p \succ q) \Rightarrow u(p) < u(q),\ v(p) < v(q),\ (q,p) \in P^2;$$

$$(**)\ u(\alpha p+(1-\alpha)q)= \alpha u(q)+(1-\alpha)u(q),$$
$$v(\alpha p+(1-\alpha)q)=\alpha v(q)+(1-\alpha)v(q), \text{ for } \alpha \in [0,1].$$

If the function $u(.)$ is constant, then the function $v(.)$ is also constant, because they are equivalent. Let us suppose that there exist single point distributions p and q, $(p, q, s) \in X^3$, $(X \subseteq P)$, for which it is fulfilled $(p \succ q)$. We put:

$$f_1(s) = \frac{u(s) - u(q)}{u(p) - u(q)},\ f_2(s) = \frac{v(s) - v(q)}{v(p) - v(q)},\ \forall s \in X.$$

Then the following three cases are possible for $s \neq p$ and $s \neq q$:

$$p \succ q \succ s \Rightarrow \exists\ \alpha \in [0,1],\ ((1-\alpha)p + \alpha s) \approx q;$$
$$p \succ s \succ q \Rightarrow \exists\ \beta \in [0,1],\ ((1-\beta)p + \beta q) \approx s;$$
$$s \succ p \succ q \Rightarrow \exists\ \gamma \in [0,1],\ ((1-\gamma)s + \gamma q) \approx p.$$

The numbers α, β, and γ are the only ones for which the above is fulfilled. From here it follows:

$$((1-\alpha) + \alpha f_i(s)) = 0,\ i = 1, 2;$$
$$f_i(s)) = 1 - \beta,\ i = 1, 2;$$
$$(1-\gamma) f_i(s)) = 1,\ i = 1, 2.$$

Therefore $f_1(s) = f_2(s)$, $s \in X$, this concludes the proof.

From the theorem follows that the *utility measurement scale* is the interval scale. The presented results are part of the so-called axiomatic or normative approach. This can be clearly seen in the previous theorem, where based on three axioms the necessary and sufficient condition for the existence of utility function is given. This approach has been widely criticized over the years, mainly because of the violation of the transitivity in the condition of weak order (Cohen, 1988; Mengov, 2010; Tversky, 1991; Kahneman, 1979; Binbaum, 2008). Here is the place of the famous example of Allais (1953). These questions were discussed in chapter 3 and we accepted the position that the violations of the transitivity may be regarded as random events with mathematical expectation zero and finite variance. Under these conditions, we can use the advancements of the stochastic programming and eliminate the transitivity violations in the weak order in the manner used to eliminate the noise in the stochastic programming. Our long-term experience and practical results have confirmed the validity of these assumptions and of this approach.

3. SOME POPULAR UTILITY FUNCTIONS: RISK AVERSION

In most cases in practice multifactor (multiattribute) problems are solved, as often times the number of factors approaches seven, which number is at the limit of human ability for simultaneous detection and correct judgment. In such case the satisfactions of several conditions, which are acceptable for a large portion of the real-world problems, allows for the decomposition of the utility functions to a more convenient for application form, uni-factor form. This theory is discussed in Kenney (1993). Here we will consider some of the more important of these conditions. Let $X = \prod_{i=1}^{n} X_i$ be the Decart product of (n) in number mono factor spaces. Let us present the set X as Decart product of two of its subspaces $X = Z \times \overline{Z}$. We will state two important from practical point of view definitions.

3.1. Definition

We say that the factor Z does not depend by preference from its complement $\overline{Z}$, if the preferences expressed with respect to Z do not depend on the particular value of $\overline{Z}$:

$$(y', \overline{\overline{y}}) \succ (y'', \overline{\overline{y}}) \Leftrightarrow (y', \overline{y}) \succ (y'', \overline{y}),$$
$$\forall (y', y'') \in Z^2, \forall (\overline{\overline{y}}, \overline{y}) \in \overline{Z}.$$

3.2. Definition

We say that the factor Z does not depend by utility from its complement $\overline{Z}$, if the preferences expressed with respect to lotteries (discrete probability distributions) over Z do not depend on the particular value of $\overline{Z}$.

The theory concerning the decomposition of multifactor value and utility functions is investigated in detail in Keeney (1993). In this chapter, we will state three fundamental theorems, providing the most general decomposition of multifactor utility to simpler utility functions.

3.3. Theorem

Let the set X is Decart product of (n) factors $X = \prod_{i=1}^{n} X_i$. We suppose that each subset from these factors does not depend by utility on its complement. Then if the conditions for existence of perfect utility function are satisfied, it has the form:

$$u(x) = \sum_{i=1}^{n} k_i u_i(x_i) + k \sum_{i=1, j>i}^{n} k_i k_j u_i(x_i) u_j(x_j) + ..$$
$$+ k^{n-1} k_1 k_2 ... k_n u_1(x_1) u_2(x_2) .. u_n(x_n),$$
$$x = (x_1, x_2, .., x_n).$$

In the above formula $k, k_1, k_2, .., k_n$ are coefficients, $u_1(x_1), u_2(x_2), .., u_n(x_n)$ are one dimensional utility functions. This result is a corollary from the fact that the utility functions are known with precision up to affine transformation. The proof of this theorem and the following two theorems may be seen in detail in Keeney (1993). Utility function of this type is called multiplicative.

3.4. Theorem

Let the set X is Cartesian product of (n) factors $X = \prod_{i=1}^{n} X_i$. We suppose that every factor X_i does not depend by utility on its complement. Then, if the conditions for the existence of perfect utility function are satisfied, it will have the form:

$$u(x) = \sum_{i=1}^{n} k_i u_i(x_i) + \sum_{i=1}^{n} \sum_{j>i} k_{ij} u_i(x_i) u_j(x_j) + ..$$
$$+ k_{12 \cdots n} u_1(x_1) u_2(x_2) .. u_n(x_n),$$
$$x = (x_1, x_2, .., x_n).$$

Utility function of this type is called polylinear.

3.5. Theorem

Let the set X is Cartesian product of (n) factors $X = \prod_{i=1}^{n} X_i$. If the preferences over the lotteries defined over $X_1, X_2, .., X_n$ depend only on the marginal distributions, then the perfect utility function has the *additive form*:

$$u(x) = \sum_{i=1}^{n} k_i u_i(x_i),\ x = (x_1, x_2, .., x_n).$$

The case of additive utility function is considered in detail in Fishburn (1970). From theorem 3.5 it is seen that the conditions for using additive function in practice requires the satisfaction of very hard conditions. They may be rephrased as non-interconnectedness of the individual factors. Whether this is so, whether, for instance, in economics the industry does not depend on the raw materials?

The usefulness of the above theorems is evident. For their construction, it is necessary to construct one dimensional utility functions and to determine their respective coefficients. If we have $u(\bar{x}_1, \bar{x}_2, .., \bar{x}_n) = 1$ and $u(\bar{\bar{x}}_1, \bar{\bar{x}}_2, .., \bar{\bar{x}}_n) = 0$, then for every function $u_i(x_i)$ has the form $u_i(x_i) = (\bar{\bar{x}}_1, \bar{\bar{x}}_2, .., x_i, .., \bar{\bar{x}}_n)$. Due to the fact that the utility functions are defined in the interval scale we can accept that $u_i(\bar{x}_i) = 1$ and $u_i(\bar{\bar{x}}_i) = 0$. After investigating the preferences, the values of the coefficients are determined for the polylinear, multiplicative and additive utility functions. These questions are discussed in detail and described in Keeney (1993).

Let us assume that we participate in game, in which at every coin toss we can bet a sum (S) and win that much money for "heads" or lose the same amount for "tails" or we can decline to participate and remain with the initial sum (C). We can describe the game through utility function $u(.)$. In participation in the game, the following cases are possible:

- We play and the win from the lottery <C-S,C+S,½> is $\frac{1}{2}u(C-S) + \frac{1}{2}u(C+S)$ (Farquhar, 1984);
- We do not play and have a fortune $u(C)$.

The following three cases are possible for the lottery participation:

$$u(C) \overset{>}{\underset{<}{=}} \left(\frac{1}{2}u(C-S) + \frac{1}{2}u(C+S)\right).$$

Let us now assume that independently from the values of (C) and (S), the inequality $u(C) > (\frac{1}{2}u(C-S) + \frac{1}{2}u(C+S))$ is always fulfilled. In other words, we always decline lottery participation. Graphically this looks like Figure 3.

This situation corresponds in life, when the utility of increase of personal fortune diminishes with the increase of the fortune. DM, who has some utility function is called over-cautious because he always declines to participate in the lottery <C-S,C+S,½>. For him/her the status quo is preferable independently of (C) and (S):

$$u(C) > \left(\frac{1}{2}u(C-S) + \frac{1}{2}u(C+S)\right).$$

Utility functions of this type (concave) are called "risk-averse." If for the DM is always in-

Figure 3. Form of utility under declination to play

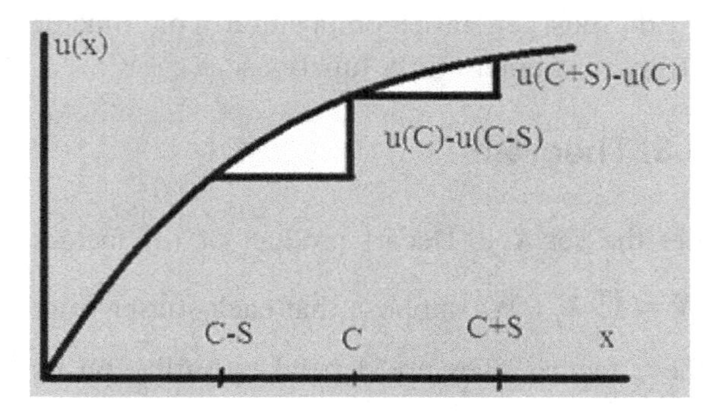

different whether he/she plays the utility function will be a straight line.

The situation changes if the function is convex. For it the DM will always participate in the lottery <C-S,C+S,½>, because the mathematical expectation of the lottery utility will always be greater than the utility of the status quo independently of (C) and (S):

$$u(\mathrm{C}) < (\frac{1}{2}u(\mathrm{C}-\mathrm{S}) + \frac{1}{2}u(\mathrm{C}+\mathrm{S}))$$

Utility functions of this type (convex) are called "risk assume." If we assume that the utility function is continuous, independently of the type of the utility function there exists numberδ, (δ<S), for which it is fulfilled:

$$u(\mathrm{C}-\delta) = (\frac{1}{2}u(\mathrm{C}-\mathrm{S}) + \frac{1}{2}u(\mathrm{C}+\mathrm{S}))\,.$$

This number δ may be called "price of the lottery" <C-S,C+S,½>. This is the price that DM is ready to pay to decline participation in the lottery. Depending on whether the price is positive or negative we have aversion or risk seeking. This quantity and other are used in (Keeney, 1993; Raiffa, 1968) to study the one-dimensional utility functions. However, in practice and from experience we know that the utility functions have inflection points, i.e. more complex form (Friedman, 1948). The methodology proposed here for the evaluation of the utility based on stochastic procedures, constructs polynomial approximation of utility independently of the type of the function (Aizerman, 1970; Pavlov, 2005, 2011). For this reason, we will stop the discussion of the utility functions of one variable here.

4. FIELD STUDIES

The utility theory is concerned with the ways in which the binary preferences may be measured in the respective scales (Litvak, 1982; Fishburn, 1970; Pfanzagl, 1971; Keeney, 1991). In complex, hard to formalize areas of application, like medicine, agriculture, ecology, etc., often times the only way to describe things is verbally. At such level of analysis and description of the objects, the main expression of knowledge is usually the explicitly stated preferences of the DM. In fact, through preferences, the empirical subjective knowledge may be included in the investigation and modeling of the processes with human participation. In such way hard to formalize empirical knowledge may be described mathematically by analytic functions and to be included in logical manner, as a part of complex system or model (Keeney, 1988; Makarov, 1987). We will state the described in this chapter mathematical result and some areas of investigation and application, through some non-standard examples of utility theory application (Pavlov, 1996, 2001).

As a first area of application and research, we will make a parallel between the fuzzy sets theory and the utility theory in an example described in detail in the help of the computational system MATLAB (MathWorks Inc.)(MATLAB-Help (MthWorks Inc.)).

The main theoretic areas, which allow such description, are the fuzzy sets, the decision-making theory and more specifically the normative decision-making theory at which core lies the utility theory (Fishburn, 1970; Pavlov, 2001). Historically the first to develop is the utility theory. Its achievements have been applied since a couple of decades ago (Keeney, 1993; Raiffa, 1968). In this period, the fuzzy theory application emerged. Its main areas quickly lead to application results. The purpose of the current investigation is a comparative analysis of the capabilities and specifics of the two scientific fields through their practical application to a widely known in the scientific

Figure 4. Example in MATLAB-fuzzy logic

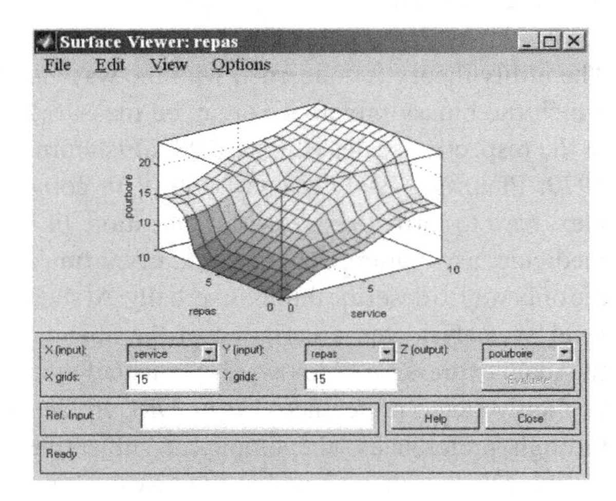

literature example (MATLAB-Help (MathWorks Inc.)). The example is of expert system and fuzzy logic described in MATLAB (MathWorks Inc.). As a demonstration of the capabilities of the fuzzy logic in the "Fuzzy logic" toolbox is proposed expert system to determine the tipping of the waiter depending on the food and the quality of service. This system is described in detail in the help of the "Fuzzy logic" toolbox and can be viewed and studied in detail in the information system MATLAB (MathWorks Inc.)(Figure 4).

This expert system will correspond maximally to its purpose if the built fuzzy membership functions correspond to the particular user. In MATLAB (MathWorks Inc.) there are options for constructing user functions of fuzzy membership, but they are based more on empirical procedures, than on mathematically established algorithms.

We will cite the production rules used in the system:

- (Bad service) or (unsatisfactory food) ⇒ (small tipping);
- (Good service) ⇒ (middle tipping);
- (Excellent service) or (excellent food) ⇒ (generous tipping).

Figure 5. Service: individual membership functions

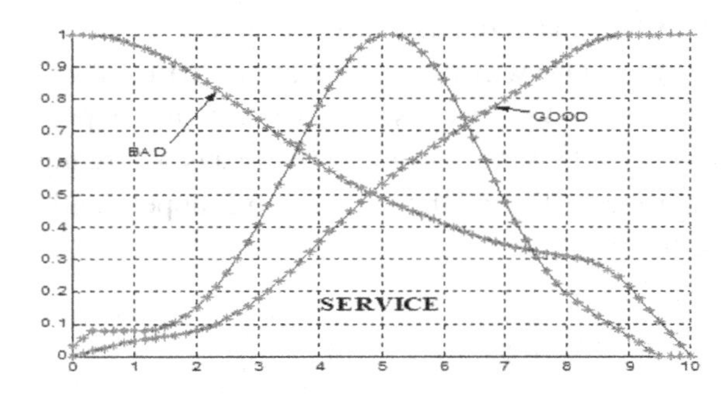

Figure 6. Food: individual membership functions

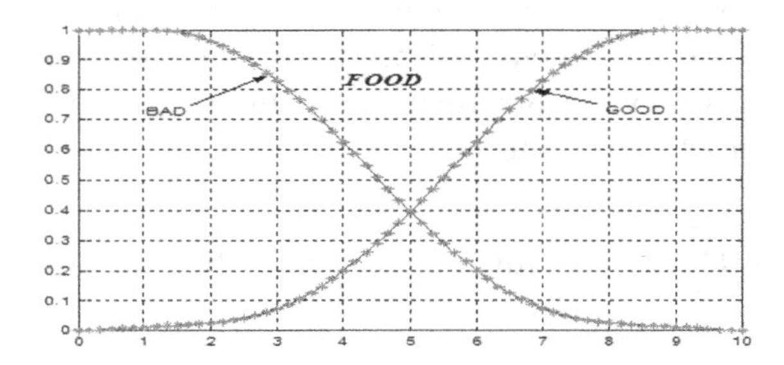

The algorithm is of Mamdani type and the defuzzification method the "centroid."

The used fuzzy functions may be seen in "Fuzzy Toolbox." In our investigation we will construct these function with the stochastic methodology proposed in this book. In the previous year, special software was developed on the base of VISUAL BASIC 6.0 by the authors. Its main part is decision support system and evaluation of the individual utility functions, aimed at the particular subject in Visual studio-Visual basic 6.0 environment. With this software are constructed own membership functions, which may be interpreted as utility functions. The constructed individual membership functions have the form shown in Figures 5-8.

There is slight change in "food." For both membership functions the interval is taken as [0, 10], which from the position of the particular subject was more sensible.

Figure 7. Tipping-membership functions

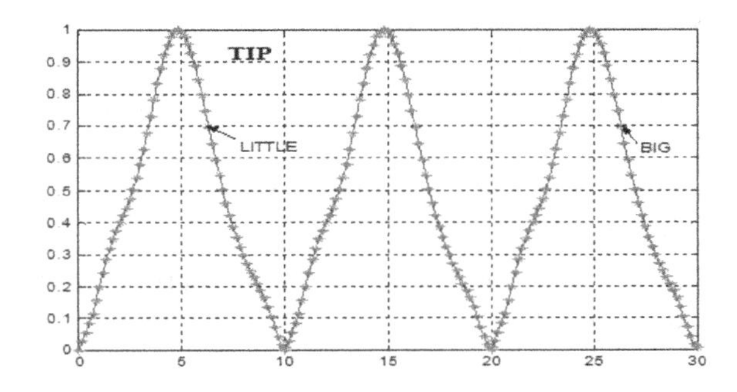

Figure 8. Defuzzification method

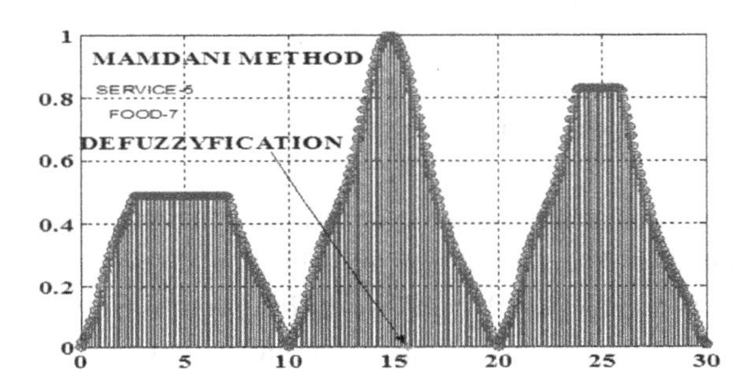

These functions are apparently different from the described in MATLAB (MathWorks Inc.), although the general scheme of the problem is preserved. It is used the same method for deffuzification—"Centroid of the area," Figure 8.

Preserving the Mamdani type algorithm and the defuzzification rule but with the individual membership functions, we have a response surface shown in Figure 9.

It differs strongly from the experimental expert system described in MATLAB (MathWorks Inc.), Figure 5. This definitely means that the final fuzzy model depends highly on the type of the membership functions. In addition, they, evidently, for different user are bound to differ.

Now we will make a comparative analysis between the fuzzy logic and the utility approach. The functional dependence "input-output" is structurally determined by Mamdani model and defuzzification rule. This part of the "fuzzy model"

Figure 9. Expert fuzzy surface, f(x,y), x-service, y-food

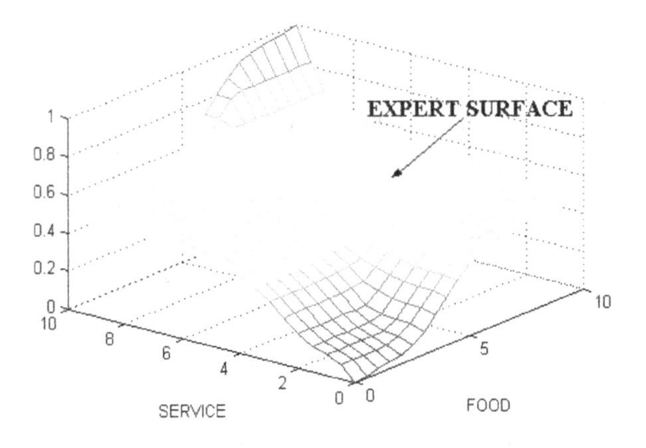

Figure 10. Utility independence

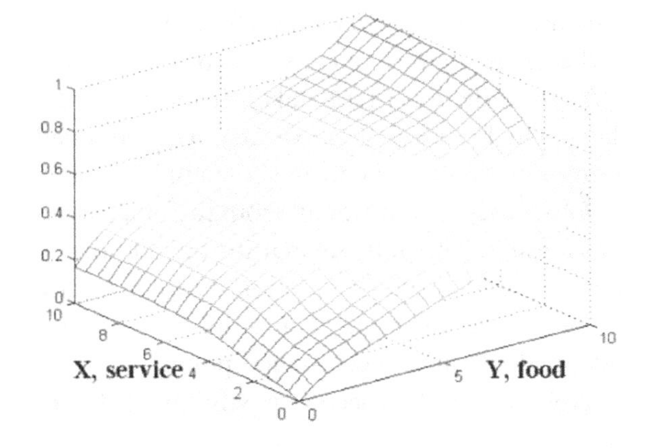

is related to the particular user with the specific production rules.

In application aspect of the utility theory, the response (utility) function is sought in dialogue with the decision maker. If we assume that the "service" and "food" are utility independent with respect to the tipping (Theorem 3.4), then the utility function will have the structured form (x-service [0, 10]; y-food [0, 10]):

$$f(x,y) = k_1 f_1(x) + k_2 f_2(y) + k_{12} f_1(x) f_2(y).$$

Figure 11. Food "utility independence"

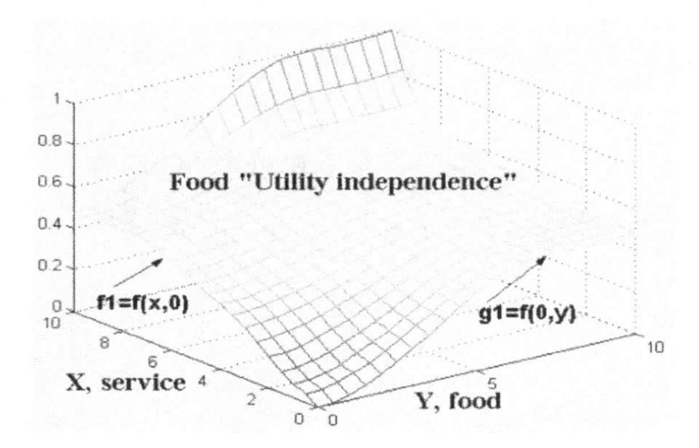

The coefficients are determined based on the expert preferences, expressed in dialogue with the decision maker and with the help of special software tool developed with VISUAL BASIC 6.0. In the same manner is constructed the function $f_2(.)$. The function $f_1(.)$ is taken from the model shown on Figure 9, $f_1(x)=f(x,0)$. The obtained utility function has the form shown in Figure 10.

Obviously, it is different from the one shown on Figure 9. This tells us that for the authors of the book, with response function shown in Figure 9, the food and service should not have "utility independent" influence on the tipping.

Now we will proceed in slightly different manner. If we assume utility independence of the influence of "the food" from the influence of "service" with respect to the tipping, the structure of the utility function is the following (Keeney, 1993):

$$f(x,y) = g_1(y) + g_2(y)f_1(x)$$

This means that the preferences with respect to the food are not affected by the service while the opposite is not true. Normalization of the partial utility functions in the general utility function is done in accordance with the normalization of the general function (between 0 and 1). The three functions in the model ($g_1(y)=f_2(y)=f(0,y)$; $f_1(x)=f(x,0)$; $g_2(y)= f(10, y)- f(0,y)$) are directly determined from the fuzzy model shown in Figure 9. The type of the model is shown in Figure 11.

The correspondence is obvious. It has to be stated that in the fuzzy set theory the inter-relations between these results are hidden and are not analyzed. At this stage, the investigation was concluded. From practical point of view for the expert system and the fuzzy model, it can be said:

- "The service" and "the food" by "tipping" relation reach in certain moment a level of saturation;
- When "food" is good, "the service" has effect only if it is very good or excellent;
- "Food" is utility independent.

These three facts determine the fuzzy model and the expert utility model. All calculations were done in MATLAB (MathWorks Inc.) or in VISUAL STUDIO-VISUAL BASIC 6.0 environment. The investigation was done iteratively with continuous transitions from the fuzzy logic to the utility theory, and vice versa, and all results were derived in accordance with the theoretical achievements of the respective theories. All partial utility functions, coefficients, fuzzy membership functions were built using the algorithms for utility function evaluation in this book. We can construct the functions ($g_1(y)=f_2(y)=f(0,y)$; $f_1(x)=f(x,0)$; $g_2(y)=f(10, y) - f(0,y)$), as utility functions. In the preference expression with respect to lotteries after a single visual comparison with the fuzzy model, almost the same model as in Figure 9 was obtained. However, if we use only our personal preference independently from the fuzzy model, the result, as we saw, can be quite different (Figures 9 and 10). On the other hand, if the structure of the utility function do not corresponds to the fuzzy model in Figure 10 the result is the same quite different. For example if we suppose utility independence between the "service" and "food" with respect to the tipping then the function has the form:

Figure 12. Utility independence determined from the fuzzy model (Figure 9)

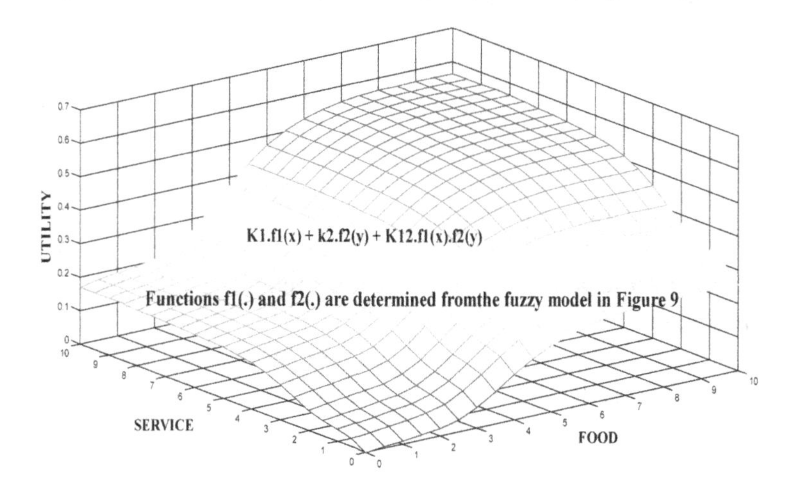

$$f(x,y) = k_1 f_1(x) + k_2 f_2(y) + k_{12} f_1(x) f_2(y).$$

If we determine the one dimensional utility functions $f_1(x)$ and $f_2(y)$ from the fuzzy model in Figure 9 the new utility function appears as the function $f(x,y)$ in Figure 12.

The difference is obvious; therefore, this structure is not appropriate.

Another non-standard example and an area of investigation is the use of the utility theory for assessment of the ecological effect of biomaterial decomposition from stockbreeding. The purpose of such decomposition is to achieve efficient and economically reasonable purification of the waters polluted by the stockbreeders.

The basis of the investigation is a subjective expert analysis of the processes of anaerobic digestion of wastewaters from calf-breeding livestock farms. The ecological problems mainly reflect the often-conflicting 'man-nature' interrelations. In essence, these are simultaneous and interacting processes, which are also understood and judged by our subjective or social conscience. Therefore, their accurate description, modeling, and control are a serious challenge faced by modern science.

The main objective of the study is "At a steady mode of operation, most reasonable methane fermentation." The main sub-objectives have been chosen to be "economic efficiency" and "ecological effect." Economic efficiency has been in turn divided into the following sub-objectives:

- Percentage of yielded gas, as an excess;
- Ability to use the remainder as a fertilizer.

Further analysis led to dividing the first objective into:

- Speed of digestion;
- Influence of ambient temperature;
- Extent of digestion of organic matter.

The extent of digestion was divided into:

- Substrate concentration; concentration in the remainder;
- Substrate type.

It was judged that from the decision maker's point of view the ability to use the remainder as a fertilizer is not essential to the established objective. On the other hand, the ecological effect is adequately described by the following sub-objectives:

- Extent of digestion of organic matter;
- Extent of inactivating of the pathogenic microorganisms.

The extent of inactivating of the pathogenic microorganisms is described by:

- Process temperature;
- Speed of digestion.

The expert analysis and structuring carried out led to accepting the following criteria, which adequately describe the main objective and are real, physically measured quantities:

- Yl: Digestion speed [10-30 g/l];
- Y2: Ambient temperature influence/fermentor volume [10-10^4 m^3];
- Y3: Substrate concentration. At the later stage of research, this interval was limited to [20-90 g/l] in order to establish a particular type of independence;
- Y4: Substrate type - % of dry substance [2.5-6.O%];
- Y5: Process temperature [25-42°C].

The "concentration in the remainder" was neglected as a criterion, since it is adequately described in the concepts of the decision maker for speed of digestion and substrate type. The following criteria were evaluated as value independent:

{Y 2} from {Y1, Y 3, Y4, Y 5},
{Y3} from {Y1, Y 2, Y 4, Y 5},
{Y4} from {Y1, Y *2, Y3,* Y5},
{Y 5} from {Y1, *Y2,* Y *3,* Y4},
{Y2, Y5} from {Y1, Y 3, Y4},
{Y1, Y2} from {Y3, Y4, Y5},
{YI, Y5} from {Y2, Y3, Y4},
{Y3, Y4} from {Y, Y2, Y5},
{Y1, Y3, Y4} from {Y2, Y5},
{Y3, Y4, Y5} from {YI, Y2},
{YI, Y2, Y3, Y5} from {Y4}.

The following criteria were evaluated as utility independent:

{Y2} from {Y1, Y3, Y4, Y5} - additive independence,
{Y1, Y5} from {Y3, Y4},{Y3} from {Y1, Y4, Y5},{Y4} from {Y1, Y3, Y5},
{Y5} from {YI, Y3, Y4},{Y2, Y5} from {Y1, Y3, Y4},{Y1, Y2, Y5} from {Y3, Y4}.

It was established that {Y3, Y5} was not utility independent of {Y1, Y4}. The following general form of the utility function U(.) was determined on the basis of the above information:

We take Y= (Y1, Y2, Y3, Y4, Y5),
$Y^o =(Y^o1, Y^o2, Y^o3, Y^o4, Y^o5) = (10, 10, 20, 2.5, 2.5)$.
$U(Y) = U(Y^o1, Y2, Y^o3, Y^o4, Y^o5)+ U(Y1, Y^o2, Y3, Y4, Y5)=K_2\ f_2(Y2) +K_1 f_{1345}(Y1, Y3, Y 4, Y5)$,
$U(Y1, Y^o2, Y3, Y4, Y5)= U(Y^o1, Y^o2, Y3, Y4, Y^o5)+ U(Y1, Y^o2, Y^o3, Y^o4, Y5)+ U(Y^o1, Y^o2, Y3, Y4, Y^o5)U(Y1, Y^o2, Y^o3, Y^o4, Y5)$.
$U(Y^o1, Y^o2, Y3, Y4, Y^o5)= U(Y^o1, Y^o2, Y3, Y^o4, Y^o5)+ U(Y^o1, Y^o2, Y^o3, Y4, Y^o5)+ U(Y^o1, Y^o2, Y3, Y^o4, Y^o5)U(Y^o1, Y^o2, Y^o3, Y4, Y^o5) = K_3\ f_3(Y3) +K_4 f_4(Y4)+ K_{34} f_3(Y3)f_4(Y4)$,

Finally, the utility function has the following form:

$$U(Y)=K_2 f_2(Y2)+K_1[\{K_3 f_3(Y3)+K_4 f_4(Y4) +K_{34}\ f_3(Y3)f_4(Y4)\}K_{134}+K_{115} f_{15}(Y1,Y5)+ K_{11345} f_{15}(Y1,Y5)\ [\ K_3\ f_3(Y3) +K_4 f_4(Y4)+ K_{34} f_3(Y3)f_4(Y4)].$$

This demonstrated function of five variables is useful for calculations and analytic representation, because it consists of functions of one and two variables and coefficients. All of them are easily evaluated with convenient for comparison lotteries of one and two variables. The concrete constructed utility functions may be viewed in detail in Pavlov (1996).

In the scientific literature, there are numerous examples of investigations and application of the theory of measurement and the utility in various areas of human activity. We used its achievements for application in the field of optimal control of bio-processes, for complete mathematical description of the complex process "technologist-process," in agriculture for modeling difficult to formalize in other way processes, etc.

We will consider one more example of application of multiattribute utility, namely the choice of optimal route in city street network. In the recent years information about researches in this area is given by Jakimavicius (2008). The main purpose is to create a model, which could evaluate urban transport system sustainability and perform rational planning of urban future development according decision support methods. The expediency of the choice of route in the city depends on the overloading of the individual street segments, the presence of speeding lanes, presence of large crossroads, current repairs, etc. By taking into account these factors, the problem becomes multi-criteria one. For analytic description of the considered problem, we will use the multiattribute utility theory (Stanoulov, 1985; Keeny, 1993).

As a main goal of the considered problem in a dialogue with the DM is defined the following goal: "The most appropriate route for moving between two points in city conditions with personal vehicle." For the determination of the sub-goals and the structure of the problem the following factors were determined by the DM:

- X_1-factors which characterize the road network;
- X_2-other factors, e.g. other vehicle, etc.
- To X_1 are assigned the following factors:
- X_{11}-condition of the road covering;
- X_{12}-length of the route;
- X_{13}-number of compulsory stops (intersections, traffic lights, etc.).
- To X_2 assigned only:
- X_2-traffic jam.
- By the opinion of the DM factors, like fuel savings are irrelevant in urban conditions and are thus neglected. The individual factors are quantitatively evaluated with the following criteria:
- X_{11}-length of the road with deteriorated surface in per cent [0, 100] %;
- X_{12}-length of the route in kilometers [0.1, 10] km;
- X_{13}-number of compulsory stops per 1km (presence of intersections, traffic lights, etc.) [0, 10];
- X_2-overload of the traffic, number of vehicles beyond the road segment's capacity in per cent [0, 100] %.
- In the process of investigation independence by utility was found by the DM between the following factors:
- $X_1=\{X_{11}, X_{12}, X_{13}\}$ from X_2;
- X_2 from X_1;
- $\{X_{13}, X_2\}$ from $\{X_{11}, X_{12}\}$.

Conditional utility independence of X_{11} from X_{12} for $X_{13}=0$ and $X_2=0$ was established. Using the theory for decomposition of multiattribute utility to simpler functions given in Keeney (1993), we determine the following structure:

$$\begin{aligned}\mathbf{k'u(X)}+1 &= \left[\mathbf{k'u}(0;0.1;\mathbf{X}_{13};0)+1\right]\\ &\times\left[\mathbf{k'u}(0;0.1;0;\mathbf{X}_2)+1\right]\\ &\times\{\mathbf{k'}[\mathbf{u}(0;\mathbf{X}_{12};0;0)\\ &\times\left(1\text{-}\mathbf{u}(\mathbf{X}_{11};0.1;0;0)\right)\\ &+\mathbf{u}(100;\mathbf{X}_{12};0;0)\mathbf{u}(\mathbf{X}_{11};0.1;0;0)]+1\}.\end{aligned}$$

With certain approximation for the partial utility functions, a constant risk propensity was found. Under these conditions, these functions have the general form $\left(a_i e^{c_i x} - b_i\right)$, where a_i, c_i, b_i are constants. In the investigation, the coefficients were determined based on the gambling approach by using comparisons of the type "equivalence" (Stanoulov, 1985; Keeny, 1993).

REFERENCES

Aizerman, M., Braverman, E., & Rozonoer, L. (1970). *Potential function method in the theory of machine learning*. Moscow, Russia: Nauka.

Allais, M. (1953). Le comportement do l'homme rationel devant le risque: Critique des postulats et axiomes de l'ecole americaine. *Econometrica, 21*, 503–546. doi:10.2307/1907921

Birnbaum, M. (2008). New paradoxes of risky decision-making. *Psychological Review, 115*(2), 463–501. doi:10.1037/0033-295X.115.2.463

Cohen, M., & Jaffray, J.-Y. (1988). Certainty effect versus probability distortion: An experimental analysis of decision-making under risk. *Journal of Experimental Psychology. Human Perception and Performance, 14*(4), 554–560. doi:10.1037/0096-1523.14.4.554

Ekeland, I. (1979). *Elements d'economie mathematique*. Paris, France: Hermann.

Farquhar, P. (1984). Utility assessment methods. *Management Science, 30*, 1283–1300. doi:10.1287/mnsc.30.11.1283

Fishburn, P. (1970). *Utility theory for decision-making*. New York, NY: Wiley.

Fishburn, P. (1988). Context-dependent choice with nonlinear and nontransitive preferences. *Econometrica, 56*(2), 1221–1239. doi:10.2307/1911365

Friedman, M., & Savage, L. (1948). Utility analysis of choices involving risk. *The Journal of Political Economy, 56*(4), 279–304. doi:10.1086/256692

Jakimavicius, M. (2008). *Multi-criteria assessment of urban areas transports systems development according to sustainability*. (Doctoral Dissertation). Vilnius-Technika. Retrieved from http://vddb.laba.lt/fedora/get/LT-eLABa-0001:E.02~2008~D_20090119_094516-74428/DS.005.1.02.ETD

Kahneman, D., & Tversky, A. (1979). Prospect theory: An analysis of decision under risk. *Econometrica, 47*, 263–291. doi:10.2307/1914185

Keeney, R. (1988). Value-driven expert systems for decision support. *Decision Support Systems, 4*(4), 405–412. doi:10.1016/0167-9236(88)90003-6

Keeney, R., & Raiffa, H. (1993). *Decision with multiple objectives: Preferences and value tradeoffs* (2nd ed.). Cambridge, UK: Cambridge University Press. doi:10.1109/TSMC.1979.4310245

Litvak, B. (1982). *Expert information, analysis and methods*. Moscow, Russia: Nauka.

Machina, M. (1982). Expected utility analysis without the independence axiom. *Econometrica, 50*(2), 277–323. doi:10.2307/1912631

Machina, M. (1987). Choice under uncertainty: Problems solved and unsolved. *The Journal of Economic Perspectives, 1*, 121–154. doi:10.1257/jep.1.1.121

Makarov, I., et al. (1987). *Theory of choice and decision-making*. Moscow, Russia: Mir.

Mengov, G. (2010). *Decision-making under risk and uncertainty*. Sofia, Bulgaria: Publishing house JANET 45.

Pavlov, Y. (1989). Recurrent algorithm for value function construction. *Proceedings of Bulgarian Academy of Sciences, 7*, 41–42.

Pavlov, Y. (2005). Subjective preferences, values and decisions: Stochastic approximation approach. *Proceedings of the Bulgarian Academy of Sciences, 58*(4), 367–372.

Pavlov, Y. (2011). Preferences based stochastic value and utility function evaluation. In *Proceedings of the InSITE 2011*. Novi Sad, Serbia: InSITE.

Pavlov, Y., Grancharov, D., & Momchev, V. (1996). Economical and ecological utility oriented analysis of the process of anaerobic digestion of waste waters. *European Journal of Operational Research*, *88*(2), 251–256. doi:10.1016/0377-2217(94)00193-6

Pavlov, Y., & Lyakova, K. (2001). Fuzzy logic and utility theory: An expert system based comparison. [Sofia, Bulgaria: BIOPS.]. *Proceedings of the, BIOPS-2001*, 37–40.

Pfanzagl, J. (1971). *Theory of measurement*. Wurzburg, Germany: Physical-Verlag.

Raiffa, H. (1968). *Decision analysis*. Reading, MA: Addison-Wesley.

Roberts, F. (1976). *Discrete mathematical models with application to social, biological and environmental problems*. Englewood Cliffs, NJ: Prentice-Hall, Inc.

Stanoulov, N., Pavlov, Y., & Tonev, M. (1985). Multi-criteria routes selection in urban conditions. In I. Popchev & P. Petrov (Eds.), *Optimization, Decision-Making, Microprocessors systems, Sofia 1983- Eight Bulgarian-Polish Symposium*. Sofia, Bulgaria: Publishing House of Bulgarian Academy of Sciences.

Tenekedjiev, K. (2004). *Quantitative decision analyze: Utility theory and subjective statistics*. Sofia, Bulgaria: Marin Drinov.

Terzieva, V., Pavlov, Y., & Andreev, R. (2007). E-learning usability: A learner-adapted approach based on the evaluation of learner's preferences. In *E-Learning III and the Knowledge Society*. Brussels, Belgium: Academic Press.

Tversky, A., & Kahneman, D. (1991). Loss aversion in riskless choice: A reference dependent model. *The Quarterly Journal of Economics*, *106*, 1039–1061. doi:10.2307/2937956

Vilkas, E. (1990). *Optimum in games and decisions*. Moscow, Russia: Nauka.

Chapter 5
Elements of Stochastic Programming

ABSTRACT

The previous chapters showed that the human preferences, which by theoretical formulation are fundamental for the determination of the value and utility functions, are characterized in their explicit expression by uncertainty of a stochastic nature. By exposing our preferences in the area of the specific complex problem to be solved, one gradually provides increasing details for her/his attitude in a recurrent manner. Because of this recurrent extraction of the preferences and the uncertainty in the explicit expression, the methods of stochastic programming of recurrent nature are of great interest for the evaluation of value and utility functions.

The authors start the chapter with one important for the stochastic recurrent procedures concept, namely with the so-called quasi-Feyer sequences. Based on this theory, they prove some significant stochastic approximation theorems and the convergence of the Robbins-Monro method. This approach is used for assessment of the subjective probabilities and for assessment of the coefficients in multiattribute utility decomposition.

In the second part of the chapter is discussed the "potential functions method," a very fruitful area of the stochastic programming. Even though it is called method, this field is actually a mathematical theory whose practical results are a large number of stochastic recurrent procedures for pattern recognition, approximating algorithms for functions in noisy conditions, development of unified mathematical approach for machine learning on the basis of human preference, proof of the perceptron theory, etc. The stochastic algorithms based on the "potential functions method" have stable convergence and flexibility, and these properties permit fruitful application in utility and value function evaluations and polynomial approximations. The last part of the chapter gives an example of pattern recognition of two sets of positive and negative answers as machine learning procedure.

DOI: 10.4018/978-1-4666-2967-7.ch005

1. BEGINNINGS: STOCHASTIC PROGRAMMING

The stochastic programming is a field in the theory of optimal decisions, in which problems for determining optimal choice in conditions described with random functions are studied (Aizerman, 1970; Ermolev, 1976), or in other words, this is a theory for solving extremal problems of stochastic nature. In this book, we will use mathematical techniques for solving problems from the decision-making theory and analytical approximation of value, utility functions and subjective probabilities, whose existence in axiomatic aspect is analyzed in the theory of measurement and the utility theory (Pfanzagl, 1971; Keneey, 1993). In the previous two chapters, we looked into some axiomatic systems and existence theorems regarding these functions and we discussed the uncertainty, which occurs in the practical aspects of the real-world decision-making problems on the basis of the normative approach. We saw that the human preferences, which by theoretical formulation are fundamental for the determination of the value and utility functions, are characterized in their explicit expression by uncertainty of stochastic nature (Keneey, 1993; Mengov, 2010). Consecutively exposing our preferences in the area of the specific problem to be solved, we gradually provide increasing details for our attitude in recurrent manner. Because of this, the problems and methods of stochastic programming of recurrent nature are of great interest to us (Pavlov, 1989, 2003, 2011). Such are the quasi-gradient recurrent methods to which we will limit our considerations and which we will discuss in more details for the purpose of this book (Robbins, 1951; Aizerman, 1970; Ermolev, 1976).

The gradient of a nonlinear function $F(x_1, x_2, \dots, x_n)$, $F_x(\mathbf{x}) = (\frac{\partial F}{\partial x_1}, \frac{\partial F}{\partial x_2}, \dots, \frac{\partial F}{\partial x_n})$ cannot be precisely determined in the stochastic programming, when the function is given algorithmically. Because of this in the stochastic programming problems a quasi-gradient is introduced, a random vector whose mathematical expectation is close to the gradient or to the generalized gradient of the investigated functions (see paragraph 2.2). Let us want to minimize the convex function:

$$F(x_1, x_2, \dots, x_n), (x_1, x_2, \dots, x_n) \in \mathbf{X}.$$

In the problems, we suppose that the set $\boldsymbol{X}$ is convex and closed. We already mentioned that with respect to the generalized gradient we can only have statistical estimates of random vectors due to the algorithmic determination of the objective function. When the set $\boldsymbol{X}$ is bounded, in the stochastic recurrent quasi-gradient procedures the prjection $\pi(z)$ may be used. For every point $z, z \in \boldsymbol{R}^n$, the point $\pi_X(z)$ belongs to $\boldsymbol{X}$, $\pi_X(z) \in \boldsymbol{X}$ and hence:

$$\| y - \pi_X(z) \| \leq \| y - z \|, \forall y \in \boldsymbol{X}.$$

Let $(\boldsymbol{\Theta}, \boldsymbol{I}, \boldsymbol{P})$ be the initial probability space (Parthasarathy, 1978; Shiryaev, 1989). We consider the sequence of random points x_s, s=0, 1, 2, 3.... Then by analogy with the gradient methods of the non-linear programming, we consider the quasi-gradient procedure:

$$x_{s+1} = \pi_X (x_s - \rho_s \gamma_s \xi_s), s=0,1,2,3 \dots\dots\dots$$

In this case, x_0 is an arbitrary initial point, ρ_s is the magnitude of the step, and γ_s is a norming coefficient. The interesting part is related to the random variable (ξ_s). Its conditional mathematical expectation is related to the gradient or the generalized gradient of the function $F(x_1, x_2, \dots, x_n)$ through the formula:

$$M(\xi_s / x_1, x_2, \dots, x_s) = a_s F_x(x_s) + \mathbf{b}^s, s=0,1,2,3 \dots\dots\dots$$

In the above formula (a_s) are non-negative quantities, and $(\mathbf{b}^s) = (b^s_1, b^s_2, \dots, b^s_n)$ are sequence of random vectors, dependent on the sequence (x_1, $x_2, \dots, x_s$) until the current moment. Actually it is assumed that (a_s) and ($\mathbf{b}^s$) are measurable with respect to the σ–algebra $\Im_s$, generated by the random variables $(x_1, x_2,\ldots, x_s)$. The variables ρ_s and γ_s are also measurable with respect to the σ–algebra $\Im_s$ (Parthasarathy, 1978; Eliott, 1982). In the above formula $F_x(x_s)$ is a generalized gradient, i.e.

$(F(z) - F(x_s)) > (F_x(x_s), z - x_s)$.

As a simple example of random vector of this kindletusconsiderthecasewhen $F(\mathbf{x}) = \sum_{i=1}^{N} f^i(x)$. In this formula $f^i(x)$, i=1,2,..N, are continuously differentiable functions and $f^i_x(x)$ is gradient of the function $f^i(x)$ (Ermolev, 1976). Let for (x_s) with probability (1/N) we choose one of the numbers (1,2,..N) and denote it by (i_s). Then for the vector $(\xi_s)=(f^{is}_x(x))$ it is fulfilled:

$$\mathbf{M}(\xi_s/x_1,x_2,..,x_s) = \frac{1}{\mathbf{N}}\sum_{i=1}^{\mathbf{N}} f^i_x(x_s) = \frac{1}{\mathbf{N}} F_x(x_s).$$

In the above formula we have $(a_s=1/N)$ and $(\mathbf{b}^s=0)$.

The convergence of the sequence (x_s) to the set of optimal solutions X^* may be in different aspects: by probability, almost everywhere or in quadratic sense (Shiryaev, 1989). Practical application has mainly the almost everywhere convergence or which is equivalent with probability one. The stochastic quasi-gradient analogues of theorems 2.2.14 and 2.2.15 from Chapter 2 remain valid. We denote by X^* the set of points, in which the continuous function $F(x)$ takes minimal value. It is supposed that there exists $x \in X^*$, such that $\|x\| \leq$ cte. Then the following theorem is valid (Ermolev, 1976).

1.1. Theorem

If for every number L there exists such a number K_L, that

$$\mathbf{M}(\|\xi_s\|^2/x_1,x_2,..,x_s) \leq \eta_s^2 \leq \mathbf{K}_L,$$

for

$$\|\mathbf{x}^k\| \leq \mathbf{L},\ k = 0,1,2,\ldots,s;$$

$$\gamma_s > 0, \quad \gamma_s(\eta_s + \tau_s\|x^s\|) \leq \mathbf{cte},$$

$$(\tau_s = 1) \text{ if } \|\mathbf{b}_s\| \neq 0, (\tau_s = 0) \text{ if } \|\mathbf{b}_s\| = 0;$$

$$\rho_s \geq 0, \quad \rho_s \to 0, \sum_{i=0}^{\infty} a_i\rho_i = \infty, \quad \sum_{i=0}^{\infty} \mathbf{M}(\|b_i\|\rho_i) < \infty,$$

then there exists subsequence $F(x_{ik})$ of the sequence $F(x_i)$, such that

$$\lim F(x_{ik}) = F(x^*),\ k \to \infty,$$
$$i.e. \lim_{i\to\infty} \min_{k<i} F(x_k) = F(x^*).$$

Now we will consider the case, when the set X^* of extremal points is bounded. Then the following theorem is valid (Ermolev, 1976).

1.2. Theorem

Let in addition to the previous theorem we assume that the set of extremal points is bounded. Then is fulfilled the following convergence $\inf_{\mathbf{x}^* \in \mathbf{X}^*} \mathbf{M}\|\mathbf{x}^* - \mathbf{x}_s\|^2 \to 0$.

With these definitions and theorems, we revealed main specificities of the stochastic-recurrent quasi-gradient procedures, which distinguish them from the problems of the nonlinear programming. We also revealed some largely used techniques, providing convergence of the recurrent stochastic procedures. Now we will look in more detail in some of the stochastic recurrent procedures.

2. ROBBINS-MONRO CLASSICAL METHOD

Now we will look into one important for the stochastic recurrent procedures concept, namely with the so-called quasi-Feyer sequences (Ermolev, 1976). Let ($\boldsymbol{\Theta}$, I, $\boldsymbol{P}$) be the initial probability space (Shiryaev, 1989). Let the set $\boldsymbol{X}$ is a subset of the n-dimensional space of the real numbers $\boldsymbol{R}^n$. The sequence of random vectors (x_s), s=0, 1, 2, 3... in $\boldsymbol{R}^n$ we will call quasi-Feyer sequence with respect to the set $\boldsymbol{X}$, if for an arbitrary point z, $z \in \boldsymbol{X}$, it is fulfilled:

$$\mathbf{M}(\left\|\mathbf{z}-\mathbf{x}_{s+1}\right\|^2 / \mathbf{x}_0, \mathbf{x}_1, .., \mathbf{x}_s) \\ \leq \left\|\mathbf{z}-\mathbf{x}_s\right\| + \mathbf{W}_s,\ \mathbf{s}=0,1,2,...$$

In this case the random variables $\mathbf{W}_s(\mathbf{w}) \geq 0$, $\mathbf{s}=0,1,2,...$ are measurable with respect to the σ–algebra $\mathfrak{I}_s$, generated by the random variables (x_1, x_2, ...,x_s) and satisfy the condition:

$$\sum_{s=0}^{\infty} \mathbf{W}_s(\mathbf{w}) < \infty.$$

For the mathematical expectation to exist in the above two formulas it is sufficient that $\mathbf{M}(\left\|\mathbf{x}_0\right\|^2) < \infty$. The following theorem is valid (Ermolev, 1976):

2.1. Theorem

If the sequence of random variables (x_s), s=0, 1, 2, 3,.. is quasi-Feyer with respect to the set $\boldsymbol{X}^*$, the random variable $\left\|\mathbf{z}-\mathbf{x}_s\right\|$, $\forall \mathbf{z} \in \mathbf{X}^*$, is convergent almost everywhere. The set of limit points of (x_s), s=0,1, 2, 3,.. is non-empty for almost every w$\in\boldsymbol{\Theta}$. If there exist two different limit points x'(w) and x''(w) of the sequence (x_s), s=0,1, 2, 3..., which to not belong to $\boldsymbol{X}^*$, then the set $\boldsymbol{X}^*$ lies in hyperspace, which is equidistant for the points.

Proof

The proof is a direct corollary from theorem 2.1.12 of Chapter 2, with respect to the sequence of random variables $\left\|\mathbf{z}-\mathbf{x}_s\right\|$, $\mathbf{s}=1,2,3..$, $\forall \mathbf{z} \in \mathbf{X}^*$. All limit points lie on a sphere with radius $lim\left\|\mathbf{z}-\mathbf{x}_s\right\|$, $\forall \mathbf{z} \in \mathbf{X}^*$. From here follows the rest of the theorem.

The proved theorem determines quite general conditions for convergence almost everywhere to points from the set $\boldsymbol{X}^*$. If the dimensionality of the set $\boldsymbol{X}^*$ is equal to the dimensionality of the main space $\boldsymbol{R}^n$, then the sequence (x_s), s=0, 1, 2, 3,... has only one limit point.

Now we will prove some important theorems concerning the convergence of the stochastic quasi-gradient procedure, discussed in the beginning of this Chapter (Aizerman, 1970; Ermolev, 1976):

$$x_{s+1}=\pi_X(x_s-\rho_s\gamma_s\xi_s),\ s=0,1,2,3..........,$$

$$\mathrm{M}(\xi_s/x_0, x_1, x_2, \ldots, x_s)=a_s F_x(x_s)+\mathbf{b}^s. \tag{5.2.1}$$

We preserve the same denotations and we put the set $\boldsymbol{X}^*$, as the set of minima of the convex function $F(x_1, x_2,\ldots, x_n)$, $(x_1, x_2,\ldots, x_n) \in \mathbf{X}$. The following theorem is valid (Ermolev, 1976).

2.2. Theorem

Let the random variable $\eta_s(\mathbf{w})$ is measurable with respect to the σ–algebra $\mathfrak{I}_s$, generated by the random variables (x_0, x_1, x_2, ...,x_s). Let for every constant L there exist a constant $\mathbf{C}_L$, for which the following condition is satisfied:

$$\mathbf{M}(\left\|\xi_s\right\|^2/x_1,x_2,..,x_s) \leq \eta_s^2 \leq \mathbf{C}_L, \\ \text{if } \left\|\mathbf{x}_k\right\| < \mathrm{L}, \mathbf{k}=1,2,3,..\mathbf{s};$$

There exist constants $\underline{\gamma}$ and $\bar{\gamma}$, such that with respect to the norming coefficient γ_s it is fulfilled:

$$\gamma_s > 0, \quad 0 < \underline{\gamma} \leq \gamma_s(\eta_s + \tau_s \|x^s\|) \leq \bar{\gamma} < \infty,$$

$$(\tau_s = 1) \text{ if } \|\mathbf{b}_s\| \neq 0, (\tau_s = 0) \text{ if } \|\mathbf{b}_s\| = 0;$$

Concerning the variables ρ_s, a_s, $\mathbf{b}_s$ it is fulfilled:

$$\rho_s \geq 0, \quad a_s > 0, \quad \sum_{s=0}^{\infty} \mathbf{M}(\|\mathbf{b}_s\| \rho_s + \rho_s^2) < \infty;$$

Then the sequence of points (x_s), s=0, 1, 2, 3,.., determined according to the stochastic quasi-gradient procedure (5.2.1) is quasi-Feyer with respect to the set X^*. If in addition with probability one it is fulfilled:

$$\sum_{s=0}^{\infty} a_s \rho_s = \infty,$$

then for almost every w∈Θ, the sequence $(x_s(w))$, s=0, 1, 2, 3,.., is convergent to an element x*(w), x*(w)∈X^*.

Proof

Let x*∈X^*. From procedure (5.2.1) follows:

$$\|\mathbf{x}^* - \mathbf{x}_{s+1}\|^2 \leq \|\mathbf{x}^* - \mathbf{x}_s + \rho_s \gamma_s \xi_s\|^2 = \|\mathbf{x}^* - \mathbf{x}_s\|^2 + 2\rho_s \gamma_s (\xi_s, \mathbf{x}^* - \mathbf{x}_s) + \rho_s^2 \gamma_s^2 \|\xi_s\|^2.$$

We take the conditional mathematical expectations from both sides of the equation:

$$\mathbf{M}(\|\mathbf{x}^* - \mathbf{x}_{s+1}\|^2 / \Im_s) \leq \|\mathbf{x}^* - \mathbf{x}_s\|^2 + 2\rho_s \gamma_s a_s (F_x(\mathbf{x}_s), \mathbf{x}^* - \mathbf{x}_s) + 2\rho_s \gamma_s (\mathbf{b}_s, \mathbf{x}^* - \mathbf{x}_s) + \rho_s^2 \gamma_s^2 \mathbf{M}(\|\xi_s\|^2 / \Im_s).$$

Since $(0 > (F_x(x_s), x^* - x_s))$, then from the Cauchy–Bunyakovsky inequality and from the fact that we can always put $\gamma_s \leq \gamma^* < \infty$, for some number γ^*, it follows:

$$\mathbf{M}(\|\mathbf{x}^* - \mathbf{x}_{s+1}\|^2 / \Im_s) \leq \|\mathbf{x}^* - \mathbf{x}_s\|^2 + 2\rho_s \|\mathbf{b}_s\| \gamma_s (\gamma^* \|\mathbf{x}^*\| + \gamma^*) + \rho_s^2 \gamma^{*2}.$$

In the above formula we assumed $0 < \varepsilon \leq \eta_s$, which is feasible for the theorem's conditions. From here and from the condition of the theorem, it follows that (x_s), s=0, 1, 2, 3,.. is quasi-Feyer sequence.

Now we will prove the second part of the theorem under the condition:

$$\sum_{s=0}^{\infty} a_s \rho_s = \infty.$$

From the previous equation it follows:

$$\|\mathbf{x}^* - \mathbf{x}_{s+1}\|^2 \leq \|\mathbf{x}^* - \mathbf{x}_0\|^2 + \sum_{k=0}^{s} 2\rho_k \gamma_k (\xi_k, \mathbf{x}^* - \mathbf{x}_k) + \sum_{k=0}^{s} \rho_k^2 \gamma_k^2 \|\xi_k\|^2.$$

We take the mathematical expectation from both sides of the inequality and having in mind the formula $\mathbf{M}(\mathbf{g}(\mathbf{k})) = \mathbf{MM}(\mathbf{g}(\mathbf{k}) / \Im_s)$ we obtain:

$$\mathbf{M}(\|\mathbf{x}^* - \mathbf{x}_{s+1}\|^2) \leq \mathbf{M}(\|\mathbf{x}^* - \mathbf{x}_s\|^2) + 2\mathbf{M}(\sum_{k=0}^{\infty} \rho_k \gamma_k a_k (F_x(\mathbf{x}_k), \mathbf{x}^* - \mathbf{x}_k)) + 2(\gamma^* \|\mathbf{x}^*\| + \gamma^*) \sum_{s=0}^{\infty} \rho_s \|\mathbf{b}_s\| + \gamma^* \sum_{s=0}^{\infty} \mathbf{M}(\rho_s^2).$$

From the condition of the theorem and from the fact that the norm is always positive it follows that:

$$-\infty < \sum_{k=0}^{\infty} \rho_k \gamma_k a_k (F_x(\mathbf{x}_k), \mathbf{x}^* - \mathbf{x}_k).$$

From the formula $\sum_{s=0}^{\infty} a_s\rho_s = \infty$ with probability one it follows that:

$$\gamma_k(F_x(x_k), x^* - x_k) \xrightarrow[k\to\infty]{} 0.$$

From the condition of the theorem it follows that $0 < \varepsilon \leq \gamma_s$ almost everywhere. Therefore, the following is fulfilled for some subsequence (x_{sK}) of (x_s), s=0, 1, 2, 3,..:

$$(F_x(x_{s_k}), x^* - x_{s_k}) \xrightarrow[k\to\infty]{} 0.$$

Since $F_x(x_k)$ is a generalized gradient and since x^* is extremal point, it follows:

$$(F(x^*) - F(x_{s_k})) \geq (F_x(x_{s_k}), x^* - x_{s_k}) \xrightarrow[k\to\infty]{} 0.$$

Therefore, at least one limit point x*(w) of the sequence $(x_s(w))$, s=0, 1, 2, 3,.. belongs to the extremal set X^*. From this and the previous theorem it follows that every limit point belongs to set X^*. If in addition, the domain X, for which we take the projection $\pi_X(.)$ is bounded, then we can assume γ_s as a constant. This concludes the proof of the theorem.

Now we will consider the case of an arbitrary continuously differentiable function $F(x_1, x_2, \ldots, x_n)$ for $X = R^n$. Without loss of generality we suppose, that in the following two theorems the function $F(x_1, x_2, \ldots, x_n)$ is positive. From these conditions, we will seek local convergence of the recurrent quasi-gradient procedures. Now we will prove two important theorems concerning the convergence of the stochastic quasi-gradient procedures in the new conditions:

$$x_{s+1} = (x_s - \rho_s\gamma_s\xi_s),\ s = 0, 1, 2, 3, \ldots,$$

$$M(\xi_s / x_0, x_1, x_2, \ldots, x_s) = a_s F_x(x_s) + b^s. \qquad (5.2.2)$$

We preserve the same denotation and put the set X^*, as the set of minima of the continuous function $F(x_1, x_2, \ldots, x_n)$, $(x_1, x_2, \ldots, x_n) \in \mathbf{X}$. The following is fulfilled for $x^* \in \mathbf{X}^*$:

$$0 \geq (F_x(x_s), x^* - x_s).$$

For the mathematical expectation to exist, it is sufficient that $M(\|x_0\|^2) < \infty$. In addition we suppose that for each constant L there exist constant C_L, for which the following condition is fulfilled:

$$\|F_x(x)\| \leq C_L, \text{ if } \|F(x)\| \leq L.$$

We also suppose that the gradient $F_x(x)$ satisfies uniformly the Lipschitz condition (Rudin, 1976):

$$\|F_x(x) - F_x(y)\| \leq C\,\|x - y\|,$$

where C is constant.

Under these conditions the following theorem is valid (Ermolev, 1976).

2.3. Theorem

Let the random variable $\eta_s(w)$ is measurable with respect to the σ–algebra $\Im_s$, generated by the random variables $(x_0, x_1, x_2, \ldots, x_s)$. Let for every constant L there exists a constant C_L, C_L, for which the following condition is fulfilled:

$$M(\|\xi_s\|^2 / x_0, x_1, x_2, .., x_s) \leq \eta_s^2 \leq C_L,$$
$$\text{if } \|F(x_k)\| < L, k = 0, 1, 2, 3, ..s;$$

There exist a constant $\underline{\gamma}$, such that with respect to the norming coefficient γ_s it is fulfilled: $\gamma_s > 0,\quad 0 < \underline{\gamma} \leq \gamma_s\eta_s \leq r_s < \infty;$

The Lipschitz condition is fulfilled. The variables ρ_s, a_s, r_s, $\mathbf{b}_s$ are measurable with respect to the σ–algebra $\Im_s$, generated by the random variables $(x_0, x_1, x_2, \ldots, x_s)$ and in addition it is fulfilled almost everywhere.

$$\rho_s \geq 0, \quad a_s \geq 0, \quad \frac{\|\mathbf{b}_s\|}{a_s} \xrightarrow{\text{uniformly with probability 1}} 0,$$

$$\sum_{s=0}^{\infty} \mathbf{M}(\|\mathbf{b}_s\| \rho_s + \rho_s^2 r_s^2) < \infty;$$

$$\sum_{s=0}^{\infty} a_s \rho_s = \infty,$$

Then the sequence of points (x_s), s=0,1,2,3,.., determined according to the stochastic quasi-gradient procedure (2) is such that $F(x_s)$ is convergent almost everywhere and it is fulfilled $\|F_x(\mathbf{x}_{s_k})\| \xrightarrow[k \to \infty]{} 0$ almost everywhere for some subsequence (x_{sk}) of (x_s), s=0,1,2,3,...

Proof

$$F(\mathbf{x}_{s+1}) - F(\mathbf{x}_s) = \int_0^1 F'_\alpha(\mathbf{x}_s - \alpha\rho_s\gamma_s\xi_s)d\alpha$$

$$= -\rho_s\gamma_s \int_0^1 (F_x(\mathbf{x}_s), \xi_s)d\alpha$$

$$+\rho_s\gamma_s \int_0^1 (F_x(\mathbf{x}_s) - F_x(\mathbf{x}_s - \alpha\rho_s\gamma_s\xi_s), \xi_s)d\alpha$$

$$\leq -\rho_s\gamma_s(F_x(\mathbf{x}_s), \xi_s) + \rho_s^2\gamma_s^2 C \|\xi_s\|^2.$$

We take the mathematical expectation from both sides and we rewrite the above formula in a new way as in Box 1.

The coefficient β_s is such that the following condition is satisfied:

$$\beta_s = \begin{Bmatrix} 1, & \|F_x(\mathbf{x}_s)\| \geq Q \\ 0, & \|F_x(\mathbf{x}_s)\| < Q \end{Bmatrix}.$$

In the formula Q is constant. We note that $\gamma_s < \bar{\gamma}$ for some constant $\bar{\gamma}$, which follows from the problem formulation and that $\frac{\|\mathbf{b}_s\|}{a_s} \xrightarrow{\text{uniformly with probability 1}} 0$. Then there exists a number N, such that for (s>N) the following inequality is fulfilled:

$$\mathbf{M}(F(\mathbf{x}_{s+1})/\Im_s) \leq F(\mathbf{x}_s)$$
$$+\rho_s\gamma_s\, a_s(1-\beta_s)\left[-a_s\|F_x(\mathbf{x}_s)\|^2 + \|\mathbf{b}_s\|\|F_x(\mathbf{x}_s)\|\right]$$
$$+\rho_s^2 \mathbf{r}_s^2 C \leq F(\mathbf{x}_s) + Q\rho_s\bar{\gamma}\|\mathbf{b}_s\| + \rho_s^2 r_s^2 C.$$

From this inequality and from theorem 2.1.12 follows the convergence of the sequence $\mathbf{F}(\mathbf{x}_s)$.

Now we will prove the last part of the theorem. From the following inequality:

Box 1.

$$\mathbf{M}(F(\mathbf{x}_{s+1})/\Im_s) \leq F(\mathbf{x}_s) + \rho_s\gamma_s\, a_s\left[-\|F_x(\mathbf{x}_s)\|^2 + \frac{\|\mathbf{b}_s\|}{a_s}\|F_x(\mathbf{x}_s)\|\right] + \rho_s^2 \mathbf{r}_s^2 C$$
$$= F(\mathbf{x}_s) + \rho_s\gamma_s\, a_s\beta_s\left[-\|F_x(\mathbf{x}_s)\|^2 + \frac{\|\mathbf{b}_s\|}{a_s}\|F_x(\mathbf{x}_s)\|\right]$$
$$+ \rho_s\gamma_s\, a_s(1-\beta_s)\left[-a_s\|F_x(\mathbf{x}_s)\|^2 + \|\mathbf{b}_s\|\|F_x(\mathbf{x}_s)\|\right] + \rho_s^2 \mathbf{r}_s^2 C.$$

$$F(\mathbf{x}_{s+1})\text{-}F(\mathbf{x}_s) = \int_0^1 F'_\alpha(\mathbf{x}_s - \alpha\rho_s\gamma_s\xi_s)d\alpha$$
$$= -\rho_s\gamma_s\int_0^1 (F_x(\mathbf{x}_s), \xi_s)d\alpha$$
$$+\rho_s\gamma_s\int_0^1 (F_x(\mathbf{x}_s)\text{-}F_x(\mathbf{x}_s - \alpha\rho_s\gamma_s\xi_s), \xi_s)d\alpha$$
$$\leq -\rho_s\gamma_s(F_x(\mathbf{x}_s), \xi_s) + \rho_s^{\,2}\gamma_s^{\,2}C\left\|\xi_s\right\|^2,$$

it follows that it is valid:

$$\mathbf{M}(F(\mathbf{x}_{s+1})) \leq \mathbf{M}(F(\mathbf{x}_0))$$
$$-\mathbf{M}(\sum_{k=0}^{s}\rho_k\gamma_k\, a_k\left\|F_x(\mathbf{x}_k)\right\|^2)$$
$$+\sum_{k=0}^{s}\mathbf{M}(\rho_k\gamma_k\left\|\mathbf{b}_k\right\|\left\|F_x(\mathbf{x}_k)\right\| + C\sum_{k=0}^{s}\mathbf{M}(\rho_k^{\,2}\,\mathbf{r}_k^{\,2}).$$

From the convergence of the sequence $F(\mathbf{x}_s)$ and from the condition of the theorem it follows:

$$\mathbf{M}(\sum_{k=0}^{s}\rho_k\gamma_k\, a_k\left\|F_x(\mathbf{x}_k)\right\|^2) < \infty.$$

The variable γ_s is uniformly bound from below and $\sum_{s=0}^{\infty} a_s\rho_s = \infty$, with probability 1. Because of this for some subsequence (x_{sk}) of (x_s), s=0, 1, 2, 3,... it follows:

$$\left\|F_x(\mathbf{x}_{s_k})\right\| \xrightarrow[k\to\infty]{} 0\,.$$

This completes the proof.

The following theorem is also useful (Ermolev, 1976).

2.4. Theorem

Let $a_s = 1$ and the random variable $\mathbf{d}_s(\mathbf{w})$ is measurable with respect to the σ–algebra $\mathfrak{I}_s$, generated by the random variables $(x_0, x_1, x_2, \ldots, x_s)$. Let for every constant L there exist a constant $\mathbf{C}_L$, for which the following condition is fulfilled:

$$\sum_{j=1}^{n}\mathbf{D}(\xi_{js}/x_0,x_1,x_2,..,x_s) \leq \mathbf{d}_s^2 \leq \mathbf{C}_L,$$
$$\text{if } \left\|F(\mathbf{x}_k)\right\| < L, \mathbf{k} = 0,1,2,3,..\mathbf{s};$$

We note by D the variance. There exists a constant $\underline{\gamma}$, such that with respect to the norming coefficient γ_s it is fulfilled:

$$\gamma_s > 0, \quad 0 < \underline{\gamma} \leq \gamma_s\mathbf{d}_s \leq \mathbf{r}_s < \infty;$$

The Lipschitz condition is fulfilled. The variables ρ_s, r_s, $\mathbf{b}_s$ are measurable with respect to the σ–algebra $\mathfrak{I}_s$, generated by the random variables $(x_0, x_1, x_2, \ldots, x_s)$ and in addition it is fulfilled:

$$\rho_s \geq 0, \quad \rho_s \to 0, \quad \left\|\mathbf{b}_s\right\| \xrightarrow{\text{uniformly with probability 1}} 0,$$

$$\sum_{s=0}^{\infty}\mathbf{M}(\left\|\mathbf{b}_s\right\|\rho_s + \rho_s^2 r_s^{\,2}) < \infty$$

almost everywhere;

$$\sum_{s=0}^{\infty}\rho_s = \infty,$$

almost everywhere (e.w. or with probability 1).

Then the sequence of points (x_s), s=0, 1, 2, 3,.. determined according to the stochastic quasi-gradient procedure (5.2.2) is such that $F(x_s)$ is convergent almost everywhere and in addition $\left\|F_x(\mathbf{x}_{s_k})\right\| \xrightarrow[k\to\infty]{} 0$ uniformly almost everywhere for some subsequence (x_{sk}) of (x_s), s=0,1,2,3,...

Proof

$$F(\mathbf{x}_{s+1})\text{-}F(\mathbf{x}_s) = \int_0^1 F'_\alpha(\mathbf{x}_s - \alpha\rho_s\gamma_s\xi_s)d\alpha$$
$$= -\rho_s\gamma_s\int_0^1 (F_x(\mathbf{x}_s), \xi_s)d\alpha$$
$$+\rho_s\gamma_s\int_0^1 (F_x(\mathbf{x}_s)\text{-}F_x(\mathbf{x}_s - \alpha\rho_s\gamma_s\xi_s), \xi_s)d\alpha$$
$$\leq -\rho_s\gamma_s(F_x(\mathbf{x}_s), \xi_s) + \rho_s^2\gamma_s^2 C\left\|\xi_s\right\|^2.$$

We take the mathematical expectation from both sides and we rewrite the above formula in new way as in Box 2.

We use the coefficient β_s, β_s, defined in the previous theorem. Then Box 2 inequality is transformed into Box 3.

We note that $\gamma_s < \bar{\gamma}$ for some constant $\bar{\gamma}$, which follows from the problem formulation and that $\left\|\mathbf{b}_s\right\| \xrightarrow{\text{uniformly with probability 1}} 0$. Then there exists a number N, such that for (s>N) is fulfilled the inequality:

$$\mathbf{M}(F(\mathbf{x}_{s+1})/\Im_s) \leq F(\mathbf{x}_s)$$
$$+\rho_s\gamma_s(1\text{-}\beta_s)\left[Q\left\|\mathbf{b}_s\right\| + \left\|\mathbf{b}_s\right\|^2\right] + \rho_s^2 r_s^2.$$

From this inequality and from theorem 2.1.12 it follows the convergence of the sequence $F(\mathbf{x}_s)$.

Now we will prove the last part of the theorem. From the inequality:

$$F(\mathbf{x}_{s+1})\text{-}F(\mathbf{x}_s) = \int_0^1 F'_\alpha(\mathbf{x}_s - \alpha\rho_s\gamma_s\xi_s)d\alpha$$
$$= -\rho_s\gamma_s\int_0^1 (F_x(\mathbf{x}_s), \xi_s)d\alpha$$
$$+\rho_s\gamma_s\int_0^1 (F_x(\mathbf{x}_s)\text{-}F_x(\mathbf{x}_s - \alpha\rho_s\gamma_s\xi_s), \xi_s)d\alpha$$
$$\leq -\rho_s\gamma_s(F_x(\mathbf{x}_s), \xi_s) + \rho_s^2\gamma_s^2 C\left\|\xi_s\right\|^2,$$

it follows that it is valid:

$$F(\mathbf{x}_{s+1}) \leq \mathbf{M}F(\mathbf{x}_0) - \sum_{k=0}^{s}\rho_k\gamma_k(F_x(\mathbf{x}_k), \xi_k) + C\sum_{k=0}^{s}\rho_k^2 r_k^2\left\|\xi_k\right\|^2.$$

Using the mathematical expectation, we obtain:

$$\mathbf{M}(F(\mathbf{x}_{s+1})) \leq \mathbf{M}(F(\mathbf{x}_0))$$
$$-\mathbf{M}(\sum_{k=0}^{s}\rho_k\gamma_k\left\|F_x(\mathbf{x}_k)\right\|^2) +$$
$$+\sum_{k=0}^{s}\mathbf{M}[\rho_k\gamma_k\left\|\mathbf{b}_k\right\|\left\|F_x(\mathbf{x}_k)\right\| + \rho_k^2\gamma_k^2\left\|F_x(\mathbf{x}_k)\right\|^2$$
$$+\rho_k^2\gamma_k^2\left\|\mathbf{b}_k\right\|\left\|F_x(\mathbf{x}_k)\right\| + \rho_k^2\gamma_k^2\left\|\mathbf{b}_k\right\|^2 + \rho_k^2 r_k^2].$$

In the above formula we use Box 4.

From the convergence of the sequence $F(\mathbf{x}_s)$ and from the condition of the theorem it follows:

$$\mathbf{M}(\sum_{k=0}^{s}\rho_k\gamma_k\left\|F_x(\mathbf{x}_k)\right\|^2) < \infty.$$

The variable γ_s is uniformly bound from below and $\sum_{s=0}^{\infty}\rho_s = \infty$, with probability 1. Because of

Box 2.

$$\mathbf{M}(F(\mathbf{x}_{s+1})/\Im_s) \leq F(\mathbf{x}_s) - \rho_s\gamma_s\left[\left\|F_x(\mathbf{x}_s)\right\|^2 - \left\|(F_x(\mathbf{x}_s),\mathbf{b}_s)\right\|\right]$$
$$+ \rho_s^2\mathbf{r}_s^2\left[\sum_{j=1}^{n}\mathbf{D}(\xi_{js}/x_0,x_1,x_2,..,x_s) + \left\|F_x(\mathbf{x}_s)\right\|^2 + 2\left\|(F_x(\mathbf{x}_s),\mathbf{b}_s)\right\| + \left\|\mathbf{b}_s\right\|^2\right].$$

Box 3.

$$\mathbf{M}(F(\mathbf{x}_{s+1})/\Im_s) \leq F(\mathbf{x}_s) +$$
$$+ \rho_s\gamma_s\beta_s\Big[(-\left\|F_x(\mathbf{x}_s)\right\| + \left\|\mathbf{b}_s\right\|)\left\|F_x(\mathbf{x}_s)\right\| + \rho_s\gamma_s\left\|F_x(\mathbf{x}_s)\right\|(\left\|F_x(\mathbf{x}_s)\right\| + 2\left\|\mathbf{b}_s\right\|) + \rho_s\gamma_s\left\|\mathbf{b}_s\right\|^2\Big] +$$
$$+ \rho_s\gamma_s(1-\beta_s)\Big[(-\left\|F_x(\mathbf{x}_s)\right\|^2 + \left\|F_x(\mathbf{x}_s)\right\|\left\|\mathbf{b}_s\right\|) + \rho_s\gamma_s(\left\|F_x(\mathbf{x}_s)\right\|^2 + 2\left\|F_x(\mathbf{x}_s)\right\|\left\|\mathbf{b}_s\right\| + \left\|\mathbf{b}_s\right\|^2)\Big] +$$
$$+ \rho_s^2 r_s^2 \mathbf{d}_s,$$
$$\beta_s = \begin{cases} 1, & \left\|F_x(x_s)\right\| \geq Q \\ 0, & \left\|F_x(x_s)\right\| < Q \end{cases}.$$

Box 4.

$$\mathbf{M}(\left\|\xi_s\right\|^2/\Im_s) = \left[\sum_{j=1}^{n} \mathbf{D}(\xi_{js}/x_0,x_1,x_2,..,x_s) + \left\|F_x(\mathbf{x}_s)\right\|^2 + 2\left\|(F_x(\mathbf{x}_s),\mathbf{b}_s)\right\| + \left\|\mathbf{b}_s\right\|^2\right].$$

this for some subsequence (x_{sk}) of (x_s), $s=0, 1, 2, 3,...$ it follows:

$$\left\|F_x(\mathbf{x}_{s_k})\right\| \xrightarrow[k\to\infty]{} 0.$$

This completes the proof.

Now we will prove the convergence of the classical stochastic recurrent procedure of Robins-Monro, which is the final goal of this section. Let us consider the function $F(x_1, x_2, \ldots, x_n)$, $\mathbf{x}=(x_1, x_2, \ldots, x_n)$, $\mathbf{x}\in\mathbf{X}\subseteq\boldsymbol{R}^n$, $F(\mathbf{x})=\mathrm{M}\left\|\mu - \mathbf{x}\right\|^2$, where μ is random vector with values in $\boldsymbol{R}^n$, whose mathematical expectation we aim to determine. Obviously the minimum of the function $F(\mathbf{x})$ is mathematical expectation of the random variable μ. If we assume that the set $\mathbf{X}$ is bounded, then by using theorem 4.2.2 and procedure (5.2.1) we obtain the Robbins-Monro procedure in the form (Robbin, 1951; Aizerman, 1970; Ermolev, 1976):

$$\mathbf{x}_{s+1} = \pi_X(\mathbf{x}_s + \frac{1}{(s+1)}(\mu_s - \mathbf{x}_s)).$$

If $\mathbf{X}= \boldsymbol{R}^n$, we obtain the Robins-Monro procedure in its classic form:

$$\mathbf{x}_{s+1} = \mathbf{x}_s + \frac{1}{(s+1)}(\mu_s - \mathbf{x}_s)) = \frac{1}{(s+1)}\sum_{k=1}^{s+1}\mu_k.$$

In these formulas the random variable ξ_s, its mathematical expectation and the step have the form:

$$\xi_s = -2(\mu_{s+1} - \mathbf{x}_s), \quad M(\xi_s/\Im_s)$$
$$= F_x(x_s),\ \rho_s = \frac{1}{2(s+1)}.$$

This procedure will be very helpful in the following chapters of the book. Now we will look at a different stochastic methods, namely the theory of "Potential functions method."

3. POTENTIAL FUNCTION METHOD AND PATTERN RECOGNITION

Now we proceed to a different and very fruitful area of the stochastic programming, namely the "*Potential functions method*." Even though it is called method, this field is actually a strict and mathematically sound theory whose practical results are a large number of stochastic recurrent procedures for pattern recognition, approximating algorithms for functions in noisy conditions, development of unified mathematical approach for *machine learning* on the base of human preference, proof of the perceptron theory, etc. (Aizerman, 1964, 1970; Pavlov, 1989, 2005; Herbrich, 1988, 1999). In some of its aspects the "Potential functions method" can also be regarded as stochastic gradient procedure or more precisely as a realization of the extremal approach in the stochastic programming. Its main difference from the methods in the previous chapters is in the way of determination of the random variable ξ_s from procedures (5.2.1) and (5.2.2) in the section 5.2. The idea that is used is that in pattern recognition, the elements of the different clusters must be grouped in sets, separated by regions where there are no points from them. Then if a potential is used (unimodal positive function, Figure 1), which is accumulated with sign plus over one of the sets and with sign minus over the other set for every training point the sets are separated, the clusters are discerned.

By a sequence of training samples, recurrently and cumulatively is constructed a function which is positive over one of the sets and negative over the other of the sets and the regions where this function is zero separate or "recognize" the two sets.

We will consider and acquaint the reader with some sections of the "Potential functions method" from the position of application in the decision-making theory and more specifically from the point of view of approximating value functions

Figure 1. Potential function

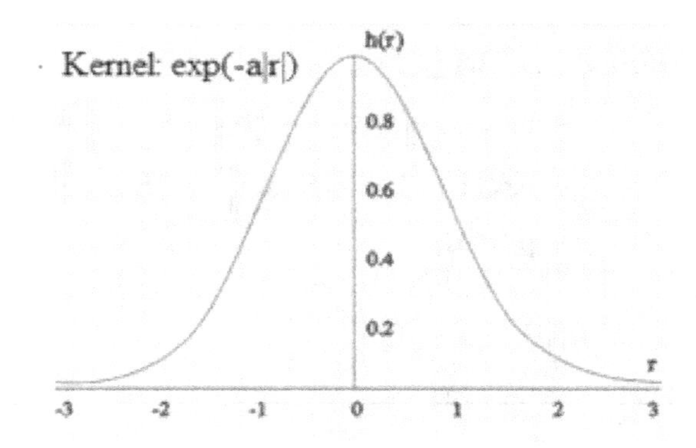

in uncertainty conditions, as well as in conditions of precise separation of the sets of positive and negative answers.

We will start with a theorem from that area, which will be needed in the proofs. Its proof may be seen in Aiserman (1970).

3.1. Theorem

Let there exist two positive functions $\mathbf{U}(y_1,y_2,..,y_n)$ and $\mathbf{V}(y_1,y_2,..,y_n)$ from generally speaking, increasing with (n) in number arguments of finite or infinite in dimension vectors $y_1,y_2,..,y_n,..$ Let the mathematical expectations $\mathbf{M}(\mathbf{U}(y_1))$ and $\mathbf{M}(\mathbf{V}(y_1))$ exist and satisfy the following conditions:

$$\mathbf{M}(\mathbf{U}(y_1,y_2,..,y_{n+1})/y_1,y_2,..,y_n)$$
$$\leq (1+\mu_n)\mathbf{U}(y_1,y_2,..,y_n)-\gamma_n\mathbf{V}(y_1,y_2,..,y_n)+\xi_n,$$
$$\mathbf{n}=1,2,3,......$$

In this case $\gamma_n \geq 0$, and the sequence μ_n satisfies:

$$\gamma_n = \mathbf{a} > 0;$$

$$\sum_1^{\infty}\left|\mu_n\right| < \infty;$$

The variables ξ_n are positive $\xi_n \geq 0$ or are sequence of functions $\xi_n(y_1,y_2,..,y_n)$ with random arguments $y_1,y_2,..,y_n,..$ and $\sum_1^{\infty} \mathbf{M}(\xi_n) < \infty$; then $\lim_{n\to\infty} \mathbf{M}(\mathbf{V}(y_1,y_2,..,y_n)) = 0$ and in addition the random variables $\mathbf{V}(y_1,y_2,..,y_n)$ tend to zero almost everywhere for $n \to \infty$.

Now we will lay out the general form of the recurrent procedures of the potential functions method:

$$\mathbf{y}_{i_{n+1}} = \mathbf{y}_{i_n} + \gamma_n \mathbf{r}^n \Phi_i(\mathbf{y}_n, \mathbf{x}_n), \quad i = 1,2,3,... \tag{5.3.3}$$

In this case $y_n = (y_{1n}, y_{2n},..)$ are finite or infinite in dimension vectors, and x_n are random vectors which appear independently at each step, with some possibly unknown conditional probability distribution $\mathbf{p}(x_n/y_n)$, which does not depend on (n) and on the function $\Phi_i(\mathbf{y}_n,\mathbf{x}_n)$. We see that for vectors of finite dimension $y_n = (y_{1n}, y_{2n},..)$ and under certain choice of the step γ_n, $\mathbf{r}^n = 1$, the above recurrent stochastic procedure degenerates to Robins-Monro procedure. The following theorem is valid (Aizerman, 1970).

3.2. Theorem

Let in the recurrent procedure (5.3.3) the vector $\Phi(\mathbf{y},x)$ is such that the function:

$\mathbf{R}(y) = -\mathbf{M}_x(\Phi(\mathbf{y},x))$ is monotonous.

We also assume that the integral in the following formula does not depend on the path:

$$\mathbf{J}(y) = \int_0^{\mathbf{y}} \mathbf{R}(y)dy.$$

We also assume the following three conditions:

$$\inf_{\|\mathbf{y}\|<\infty} J(y) = J_{\min} > -\infty;$$

$$\mathbf{M}_x(\Phi^2(\mathbf{y},x)) \leq \mathbf{a} + \mathbf{b}(\mathbf{J}(y) - \mathbf{J}_{min}), \ \mathbf{a} > 0, \ \mathbf{b} > 0;$$

$$\sum_1^{\infty} \gamma_n = \infty, \ \sum_1^{\infty} \gamma_n^2 < \infty.$$

Then from the recurrent procedure (5.3.3) follows

$$\mathbf{J}(y_n) \xrightarrow{\text{e.w.}} \mathbf{J}_{\min}.$$

Now we will consider an interesting from the practical point of view algorithm (Aizerman, 1970):

$$\mathbf{f}^{n+1}(\mathbf{x}) = \mathbf{f}^n(\mathbf{x}) + \mathbf{sign}(\mathbf{sign}(\mathbf{f}^*(\mathbf{x}_{n+1})) - \mathbf{sign}(\mathbf{f}^n(\mathbf{x}_{n+1})))\mathbf{K}(\mathbf{x}, \mathbf{x}_{n+1}).$$

Let us assume that there exist two sets $\mathbf{A}$ and $\mathbf{B}$, $\mathbf{A} \cap \mathbf{B} = \varnothing$. We assume that there exists function $\mathbf{f}^*(\mathbf{x})$ separating the sets completely, i.e. is positive over $\mathbf{A}$ and negative over $\mathbf{B}$ and its absolute value over $\mathbf{A} \cup \mathbf{B}$ is greater from a given positive number ε. We assume that the training points $\mathbf{x}$ over $\mathbf{A} \cup \mathbf{B}$ appear independently of one another and the probability for appearance of training point under incomplete recognition is different from zero. The conditions for the function $\mathbf{f}^*(\mathbf{x})$ and for the complete recognition of the sets $\mathbf{A}$ and $\mathbf{B}$ have the form:

$$\mathbf{f}^*(\mathbf{x}) \begin{cases} \geq \varepsilon, \ \mathbf{x} \in \mathbf{A} \\ \leq -\varepsilon, \ \mathbf{x} \in \mathbf{B} \end{cases}, \ \varepsilon > 0.$$

Regarding the potential function $\mathbf{K}(\mathbf{x},\mathbf{y})$ we consider the following general conditions:

$$\mathbf{K}(\mathbf{x},\mathbf{y}) = \sum_1^{\infty} \lambda_i^2 \Theta_i(\mathbf{x})\Theta_i(\mathbf{y}) \geq 0, \ \sum_1^{\infty} \lambda_i^2 < \infty,$$
$$\mathbf{K}(\mathbf{x},\mathbf{y}) = \mathbf{K}(\mathbf{y},\mathbf{x}), \ \mathbf{K}(\mathbf{x},\mathbf{y}) < (\mathbf{K}(\mathbf{x},\mathbf{x})) < \mathbf{C},$$
$$\mathbf{C} = \mathbf{const}.$$

In this case $(\Theta_i(x))$ is a family of functions, like the Lagrange functions (Aizernan, 1970). Regarding the function $f^*(x)$ we assume that it is fulfilled:

$$f^*(x) = \sum_1^\infty c_i^* \Theta_i(x) \ , \sum_1^\infty \left(\frac{c_i^*}{\lambda_i}\right)^2 < \infty.$$

In Novikof (1963) and Aizerman (1970), it is proved that under the above conditions, the algorithm construct in finite number of steps a function $f^n(x)$, which recognizes completely the sets A and B, $A \cap B = \varnothing$. The algorithm is convenient for evaluating value functions because in the "training" "yes" and "no" answer are required for the belongingness of the points x to the sets A and B, $A \cap B = \varnothing$.

In the expression of the preferences of the decision maker, two sets are formed A and B corresponding to positive preferences and negative preferences, in the case when $f^*(.)$ is a value function (Pavlov, 1989, 2005, 2011):

$$A = \{(x, y) \,/\, (x, y) \in R^{2m} \wedge (f^*(x) > f^*(y))\},$$
$$B = \{(x, y) \,/\, (x, y) \in R^{2m} \wedge (f^*(x) < f^*(y))\}.$$

In practice, the training answers as expressed preferences for the points $(x, y), (x, y) \in R^{2m}$ are always finite in number. Because of this for a smaller number of points it can be expected that for A and B, it is fulfilled $A \cap B = \varnothing$. From the previous description we see, that the separating value function for the preferences has the form $\Phi(x, y) = (f^*(x) - f^*(y))$. By taking into account the decomposition $f^*(x) = \sum_1^\infty c_i^* \Theta_i(x)$ of the function $f^*(x)$ we reach the decomposition

$$\Phi^*(x, y) = (f^*(x) - f^*(y))$$
$$= \sum_1^\infty c_i^* (\Theta_i(x) - \Theta_i(y)).$$

Therefore, if we can approximate $\Phi^*(x, y)$ under the conditions of the above algorithm, by a function of the same type, this will be a value function. Thus we change the algorithm to the following form, preserving the old denotations:

$$\Phi^{n+1}(x, y) = \Phi^n(x, y) + \mathrm{sign}(\mathrm{sign}(f^*(x_{n+1}) - f^*(y_{n+1})) - \mathrm{sign}(f^n(x_{n+1}) - f^n(x_{n+1})))\Lambda((x, y), (x_{n+1}, y_{n+1})), \Lambda((x, y), (x_{n+1}, y_{n+1})) = K(x, x_{n+1}) + K(y, y_{n+1}).$$

Under these conditions the following theorem is valid (Pavlov, 1988):

3.3. Theorem

We consider the recurrent procedure:

$$\Phi^{n+1}(x, y) = \Phi^n(x, y) + \mathrm{sign}(\mathrm{sign}(f^*(x_{n+1}) - f^*(y_{n+1})) \mathrm{sign}(f^n(x_{n+1}) - f^n(x_{n+1})))\Lambda((x, y), (x_{n+1}, y_{n+1})).$$

If the point (x_n, y_n) is training for this procedure $(\mathrm{sign}\Phi^n(x_n, y_n) \neq (\mathrm{sign}(f^*(x_n) - f^*(y_n)))$, such will also be (y_n, x_n) for the next step. Then on each even step, under the condition $(x_{n+1}, y_{n+1}) = (y_n, x_n)$, will be valid the representation:

$$\Phi^{2n}(x, y) = f^{2n}(x) - f^{2n}(y).$$

Proof

Let $\Phi^0(x, y) = 0$. Then the first training point (x_1, y_1) with expressed preference

$$(x_1 \succ y_1) \equiv (f^*(x_1) > f^*(y_1))$$

for example, determines through the procedure:

$$\Phi^1(z, t) = \Phi^0(z, t) + \Lambda((z, t), (x_1, y_1)) = K(x_1, z) + K(t, y_1).$$

From the condition of the theorem, it follows $\Phi^1(x_1, y_1) > 0$, while for the training point $(x_2, y_2) = (y_1, x_1)$ is valid

$$(x_2 \prec y_2) \equiv ((f^*(y_1) - f^*(x_1)) < 0).$$

Then through the procedure we obtain the second approximation:

$$\begin{aligned}\Phi^2(x,y) &= \Phi^1(x,y) - \Lambda((x,y),(y_1,x_1)) \\ &= K(x_1,x) + K(y_1,y) - K(y_1,x) - K(x_1,y) \\ &= (K(x_1,x) - K(y_1,x)) - (K(x_1,y) - K(y_1,y)) \\ &= f^2(x) - f^2(y).\end{aligned}$$

We assume, that the above decomposition is true for (2n-2) step and let for the first training n^{-th} in order point it is fulfilled for example $(x_n \prec y_n) \equiv (f^*(x_n) < f^*(y_n))$. This means that it is fulfilled

$$(\Phi^{2n-2}(x_n, y_n) = f^{2n-2}(x_n) - f^{2n-2}(y_n) > 0),$$

while $(f^*(x_n) - f^*(y_n)) < 0$. Applying the recurrent procedure for the (2n-1) step we obtain:

$$\begin{aligned}\Phi^{2n-1}(x,y) &= \Phi^{2n-2}(x,y) - \Lambda((x_n,y_n),(x,y)) \\ &= \Phi^{2n-2}(x,y) - K(x_n,x) - K(y_n,y).\end{aligned}$$

On the next step (2n) for the procedure, by condition, training is (y_n, x_n). It follows from $(K(x,y) > 0)$ by condition and from the fact that $\Phi^{2n-2}(x_n, y_n) = -\Phi^{2n-2}(y_n, x_n)$ by recurrent assumption. Then on (2n) step, we obtain from the recurrent procedure in Box 5.

This completes the proof.

Hence this procedure allows for a recurrent construction of functions $f^n(x)$, which, if they recognize correctly the sets A and B, $A \cap B = \varnothing$ are actually value functions. Now we have to prove the convergence of the considered procedure. This idea will be used in the next two theorems. The following theorem is valid (Aizerman, 1970; Pavlov, 1989, 2011).

3.4. Theorem

We consider again the recurrent procedure with the same denotation as before:

$$\begin{aligned}\Phi^{n+1}(x,y) &= \Phi^n(x,y) \\ &+ sign(sign(f^*(x_{n+1}) - f^*(y_{n+1})) \\ &- sign(f^n(x_{n+1}) - f^n(x_{n+1})))\Omega((x,y),(x_{n+1}, y_{n+1})), \\ \Omega((x,y),(x_{n+1},y_{n+1})) &= K(x, x_{n+1}) \\ &+ K(y, y_{n+1}) - K(x, y_{n+1}) - K(y, x_{n+1}).\end{aligned}$$

Let the empirical value function, separating the sets A and B, $A \cap B = \varnothing$, have the form:

$$\Phi^*(x,y) = (f^*(x) - f^*(y)) = \sum_1^\infty c_i^*(\Theta_i(x) - \Theta_i(y)),$$

$$f^*(x) = \sum_1^\infty c_i^* \Theta_i(x), \quad \sum_1^\infty \left(\frac{c_i^*}{\lambda_i}\right)^2 < \infty.$$

Box 5.

$$\begin{aligned}\Phi^{2n}(x,y) &= \Phi^{2n-1}(x,y) + \Lambda((y_n, x_n,),(x,y)) = \Phi^{2n-1}(x,y) + K(y_n,x) + K(x_n,y) = \\ &= f^{2n-2}(x) - f^{2n-2}(y) + K(y_n,x) + K(x_n,y) - K(x_n,x) - K(y_n,y) = \\ &= f^{2n-2}(x) - f^{2n-2}(y) + (K(y_n,x) - K(x_n,x)) - (K(y_n,y) - K(x_n,y)) = \\ &= f^{2n}(x) - f^{2n}(y).\end{aligned}$$

Box 6.

$$M_{(x,y)}\left(\left|\mathrm{sign}(f*(x_n)-f*(y_n))-\mathrm{sign}(\Phi^n(x,y))\right|\right)\xrightarrow[n\to\infty]{a.e.}0.$$

Box 7.

$$\begin{aligned}&\mathrm{sign}(\mathrm{sign}(\Phi*(x,y))-\mathrm{sign}(\Phi^n(x,y))((f*(x)-f*(y))-\Phi^n(x,y))=\\&=r^n(x,y)((f*(x)-f*(y))-\Phi^n(x,y))=\\&=\begin{cases}\left|\sum_1^\infty \Delta c_i^n(\Theta_i(x)-\Theta_i(y))\right| & \text{, if } \mathrm{sign}(f*(x)-f*(y))\neq \mathrm{sign}(\Phi^n(x,y))\\ 0 & \text{, if } \mathrm{sign}(f*(x)-f*(y))=\mathrm{sign}(\Phi^n(x,y))\end{cases}\end{aligned}$$

We assume that it is fulfilled:

$$\Phi*(x,y)\begin{Bmatrix}\geq \varepsilon,\ (x,y)\in A\\ \leq -\varepsilon,\ (x,y)\in B\end{Bmatrix},\ \varepsilon>0.$$

We assume that the training points (x,y) appear independently, with the same distribution and the probability of having training point under incomplete recognition of the sets A and B, $A\cap B=\varnothing$, is always different than zero.

Then, due the recurrent algorithm it is valid as shown in Box 6.

Proof

We choose the separating function, so that the following condition is valid:

$$\inf_{A\cup B}\Phi*(x,y)=\sup_{A\cup B}\Omega((x,y),(x,y)).$$

This is possible due to the assumption:

$$\Phi*(x,y)\begin{Bmatrix}\geq \varepsilon,\ (x,y)\in A\\ \leq -\varepsilon,\ (x,y)\in B\end{Bmatrix},\ \varepsilon>0.$$

From the way of construction of the function $\Phi^n(x,y)$ and from the decomposition

$$K(x,y)=\sum_1^\infty \lambda_i^2\,\Theta_i(x)\Theta_i(y),$$

it follows that the function can be presented in the following way:

$$\Phi^n(x,y)=\sum_1^\infty c_i^n(\Theta_i(x)-\Theta_i(y)).$$

We introduce the denotation

$$\Delta c_i^n=(c_i*-c_i^n),\ i=1,2,3,...$$

From the recurrent procedure it follows that it is fulfilled in Box 7.

We introduce the following notation:

$$(\Delta c^n,(\Theta(x)-\Theta(y))=\sum_1^\infty \Delta c_i^n(\Theta_i(x)-\Theta_i(y))$$

If it is fulfilled $r(x,y)\neq 0$, then the strong inequality is valid:

Box 8.

$$U_{n+1} = \sum_1^\infty (\Delta\bar{c}_i^{n+1})^2 = \sum_1^\infty (\Delta\bar{c}_i^n - r^n(x_{n+1}, y_{n+1})(\lambda_i\Theta_i(x_{n+1}) - \lambda_i\Theta_i(y_{n+1})))^2 =$$
$$= U_n - 2r^n(x_{n+1}, y_{n+1})\sum_1^\infty \Delta\bar{c}_i^n(\lambda_i\Theta_i(x) - \lambda_i\Theta_i(y)) + (r^n)^2(x_{n+1}, y_{n+1})(\lambda\Theta(x_{n+1}) - \lambda\Theta(y_{n+1}))^2) =$$
$$= U_n - 2r^n(x_{n+1}, y_{n+1})\sum_1^\infty \Delta c_i^n(\Theta_i(x) - \Theta_i(y)) + (r^n)^2(x_{n+1}, y_{n+1})\Omega((x_{n+1}, y_{n+1}), (x_{n+1}, y_{n+1}))).$$

Box 9.

$$U_{n+1} = \sum_1^\infty (\Delta\bar{c}_i^{n+1})^2 = U_n - 2\left|r^n(x_{n+1}, y_{n+1})\right| \sup_{A\cup B} \Omega((x,y),(x,y) +$$
$$+ {r^n}^2(x_{n+1}, y_{n+1})\Omega((x_{n+1}, y_{n+1}), (x_{n+1}, y_{n+1}) =$$
$$= U_n - 2\left|r^n(x_{n+1}, y_{n+1})\right| \sup_{A\cup B} \Omega((x,y),(x,y) +$$
$$+ \left|r^n(x_{n+1}, y_{n+1})\right| \Omega((x_{n+1}, y_{n+1}), (x_{n+1}, y_{n+1}) \leq$$
$$\leq U_n - \left|r^n(x_{n+1}, y_{n+1})\right| \sup \Omega((x,y),(x,y).$$

$$\left|(\Delta c^n, (\Theta(x) - \Theta(y))\right| = \left|\sum_1^\infty \Delta c_i^n(\Theta_i(x) - \Theta_i(y))\right|$$
$$= \left|((f*(x) - f*(y)) - \Phi^n(x,y))\right|$$
$$\geq \sup_{A\cup B} \Omega((x,y),(x,y).$$

This means that it is valid:

$$r^n(x,y)((f*(x) - f*(y)) - \Phi^n(x,y))$$
$$\geq \left|r^n(x,y)\right| \sup_{A\cup B} \Omega((x,y),(x,y)$$

We put

$$U_n = \sum_1^\infty \left(\frac{\Delta c_i^n}{\lambda_i}\right)^2 = \sum_1^\infty \left(\frac{c_i^* - c_i^n}{\lambda_i}\right)^2 = \sum_1^\infty (\Delta\bar{c}_i^n)^2.$$

Then, the recurrent procedure with the new denotations takes the form:

$$(c^{n+1}, (\Theta(x) - \Theta(y)) = (c^n, (\Theta(x) - \Theta(y))$$
$$+ r^n(x_{n+1}, y_{n+1})((\lambda\Theta(x_{n+1})$$
$$- \lambda\Theta(y_{n+1})), (\lambda\Theta(x) - \lambda\Theta(y))).$$

From this equality we obtain Box 8.

From the inequality

$$\left|(\Delta c^n, (\Theta(x) - \Theta(y))\right| \geq \sup_{A\cup B} \Omega((x,y),(x,y)$$

it follows in Box 9.

Going to conditional mathematical expectation, we reach the following inequality:

$$M(U_{n+1} / (x_{n+1}, y_{n+1}), .., (x_n, y_n)) \leq U_n$$
$$- M_{(x,y)}\left|r^n(x, y_n)\right| \sup_{A\cup B} \Omega((x,y),(x,y).$$

From theorem (5.3.1) it follows in Box 10.

This completes the proof.

The proof of the theorem follows an idea given in Aizerman (1970), but with potential function defined in theorem (5.3.3) and allowing

Box 10.

$$M_{(x,y)}\left|r^n(x,y_n)\right| = M_{(x,y)}\left|\text{sign}(\text{sign}(\Phi*(x,y)) - \text{sign}(\Phi^n(x,y))\right| \xrightarrow[n\to\infty]{e.w} 0,$$

$$M_{(x,y)}\left|r^n(x,y_n)\right| = M_{(x,y)}\left|\text{sign}(\text{sign}(f*(xn+1) - f*(yn+1)) - \text{sign}(f^n(xn+1) - f^n(xn+1)))\right| \xrightarrow[n\to\infty]{e.w} 0.$$

Box 11.

$$\Phi^{n+1}(x,y) = \Phi^n(x,y) + \text{sign}(\text{sign}(f*(x_{n+1}) - f*(y_{n+1})) -$$
$$-\text{sign}(f^n(x_{n+1}) - f^n(x_{n+1})))\Omega((x,y),(x_{n+1},y_{n+1})),$$
$$\Omega((x,y),(x_{n+1},y_{n+1})) = K(x,x_{n+1}) + K(y,y_{n+1}) - K(x,y_{n+1}) - K(y,x_{n+1}).$$

for an analytical representation of the expert preferences as value functions. From theorems (5.3.3) and (5.3.4) it follows, that the function $f^n(x)$ is approximation of the empiric value function $f*(x)$. Now we will prove a theorem which proves that the recognition of the sets A and B, $A \cap B = \varnothing$, with the recurrent procedure stops in finite number of steps (Novikof, 1963; Aizerman, 1970; Pavlov, 1989, 2011).

3.5. Theorem

We consider again the recurrent procedure with the already familiar denotations in Box 11.

Let the empiric value function, separating the sets A and B, $A \cap B = \varnothing$, has the form:

$$\Phi*(x,y) = (f*(x) - f*(y)) = \sum_1^\infty c_i^*(\Theta_i(x) - \Theta_i(y)),$$

$$f*(x) = \sum_1^\infty c_i^*\Theta_i(x), \sum_1^\infty \left(\frac{c_i^*}{\lambda_i}\right)^2 < \infty.$$

We assume that it is fulfilled:

$$\Phi*(x,y)\begin{Bmatrix} \geq \varepsilon, (x,y) \in A \\ \leq -\varepsilon, (x,y) \in B \end{Bmatrix}, \varepsilon > 0.$$

We assume that the training points (x_i, y_i), $i = 1,2,3,....$ appear independently, with the same distribution and the probability of having training point for an incomplete recognition of the sets A and B, $A \cap B = \varnothing$, is always different from zero.

Under these conditions, there exists a number m, independent of the particular sequence, such that the number of training points

$$(\text{sign}\Phi^n(x_i,y_i) \neq (\text{sign}(f*(x_i) - f*(y_i)))$$

does not exceed m.

Proof

We put

$$f*(x) \underset{L_2}{=} \sum_1^\infty c_i^*\Theta_i(x) = \sum_1^\infty \left(\frac{c_i^*}{\lambda_i}\right)\Psi_i(x) = \sum_1^\infty \bar{c}_i * \Psi_i(x), \quad \sum_1^\infty \left(\frac{c_i^*}{\lambda_i}\right)^2 < \infty.$$

We put

$$K(x,y)=\sum_{1}^{\infty}\lambda_i^2\,\Theta_i(x)\Theta_i(y)$$
$$=\sum_{1}^{\infty}\Psi_i(x)\Psi_i(y)\geq 0,\ \sum_{1}^{\infty}\lambda_i^2<\infty.$$

We put

$$a=\inf_{A\cup B}\frac{\left|(\bar{c}^*,\Psi(x)-\Psi(y))\right|}{\left\|\bar{c}^*\right\|},\quad \left\|\bar{c}^*\right\|=\sqrt{\sum_{1}^{\infty}(\bar{c}_i^*)^2},$$

$$b=\sup_{A\cup B}\sqrt{\sum_{1}^{\infty}(\Psi(x)-\Psi(y),\Psi(x)-\Psi(y))}\leq 2\sqrt{C}.$$

The number b is correctly defined, because $K(x,y)<\ (K(x,x))$. The following inequality is satisfied:

$$a=\inf_{A\cup B}\frac{\left|(f^*(x)-f^*(y))\right|}{\left\|\bar{c}^*\right\|}\geq\frac{\varepsilon}{\left\|\bar{c}^*\right\|}>0.$$

Therefore, $\left|(f^*(x)-f^*(y))\right|>a\left\|\bar{c}^*\right\|$.

From the recurrent procedure it follows:

$$\Phi^n(x,y)=\sum_{i=1}^{n}\theta_i(\Psi(x_i)-\Psi(y_i),\Psi(x)-\Psi(y)),$$
$$\left|\theta_i\right|=1,\ \Phi^0(x,y)=0,$$
$$\theta_i=1,\ (\bar{c}^*,\Psi(x_i)-\Psi(y_i))>0,$$
$$\theta_i=-1,\ (\bar{c}^*,\Psi(x_i)-\Psi(y_i))<0.$$

We put $c^n=\sum_{i=1}^{n}\theta_i(\Psi(x_i)-\Psi(y_i))$. From the definition of a and c^n it follows, that it is fulfilled $(\bar{c}^*,c^n)\geq na\left\|\bar{c}^*\right\|$. Therefore,

$$(\left\|\bar{c}^*\right\|\left\|c^n\right\|\geq(\bar{c}^*,c^n)\geq na\left\|\bar{c}^*\right\|)$$

and $(\left\|c^n\right\|\geq na)$.

From the recurrent procedure follows the equality:

$$c^{n+1}=c^n+\theta_{n+1}(\Psi(x_{n+1})-\Psi(y_{n+1})).$$

Therefore, the following equality is valid:

$$\left\|c^{n+1}\right\|^2=\left\|c^n\right\|^2+(\Psi(x_{n+1})-\Psi(y_{n+1}),$$
$$\Psi(x_{n+1})-\Psi(y_{n+1}))+2\theta_{n+1}(c^n,\Psi(x_{n+1})-\Psi(y_{n+1})).$$

If it is fulfilled $(\bar{c}^*,\Psi(x_{n+1})-\Psi(y_{n+1}))<0$, then $\theta_{n+1}=-1$ and since (x_{n+1},y_{n+1}) is a training point it is fulfilled:

$$((c^n,\Psi(x_{n+1})-\Psi(y_{n+1}))>0)$$
$$\Rightarrow(\theta_{n+1}(c^n,\Psi(x_{n+1})-\Psi(y_{n+1}))<0).$$

Conversely, if $(\bar{c}^*,\Psi(x_{n+1})-\Psi(y_{n+1}))>0$, then $\theta_{n+1}=1$ and since (x_{n+1},y_{n+1}) is a training point it is fulfilled:

$$((c^n,\Psi(x_{n+1})-\Psi(y_{n+1}))<0)$$
$$\Rightarrow(\theta_{n+1}(c^n,\Psi(x_{n+1})-\Psi(y_{n+1}))<0).$$

Therefore, it is fulfilled:

$$\left\|c^{n+1}\right\|^2=\left\|c^n\right\|^2+b^2+2\theta_{n+1}(c^n,\Psi(x_{n+1})$$
$$-\Psi(y_{n+1}))<\left\|c^n\right\|^2+b^2\leq(n+1)b^2.$$

Therefore, the following inequality is valid:

$$(n)^2a^2\leq\left\|c^n\right\|\leq(n)b^2,$$
$$(n)\leq\frac{b^2}{a^2}=m,$$
$$m=\frac{\sup_{A\cup B}(\Psi(x)-\Psi(y),\Psi(x)-\Psi(y))}{\inf_{A\cup B}\left|(f^*(x)-f^*(y))\right|^2}\left\|\bar{c}^*\right\|^2.$$

This completes the proof.

Theorems (5.3.3), (5.3.4), and (5.3.5) proved that the stochastic recurrent procedure constructs value function $f^n(x)$ for a finite number of steps and completely recognizes the sets A and B, $A \cap B = \varnothing$. We can say that we have trained with the function $f^n(x)$ the computer to have the same preferences as the decision maker. In Aizerman (1970) is proved a theorem, which allows for the construction of decision rule when to terminate the recurrent procedure. The theorem states the following (Aizerman, 1970; Pavlov, 2011).

3.6. Theorem

The value function $f^n(.)$ constructed with the recurrent procedure recognizes correctly (1-β)100% of the set **A∪B, A∩B=∅**, with probability grater then (1-δ), if after the ith mistake

$$(sign(u*(x_n) - u*(y_n)) \neq sign(u^{n-1}(x_n) - u^{n-1}(y_n))),$$

the next (L_0+i) points are recognized correctly

$$sign(u*(x^n) - u*(y_n)) = sign(u^{n-1}(x_n) - u^{n-1}(y_n))$$

The learning points $(x,y) \in \mathbf{R}^m x \mathbf{R}^m$ are uniformly distributed. The number L_0 fulfils the condition $L_0 \geq \dfrac{\ln(\beta.\delta)}{\ln(1-\beta)}$.

The theorems given here and the developed recurrent procedure, allow a flexible strategy and different possibilities for the construction of value functions for a finite number of expert preferences. The main limitation of this approach is the condition $A \cap B = \varnothing$. This means that the contradictions in the expert's preferences are eliminated either during the work of the algorithm or preliminary.

Now we will consider the case of expressing preferences in conditions of uncertainty $A \cap B \neq \varnothing$. We consider the general recurrent formula of the potential functions method defined in theorem (5.3.3) and with presence of uncertainty in the answers. The uncertainty is described by the presence of a random variable ξ with bounded variance and mathematical expectation zero.

3.7. Theorem

We consider the general recurrent procedure of the potential functions method in noisy conditions (Aizerman, 1970; Pavlov, 2003, 2005, 2011):

$$\Phi^{n+1}(x,y) = \Phi^n(x,y) + \gamma_n(r(\Phi^n(x_{n+1}, y_{n+1}),$$
$$\Phi*(x_{n+1}, y_{n+1}) + \xi_n)\Omega((x,y),(x_{n+1}, y_{n+1})),$$
$$\Omega((x,y),(x_{n+1}, y_{n+1})) = K(x, x_{n+1})$$
$$+K(y, y_{n+1}) - K(x, y_{n+1}) - K(y, x_{n+1}).$$

We assume that the training points (x_i, y_i), $i = 1,2,3,....$ appear independently, with the same distribution. The separation function, as presentation of the value function, is of the type $\Phi*(x,y) = (f*(x) - f*(y))$. We assume that the following conditions are true:

$$M_\xi(\xi_n / (x,y), c^n) = 0,$$

$$M_\xi((\xi_n)^2 / (x,y), c^n) \leq Q, \quad Q = \text{const},$$

$$\sum_1^\infty \gamma_n = \infty, \sum_1^\infty \gamma_n^2 < \infty.$$

The potential function satisfies the conditions:

$$\Omega((x,y),(z,t)$$
$$= \sum_{i=1}^\infty \lambda_i^2 (\Theta_i(x) - \Theta_i(y), \Theta_i(z) - \Theta_i(t))$$
$$= \sum_{i=1}^\infty (\Psi_i(x) - \Psi_i(y), \Psi_i(z) - \Psi_i(t)),$$

$$K(x,y)=\sum_1^{\infty}\lambda_i^2\,\Theta_i(x)\Theta_i(y)=\sum_1^{\infty}\Psi_i(x)\Psi_i(y)$$
$$\geq 0,\ \sum_1^{\infty}\lambda_i^2<\infty,\ K(x,x)\leq H,\ H=\text{const}.$$
$$\Omega((x,y),(x,y)=K(x,x)+K(y,y)$$
$$-2K(x,y),\ \max_{(x,y)}\Omega((x,y),(x,y)=\Psi_{max},$$
$$\Psi_{max}=\text{const}.$$

The function $r(u,\Phi*(x,y))$ satisfies the following conditions:

The function $r(u,\Phi*(x,y))$ is non-strictly decreasing by its first variable.

For each pair (x,y) and for every two u_1 and u_2 is fulfilled the following:

$$|r(u_1,\Phi*(x,y))-r(u_2,\Phi*(x,y))|$$
$$\leq a+b|u_1-u_2|,\quad a=\text{const},\ b=\text{const};$$

$$M_{(x,y)}(r^2(0,\Phi*(x,y)))<\infty;$$

The function

$$J(\Phi)=M_{(x,y)}\left(-\int_0^{\Phi(x,y)}r(u,\Phi*(x,y)))du\right)$$

is bounded from below over the set of functions:

$$\Phi(x,y)$$
$$\underset{L_2}{=}\sum_1^{\infty}c_i(\Theta_i(x)-\Theta_i(y))\underset{L_2}{=}\sum_1^{\infty}\frac{c_i}{\lambda_i}(\Psi_i(x)-\Psi_i(y))$$
$$\underset{L_2}{=}\sum_1^{\infty}\bar{c}_i(\Psi_i(x)-\Psi_i(y)),$$
$$\sum_1^{\infty}\left(\frac{c_i}{\lambda_i}\right)^2<\infty.$$

The space of square-integrable functions L_2 is defined by the probability distribution of the appearance of the points (x,y).

Then, from the recurrent procedure it follows:

$$J(\Phi^n)\xrightarrow[n\to\infty]{e.w.}J_{min},$$
$$J_{min}=\min_{\Phi(x,y)\underset{L_2}{=}\sum_1^{\infty}c_i(\Theta_i(x)-\Theta_i(y)),\ \sum_1^{\infty}\left(\frac{c_i}{\lambda_i}\right)^2<\infty}(J(\Phi)).$$

Proof

Let l_2 is the Hilbert space of square-convergent number sequences (Hutson, 1980; Trenogin, 1985). We introduce the following denotation (Aizerman, 1970; Trenogin, 1985):

$$\int_{c_1}^{c_2}R(y)dy=\int_0^1(R((1-\alpha)c_1+\alpha c_2),c_2-c_1)d\alpha,$$
$$c_1,c_2,R(y)\in l_2.$$

In the above denotation $(R((1-\alpha)c_1+\alpha c_2),c_2-c_1)$ is scalar product in l_2.

We will say that $\int_0^z R(y)dy$ does not depend on the path if it is satisfied:

$$\int_{c_1}^{c_2}R(y)dy=\int_{c_1}^{z}R(y)dy+\int_{z}^{c_2}R(y)dy,\quad z\in l_2$$

We introduce the denotation $J(y)=\int_0^y R(z)dz.$

Since this integral does not depend on the path, we can note:

$$J(c_2)-J(c_1)=\int_{c_1}^{c_2}R(y)dy.$$

We will say that $R(z)$ is monotonous if it is fulfilled:

$$(R(c_2)-R(c_1),c_2-c_1)\geq 0.$$

It is known that if $R(y)$ is monotonous function and the integral $J(y) = \int_0^y R(z)dz$ by $R(y)$ does not depend on the path, then $J(y)$ is convex function and satisfies the following (Aizerman, 1970):

$$J(y_1) - J(y_2) \leq (R(y_1), y_1 - y_2).$$

We put:

$$g(u,(x,y)) = -r(u, \Phi * (x,y)),$$

$$G((c,(x,y)) = (g((c, \Psi(x) - \Psi(y)), (x,y)) - \xi)(\Psi(x) - \Psi(y)).$$

In these formulas we used the denotations $(c, \Psi(x) - \Psi(y)) = \sum_{i=1}^{n} (c_i, \Psi_i(z) - \Psi_i(t))$ and if $\sum_1^{\infty} \lambda_i^2 < \infty$, then $G((c,(x,y)) \in l_2$.

See Box 12.

We note that the conditional mathematical expectation of the random variable ξ is zero $M_\xi(\xi_n/(x, y), c^n) = 0$. From the Fubini theorem and from $M_{(x, y)}(r^2(0, \Phi^*(x, y))) < \infty$ it follows (Partasaraty, 1978):

$$J(\bar{\bar{c}}) = \int_0^{\bar{\bar{c}}} R(\bar{c})d\bar{c} = M_{(x, y)}(\int_0^{\bar{\bar{c}}} (g((\bar{c}, \Psi(x) - \Psi(y)), (x,y)) - \xi)(\Psi(x) - \Psi(y))d\bar{c})$$

We note that the function $R(\bar{c})$ in the integral $J(\bar{\bar{c}}) = \int_0^{\bar{\bar{c}}} R(\bar{c})d\bar{c}$ is monotonous by $\bar{c}$ due to the conditions of the theorem and the properties of the function $r(u, \Phi * (x,y))$. See Box 13.

We will prove the following equality:

$$\int_0^{\bar{\bar{c}}} (g((\bar{c}, \Psi(x) - \Psi(y)), (x,y))(\Psi(x) - \Psi(y))d\bar{c} = -\int_0^{(\bar{c}, \Psi(x) - \Psi(y))} r(u, \Phi * (x,y))du.$$

The integral on the left side of the above equality by definition is equal to:

Box 12.

$$R(\bar{c}) = M_{\xi, (x, y)}(G(\bar{c}, (x,y))) = M_{\xi, (x, y)}((g((\bar{c}, \Psi(x) - \Psi(y)), (x,y)) - \xi)(\Psi(x) - \Psi(y)) =$$
$$= M_{(x, y)}(M_\xi((g((\bar{c}, \Psi(x) - \Psi(y)), (x,y)) - \xi)(\Psi(x) - \Psi(y))) =$$
$$= M_{(x, y)}(g((\bar{c}, \Psi(x) - \Psi(y)), (x,y))(\Psi(x) - \Psi(y)).$$

Box 13.

$$(g((\bar{c}_2, \Psi(x) - \Psi(y)), (x,y))(\Psi(x) - \Psi(y)) -$$
$$-g((\bar{c}_1, \Psi(x) - \Psi(y)), (x,y))(\Psi(x) - \Psi(y)), \bar{c}_2 - \bar{c}_1) =$$
$$= (g((\bar{c}_2, \Psi(x) - \Psi(y)), (x,y)) -$$
$$-g((\bar{c}_1, \Psi(x) - \Psi(y)), (x,y))(\bar{c}_2, (\Psi(x) - \Psi(y))) - (\bar{c}_1, (\Psi(x) - \Psi(y)))) \geq 0.$$

$$\int_0^{c}(g((\bar{c},\Psi(x)-\Psi(y)),(x,y))(\Psi(x)-\Psi(y))d\bar{c} =$$
$$= \int_0^1 (g((\alpha\bar{\bar{c}},\Psi(x)-\Psi(y)),(x,y))(\bar{c},\Psi(x)-\Psi(y))d\alpha =$$
$$= \int_0^{(\bar{c},\Psi(x)-\Psi(y))} g(u,(x,y))du = -\int_0^{(\bar{c},\Psi(x)-\Psi(y))} r(u,\Phi^*(x,y))du.$$

Therefore, there is an independence of the path of the integral:

$$\int_0^{\bar{\bar{c}}}(g((\bar{c},\Psi(x)-\Psi(y)),(x,y))(\Psi(x)-\Psi(y))d\bar{c}.$$

Hence, there is also independence of the path of the integral:

$$M_{(x,y)}(\int_0^{(\bar{c},\Psi(x)-\Psi(y))} g(u,(x,y))du)$$
$$= M_{(x,y)}(-\int_0^{(\bar{c},\Psi(x)-\Psi(y))} r(u,\Phi^*(x,y))du) > -\infty.$$

From the conditions of the theorem, Box 14 inequality is valid.

On page 220 on the basis of the above inequality in (Aizerman, 1970) is proven the inequality:

$$M_{\xi,(x,y)}((g((\bar{c},\Psi(x)-\Psi(y)),(x,y))$$
$$-\xi)(\Psi(x)-\Psi(y))^2$$
$$= M_{\xi,(x,y)}(G((c,(x,y)))^2 \le \chi$$
$$+\beta(J(c)-J_{min}), \chi > 0, \beta > 0.$$

Therefore, for the function $R(c) = -M_{(x,y),\xi}(G(c,(x,y)))$ are fulfilled all conditions of theorem 5.3.2. From that it follows the following almost everywhere convergence:

$$J(\bar{c}^n) = \int_0^{\bar{c}^n} R(c)dc$$
$$= M_{(x,y)}(-\int_0^{(\bar{c}^n,\Psi(x)-\Psi(y))} r(u,\Phi^*(x,y))du)$$
$$= M_{(x,y)}(-\int_0^{(\Phi^n(x,y))} r(u,\Phi^*(x,y))du) \xrightarrow[n\to\infty]{e.w.} J_{min}.$$

This concludes the proof.

In this prove we used ideas of the proofs in Aizerman (1970), but with the potential function defined by theorem 5.3.3 (Pavlov, 1989, 2011). We will note some important properties of the recurrent procedure:

$$\Phi^{n+1}(x,y) = \Phi^n(x,y)$$
$$+\gamma_n(r(\Phi^n(x_{n+1},y_{n+1}),\Phi^*(x_{n+1},y_{n+1}) + \xi_n)$$
$$\Omega((x,y),(x_{n+1},y_{n+1})),$$
$$\sum_1^\infty \gamma_n = \infty, \sum_1^\infty \gamma_n^2 < \infty, \quad \Phi^0(x,y) = 0.$$

On each step, the function $\Phi^n(x,y)$ has the form $\Phi^n(x,y) = f^n(x) - f^n(y)$. And the second important moment, proved in Aizerman (1970), is that on each step it is fulfilled:

$$\Phi^n(x,y) \underset{L_2}{=} \sum_1^\infty c_i^n(\Theta_i(x)-\Theta_i(y))$$
$$= f^n(x) - f^n(y), \sum_1^\infty (\frac{c_i^n}{\lambda_i})^2 < \infty.$$

In the above formula $(\Theta_i(x)), i = 1,2,3,..$ are families of orthogonal polynomials, e.g. the Lagrange polynomials or some, possibly finite, subset of them (Partasaraty, 1987).

We introduce the following notation:

$$\bar{f} = \begin{cases} f, & \text{if } 0 < f < 1, \\ 0, & \text{if } f = 0, \\ 1, & \text{if } 1 < f. \end{cases}$$

Box 14.

$$\begin{aligned}&M_{\xi,(x,y)}((g((\overline{c},\Psi(x)-\Psi(y)),(x,y))-\xi)(\Psi(x)-\Psi(y))^2\leq\\&\leq M_{(x,y)}((g^2((\overline{c},\Psi(x)-\Psi(y)),(x,y))\Psi^2_{max}+Q\Psi^2_{max}-\\&-2M_{(x,y)}(M_{\xi}((g((\overline{c},\Psi(x)-\Psi(y)),(x,y))(\xi)(\Psi(x)-\Psi(y),\Psi(x)-\Psi(y))=\\&=M_{(x,y)}((g^2((\overline{c},\Psi(x)-\Psi(y)),(x,y))\Psi^2_{max}+Q\Psi^2_{max}\ .\end{aligned}$$

We will assume that over the set of alternatives for comparison *X*, the empiric utility $f^*(x)$ satisfies the condition $|f^*(x)-f^*(y)|\leq 1$. Now we will consider the following procedure:

$$\Phi^{n+1}(x,y)=\Phi^n(x,y)$$
$$+\gamma_n\delta_n\Omega((x,y),(x_{n+1},y_{n+1})),$$
$$\sum_1^{\infty}\gamma_n=\infty,\sum_1^{\infty}\gamma_n^2<\infty,\quad \Phi^0(x,y)=0.$$

The numbers δ_n, $\delta_n^2=1$ are constructed in the following manner. On each (n+1)-th step with probability $\overline{\Phi^n(x_{n+1},y_{n+1})}$ the training point (x_{n+1},y_{n+1}) is assigned to the set of positive preferences

$$A=\{(x,y)\,/\,(x,y)\in R^{2m}\wedge(f^*(x)>f^*(y))\}$$

or with probability $(1-\overline{\Phi^n(x_{n+1},y_{n+1})})$ is assigned to the set of the negative preferences

$$B=\{(x,y)\,/\,(x,y)\in R^{2m}\wedge(f^*(x)<f^*(y))\}$$

The decision maker also assigns the point to one of the two sets $\mathbf{A}$ and $\mathbf{B}$, $A\cap B\neq\varnothing$. Four cases are possible:

$$\delta_n=\begin{cases}0 & A,A\\0 & B,B\\1 & A,B\\-1 & B,A\end{cases}.$$

In the scientific source (Aizerman, 1970) it is proved that in this case the random variable has the form:

$$\delta_n=\overline{(\text{sign}(|f^*(x)-f^*(y)|))^+}$$
$$-\overline{\Phi^n(x_{n+1},y_{n+1})}+\xi_{n+1},$$
$$f^+=f \text{ if } f\geq 0, f^+=0 \text{ if } f<0,$$

$$M_{\xi}(\xi_n\,/\,(x,y),c^n)=0.$$

The so constructed algorithm is a particular case of the algorithm from theorem 5.3.7.

We have the following representation:

$$r(\Phi^n(x,y),\Phi^*(x,y))$$
$$=(\text{sign}(|f^*(x)-f^*(y)|))^+-\overline{\Phi^n(x_{n+1},y_{n+1})}.$$

In this case is used the notation

$$\Phi^*(x,y)=(\text{sign}(|f^*(x)-f^*(y)|))^+\ .$$

From the previous theorem, under complete system of orthogonal polynomials $(\Theta_i(x)),i=1,2,3,...,$ it follows the almost everywhere convergence shown in Box 15.

It is known that:

$$J(\overline{c}^n)\text{-}J(\overline{c}^*)=M_{(x,y)}(-\int_{\Phi^*(x,y)}^{(\overline{c}^n,\Psi(x)-\Psi(y))}(\Phi^*(x,y)-\overline{u})du)$$
$$\geq M_{(x,y)}(\frac{1}{2}((\Phi^*(x,y)-\Phi^n(x,y))^2).$$

Box 15.

$$J(\overline{c}^{n})\text{-}J(\overline{c}^{*}) = M_{(x,y)}\Big(-\int_{\Phi^{*}(x,y)}^{(\overline{c}^{n},\Psi(x)-\Psi(y))} (\Phi^{*}(x,y) - \overline{u})du\Big) =$$

$$= M_{(x,y)}\Big(-\int_{\Phi^{*}(x,y)}^{(\Phi^{n}(x,y))} (\mathrm{sign}(|f^{*}(x) - f^{*}(y)|))^{+} - \overline{u})du\Big) \xrightarrow[n\to\infty]{e.w.} 0,$$

$$f^{*}(x) = \sum_{1}^{\infty} \overline{c}^{*}\Psi(x), \sum_{1}^{\infty} \Big(\frac{c_i^{*}}{\lambda_i}\Big)^{2} < \infty,\ M_{(x,y)}(f^{*}(x) - f^{*}(y))^{2} < \infty.$$

Therefore:

$$= M_{(x,y)}\Big(\frac{1}{2}((\Phi^{*}(x,y) - \Phi^{n}(x,y))^{2})\Big) \xrightarrow[n\to\infty]{e.w.} 0$$

In this manner, we have constructed the recurrent procedure for value function approximation in conditions of uncertainty – $A \cap B \neq \varnothing$.

4. FIELD STUDIES

The fields of investigation and application of the stochastic programming are versatile: from Control theory, pattern recognition, extremal problems in mathematical statistics, game theory, and economic planning to various optimization problems, etc. Our interests in this area are related to the decision-making theory, Bayesian approach of Professor Raiffa and with utility theory and the possibilities for evaluation of the subjective probabilities and approximation of the value and utility functions.

We start with the case of exact recognition of the set A of the positive and the set B of the negative answers $(A \cap B = \varnothing)$:

$$\begin{aligned}
\Phi^{n+1}(x,y) &= \Phi^{n}(x,y) \\
&+\mathrm{sign}(\mathrm{sign}(f^{*}(x_{n+1}) - f^{*}(y_{n+1})) \\
&-\mathrm{sign}(f^{n}(x_{n+1}) - f^{n}(x_{n+1}))) \\
&\Omega((x,y),(x_{n+1},y_{n+1})), \\
\Omega((x,y),(x_{n+1},y_{n+1})) &= K(x,x_{n+1}) \\
&+K(y,y_{n+1}) - K(x,y_{n+1}) - K(y,x_{n+1}).
\end{aligned}$$

In other denotations:

$$\begin{aligned}
f^{n}(x) - f^{n}(y) &= f^{n-1}(x) - f^{n-1}(y) \\
&+\mathrm{sign}(\mathrm{sign}(f^{*}(x_n) - f^{*}(y_n)) \\
&-\mathrm{sign}(f^{n}(x_n) - f^{n}(x_n)))(K(y_n,x) - K(x_n,x)) \\
&-(K(y_n,y) - K(x_n,y)).
\end{aligned}$$

Due to the symmetry, we can write the algorithm just as an algorithm for construction of the value function:

$$\begin{aligned}
f^{n}(y) &= f^{n-1}(x) \\
&+\mathrm{sign}(\mathrm{sign}(f^{*}(x_n) - f^{*}(y_n)) \\
&-\mathrm{sign}(f^{n}(x_n) - f^{n}(x_n)))(K(x_n,x)) - (K(y_n,x).
\end{aligned}$$

Now we will consider the recurrent stochastic procedures from the previous section under intersection of the sets A and B or in other words in conditions of uncertainty $(A \cap B \neq \varnothing)$. The algorithm from the previous section has the form:

$$\Phi^{n}(x,y) = \Phi^{n-1}(x,y) + \gamma_n\delta_n\Omega((x,y),(x_n,y_n)),$$
$$\Omega((x,y),(x_n,y_n)) = K(x,x_n) + K(y,y_n)$$
$$-K(x,y_n) - K(y,x_n),$$
$$\sum_{1}^{\infty}\gamma_n = \infty, \sum_{1}^{\infty}\gamma_n^2 < \infty, \quad \Phi^{0}(x,y) = 0.$$

The same reasoning leads us to the following algorithm for value function construction:

$$f^{n}(x) = f^{n-1}(x) + \gamma_n\delta_n(K(x,x_n)$$
$$-K(x,y_n), \sum_{1}^{\infty}\gamma_n = \infty, \sum_{1}^{\infty}\gamma_n^2 < \infty, \quad \Phi^{0}(x,y) = 0.$$

This algorithm allows for another representation, which will be discussed in the following chapters. We continue with an example of application of the potential functions method for pattern recognition. Figure 2 gives two sets A={Yes} and B={No}.

We will make the recognition of these sets with the algorithm discussed in theorems (5.3.3) and (5.3.4). We use the algorithm in the following form:

$$f^{n+1}(x) = f^{n}(x) + \text{sign}(\text{sign}(f * (x_{n+1}))$$
$$-\text{sign}(f^{n}(x_{n+1})))K(x,x_{n+1}),$$
$$x = (x_1,x_2), K(x,y) = e^{-\alpha\sum_{i=1,2}(x_i-y_i)^2}, \alpha > 0.$$

Figure 2. A={Yes}, B={No}

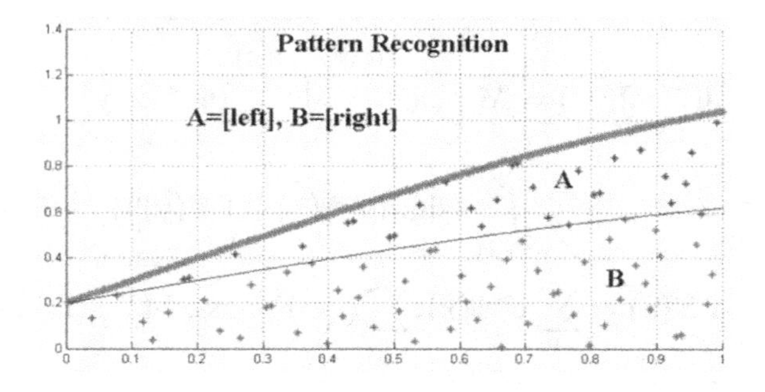

In Figures 3, 4, and 5 is shown the constructed function, which is positive over the set A={Yes} and negative over the set B={No}.

The algorithm ends in a finite number of steps for sufficiently large α. The procedures consist of cyclic rotation of the points until the two sets are completely recognized. Under this realization for every training point is formed a coefficient which accumulates at each cyclic rotation. The general form of the constructed function is:

$$f^{m}(x_1,x_2) = \sum_{j=1}^{m} D_j \exp(-\alpha\sum_{i=1}^{2}(x_i - y_i^j)^2), D_j \in R.$$

In the above formula the training points $(y_1^j, y_2^j) \in A \cup B$ are (m) number, m=**Card** $(A \cup B)$.

Figure 3. Constructed recognizing function

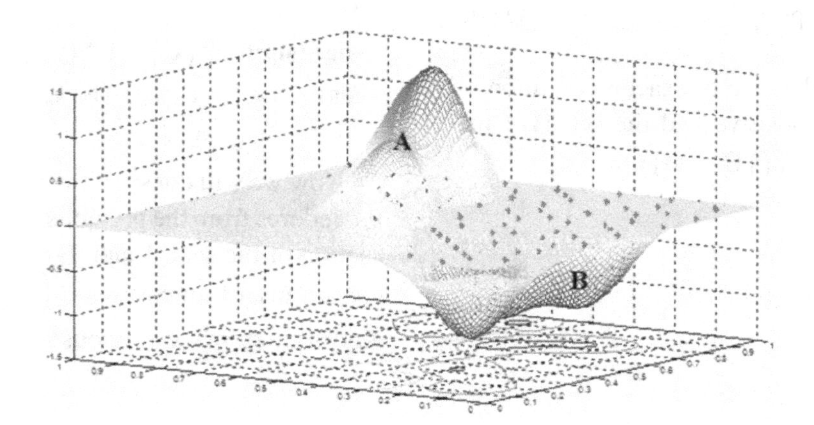

Figure 4. Constructed function and the separating line (zero)

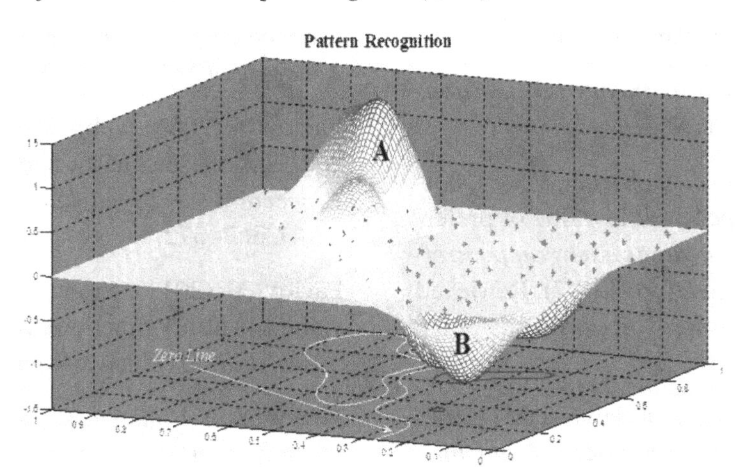

Figure 5. Constructed function: a view from above

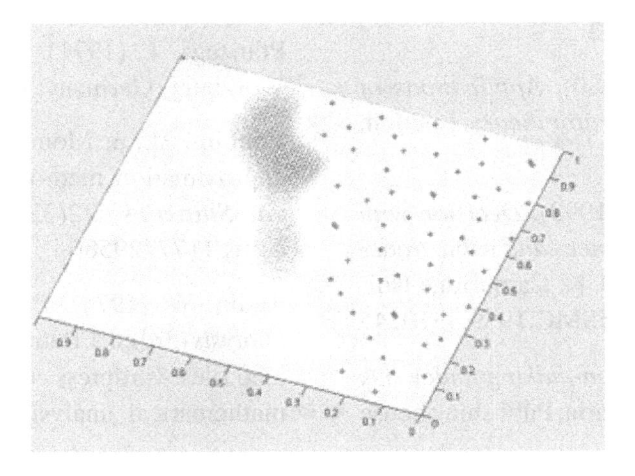

The same approach may be used for the construction of value functions with the second algorithm in conditions of uncertainty or when the sets of positive and negative answers intersect. But there the convergence does not terminate for a finite number of steps. This algorithm has another representation, which we will look into in the following chapters.

REFERENCES

Aizerman, M., Braverman, E., & Rozonoer, L. (1964). Theoretical foundations of the potential function method in pattern recognition learning. *Automation and Remote Control*, *25*, 821–837.

Aizerman, M., Braverman, E., & Rozonoer, L. (1970). *Potential function method in the theory of machine learning*. Moscow, Russia: Nauka.

Elliott, R. J. (1982). Stochastic calculus and applications. In *Applications of Mathematics*. Berlin, Germany: Springer-Verlag.

Ermolev, Y. M. (1976). *Methods of stochastic programming*. Moscow, Russia: Nauka.

Fishburn, P. (1970). *Utility theory for decision-making*. New York, NY: Wiley.

Herbrich, R., Graepel, T., Bolmann, P., & Obermayer, K. (1998). Supervised learning of preferences relations. In *Proceedings Fachgruppentreffen Maschinelles Lernen*, (pp. 43-47). Retrieved from http://research.microsoft.com/apps/pubs/default.aspx?id=65615

Herbrich, R., Graepel, T., & Obermayer, K. (1999). *Regression models for ordinal data: A machine learning approach*. TechReport TR99-3. Berlin, Germany: Technical University of Berlin. Retrieved from http://research.microsoft.com/apps/pubs/default.aspx?id=65632

Hutson, V., & Pim, J. (1980). *Applications of functional analysis and operator theory*. London, UK: Elsevier.

Keeney, R., & Raiffa, H. (1993). *Decision with multiple objectives: Preferences and value trade-offs* (2nd ed.). Cambridge, UK: Cambridge University Press. doi:10.1109/TSMC.1979.4310245

Mengov, G. (2010). *Decision-making under risk and uncertainty*. Sofia, Bulgaria: Publishing house JANET 45.

Novikof, A. (1963). A convergence proof for perceptrons. In *Symposium on Mathematical Theory of Automation*, (vol. 5, p. 12). Brookline.

Parthasarathy, K. R. (1978). *Introduction to probability and measure*. Berlin, Germany: Springer-Verlag.

Pavlov, Y. (1989). Recurrent algorithm for value function construction. *Proceedings of Bulgarian Academy of Sciences*, *7*, 41–42.

Pavlov, Y. (2005). Subjective preferences, values and decisions: Stochastic approximation approach. *Proceedings of Bulgarian Academy of Sciences*, *58*(4), 367–372.

Pavlov, Y. (2011). Preferences based stochastic value and utility function evaluation. In *Proceedings of the InSITE 2011*. Novi Sad, Serbia: InSITE.

Pavlov, Y., & Ljakova, K. (2003). Machine learning and expert utility assessment. In *Proceedings of the CompSysTech 2003*. Retrieved from http://ecet.ecs.ru.acad.bg/cst/Docs/proceedings/S3A/IIIA-5.pdf

Pfanzagl, J. (1971). *Theory of measurement*. Wurzburg, Germany: Physical-Verlag.

Robbins, H., & Monro, S. (1951). A stochastic approximation method. *Annals of Mathematical Statistics*, *22*(3), 400–407. doi:10.1214/aoms/1177729586

Rudin, W. (1974). *Principles of mathematical analysis* (3rd ed.). Retrieved from http://dangtuanhiep.files.wordpress.com/2008/09/principles_of_mathematical_analysis_walter_rudin.pdf

Shiryaev, A. N. (1989). *Probability*. New York, NY: Springer.

Trenogin, V. (1985). *Functional Analysis* (2nd ed.). Moscow, Russia: Nauka.

Chapter 6
Stochastic Utility Evaluation

ABSTRACT

The development of science and technologies accounts for the ever-growing use of modern engineering, mathematical- and computer-oriented methods and means. At the same time, the rational actions of the expert or the Decisions Maker (DM) are of great importance in the goal-seeking systems. The qualitative nature of man's evaluation and analysis is a limitation to this process. This imposes the need of developing specific approaches and methods for the analysis and evaluation of qualitative, conceptual information. Human values (utility) are part and parcel of the separate individual's process of making decisions. They are the internal motivation for determining the goal in goal-seeking systems. Unfortunately, in most scientific research studies and works the values are neither explicitly expressed nor directly oriented to the problem under consideration. In this aspect, the task of connecting the two contradicting tendencies is a topical one—the requirement for ordinal information from the mathematical and computer point of view and the cardinal nature of the knowledge of second level that is characteristic of the specific subject.

One of the possible approaches to the problem is the multiattribute utility approach. In that aspect, the program of actions of the subject towards the computer is a hierarchical decomposition of the considered goal. Preferences are expressed concerning the chosen goal and sub-objectives. Value and utility functions are made that are later used in the standard mathematical processing.

A mathematically grounded method is offered in the chapter that could be used at the stage of expressing the preferences and constructing the value and utility functions. In this case, it is necessary to take account of some important characteristics of the DM-Computer dialogue such as the conceptual, qualitative nature of the subject's thinking and the probable and subjective indefiniteness of expressing the expert's preferences. The utility functions are constructed by the means of stochastic recurrent procedures as a recognition of a set by learning the computer in the same preferences as these of the expert.

DOI: 10.4018/978-1-4666-2967-7.ch006

1. PREFERENCES RELATIONS, AXIOMS, UTILITY EXISTENCE, INTERVAL SCALE, GAMBLING APPROACH

The advances in science and technology require broader use of contemporary engineering, mathematical and computer methods and tools. On the other hand, the adequate reactions of the expert or the Decision Maker (DM) are of crucial significance in goal-oriented systems. An important limitation is the qualitative, verbal-conceptual manner of human reasoning and judgments. This imposes the use of specific approaches and evaluation methods for the conceptual empirical information (verbal expression or second order knowledge) (Larichev, 2001). Human values (utilities) are integral part of the decision-making process of the individual. They are the internal motivation for determining the main objective in the goal-oriented systems. Unfortunately, in most scientific investigations and developments, the subjective values and probabilistic expectations are not explicitly related and directly oriented towards the considered problem (Raiffa, 1968; Keeney, 1993). In this aspect, especially important is the task of connecting the two contradicting tendencies—the requirement of ordinal information from mathematical and computational point of view and the cardinal nature of the empirical (second order) knowledge, empirical ability, intuition, etc. representative for the particular subject (Pavlov, 2004; Mengov, 2010).

In scientific investigation, two main approaches are utilized, analytic, and synthetic. The analytic method is intrinsic to the human thinking, but in explicit form as main technical practice, representing the rationality of the human behavior was not introduced until 17 century. The success of this analytic approach lies not only in dividing the whole into smaller simpler parts, but also in the fact that by combining them in an appropriate goal-orientated way we obtain again another whole related to the investigated phenomena. The point of view is now bringing into line with our utility oriented position. Only in this stage can we explain the occurrence of the whole as a synchronous occurrence of its composing parts through the structure of synthesis. The above may be rephrased as "we perceive the real world through our goal-driven thinking and experience" and also as "no reality and rationality exist to us unless pierced by our intellect." We can generally accept that the analytic approach is realization of our intellect, as rationality and goal focus. In analysis, the main essential aspects of the systems are expressed as sub-objectives to the point of view of our interest towards the event, to our utility from the occurrence of the investigated event. On a following stage, inseparable from analysis, is the synthesis of the individual elements or sub-objectives in a unified structure, in which the interrelations reveal the mechanism of occurrence of the synthesized whole, but now in the direction of our main objective.

One of the possible scientific approaches to these problems is that of multiattribute utility. In this aspect, the program for action from the subject to the computer is structurization of the considered goal in the given system, fixation of scales by individual criteria, evaluating the structural sub-objectives (Raiffa, 1968; Keeney, 1993). Then comes the expression of the preferences of the decision maker and construction of the value or utility function, evaluation of the subjective probabilities and further standard mathematical processing.

With these several sentences, we sketched the mechanism of analysis and the synthesis in the system approach. Now we will outline the above through the point of view of the analytic approach in decision-making, based on the utility theory. We decompose the main objective into sub-objectives and for every sub-objective, we construct simpler, having less arguments utility functions. Depending on the revealed interrelations between the sub-objectives like "dependence by utility or value," a structure of the multaittribute utility is determined.

We determine the coefficients participating in the multaittribute utility and in conjunction with the constructed analytically simpler utility functions by the individual sub-objectives we determine the general analytical form of the multaittribute utility (Keeney, 1993). The constructed utility in conjunction with the evaluations of the subjective probabilities and the accumulated statistics regarding the objective probabilities are the fundamental analytic part of the Bayesian approach in decision-making (Raiffa, 1968).

In this manner in difficult for formalization and even verbally expressed weakly structurized problems and events we introduce the strict analytical approach, as analysis and analytically based synthesis, which allows for logically sound and mathematically precise decision formation. We achieve analytic, model description of the investigated process and the mechanisms which generate it by its reflection in our rational goal-driven thinking and we expressed the causal relations in the mechanism of the decision formations (Raiffa, 1968; Tenekedjiev, 2004; Mengov, 2008). Thus, we used the analytic approach, representing the rationality in human behavior, in the process of the decision formation. The choice of such decision is concordant with the fundamental principles of rational behavior, set in the axiomatic of the underlying mathematical apparatus. The choice is in accordance with our goal, an expression of our experience and rationality and the final decision is a mathematically optimal solution based on the knowledge, empiricism, and our experience at the moment.

In this chapter, we will lay out mathematically founded methodic for evaluation and construction of polynomial utility function approximation (Pavlov, 1999, 2005, 2011). This mathematical technique is used at the stage of preference expression and utility or value functions construction as analytic base of decision-making. We will take into account important features of the dialogue "decision maker-computer" as conceptual, qualitative nature of the subject's reasoning and probabilistic and subjective uncertainty of expressing expert preferences (Pavlov, 2003). The utility functions are constructed by stochastic recurrent procedures as a process of recognition of sets and training a computer to have the same preferences as the expert (Aizerman, 1970). We start with a description of the methodic and for starter we will repeat in short the necessary definitions and conditions for existence of utility functions given in chapter 3 and chapter 4.

Let as a result of a certain action there is a possibility of having different situations. Each of them can be assessed by the utility u(.) (Figure 1) (Pavlov, 2011).

Standard description of the situation is that an utility function ***u***(.) assesses each of this final results (x_i, i=1÷n). The DM judgment of the process behavior is measured quantitatively by the following formula

$$\mathbf{u}(\mathbf{p}) = \sum_{i=1}^{n} \mathbf{p}_i \mathbf{u}(\mathbf{x}_i),$$

where $\mathbf{p} = (p_1, p_2, \ldots, p_i, \ldots p_n)$, $\sum_{i=1}^{n} \mathbf{p}_i = 1.$

To every choice of the Decision Maker (DM) correspond (*n*) possible outcomes, (*i*=1÷*n*), each of which occurs with probability p_i, $\left(\sum_{i=1}^{n} p_i = 1\right)$.

We denote with p_i subjective or objective probabilities, which reflect the uncertainty of the final result. Following the Bayesian approach it is reasonable to choose an action that will generate result with maximal mathematical expectation $\sum_{i=1}^{n} \mathbf{p}_i \mathbf{u}(\mathbf{x}_i)$. In the formula the utility function ***u***(.) is a utility measure of the final variants ***x***, belonging to the set of final results ***X***, ***x***∈***X***. The important case for the practical needs is the case of finite set of final results ***X*** (**card**(*X*) is finite number). Now we shall give a brief mathematical description of the problem. Let ***P*** is a subset of discrete

Figure 1. Assessment

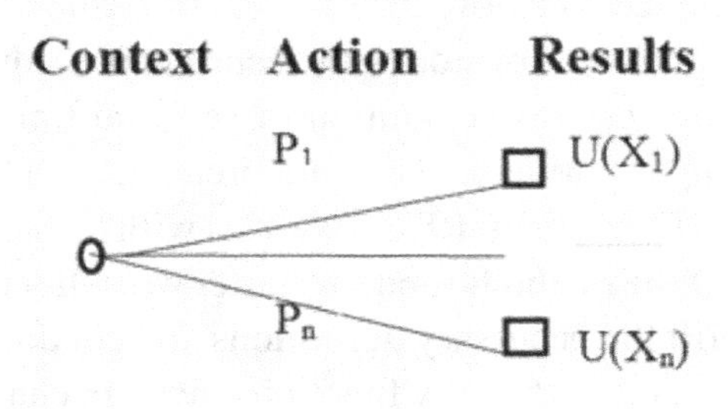

probability distributions over X. A utility function by definition is any function $\mathbf{u}(.)$ for which it is fulfilled (Fishburn, 1970):

$(p \succ q, (p,q) \in \boldsymbol{P}^2) \Leftrightarrow ((\int \boldsymbol{u}(.)dp > \int \boldsymbol{u}(.)dq), p \in \boldsymbol{P}, q \in \boldsymbol{P})$.

The utility function $\boldsymbol{u}(.)$ is linear with respect to the probability distributions over $\boldsymbol{X}$ and the mathematical expectation of $\boldsymbol{u}(.)$ is a quantitative measure concerning the expert's preferences referred to the probability distributions $\boldsymbol{P}$ over $\boldsymbol{X}$ (Fishburn, 1970; Keeney, 1993). Let the DM's preferences over $\mathbf{P}$, including those over $\boldsymbol{X}$, $(\boldsymbol{X} \subseteq \boldsymbol{P})$ are expressed by $(\succ)$. Because of the inclusion $(\boldsymbol{X} \subseteq \boldsymbol{P})$, the preferences $(\succ)$ are defined over the set of probability distributions over $\boldsymbol{X}$. The "indifference" relation $(\approx)$ is defined by $((x \approx y) \Leftrightarrow \neg((x \succ y) \vee (x \prec y)))$. There are different systems of mathematical axioms that give satisfactory conditions of a utility function existence. The most famous of them is the system of *Von Neumann and Morgenstern's Axioms* (Fishburn, 1970):

- (A.1) The *preferences* relations $(\succ)$ and $(\approx)$ are transitive, i.e. the binary preference relation $(\succ)$ is weak order;
- (A.2) *Archimedean Axiom*: for all $p,q,r \in \boldsymbol{P}$ such that $(p \succ q \succ r)$, there is an $\alpha, \beta \in (0,1)$ such that $((\alpha p + (1-\alpha)r) \succ q)$ and $(q \succ (\beta p + (1-\beta)r))$;
- (A.3) *Independence Axiom*: for all $p,q,r \in \boldsymbol{P}$ and any $\alpha \in (0, 1)$, then $(p \succ q)$ if and only if $((\alpha p + (1-\alpha)r) \succ (\alpha q + (1-\alpha)r))$.

Axioms (A1) and (A3) cannot give solution. Axioms (A1), (A2) and (A3) give solution in the interval scale (precision up to an affine transformation):

$((p \succ q) \Leftrightarrow (\int v(x)dp \succ \int v(x)dq) \Leftrightarrow (v(x) = au(x)+b, a,b \in \boldsymbol{R}, a>0))$.

The utility function $\boldsymbol{u}(.)$ is unique with precision to positive linear transformation. And vice versa, the presumption of existence of a utility function $u(.)$ leads to the "*negatively transitive*" and "*asymmetric*" relation $(\succ)$ and transitive "indifference" relation $(\approx)$ (Fishburn, 1970).

The objective of this chapter is to present a strict logical mathematical approach for modeling and estimation as utility function of the DM's preferences $(\succ)$ expressed over the set of probability distributions $\mathbf{P}$. The merger of the mathematical exactness with the empirical uncertainty in the human notions and preferences is the main challenge in the problems to solve. There are quite different utility evaluation methods and they based prevailing on the "lottery" approach (gambling approach) (Raiffa, 1968; Furqaher, 1984). In Chapter 3 and 4, we discussed the measurement scale of the utility function and we saw that the gambling approach is necessary condition for the evaluation of the utility function in the sense of von Neumann.

A "lottery" is called every discrete probability distribution over X. We mark as $<x,y,\alpha>$ the lottery: α is the probability of the appearance of the alternative x and $(1-\alpha)$—the probability of the alternative y. The gambling evaluation approach consists in comparisons of lotteries shown on Figure 2 (Raiffa, 1968; Keeney, 1993).

In the figure Θ denotes manifestation of an uncertainty event and $\neg\Theta$ denotes the conversely. We know that the weak points of this approach are violations of the transitivity of the preference relations and the indifference relation $(\approx)$. Weak points is also the determination of alternatives x (*the best*) and y (*the worst*) on condition that $(x \succ z \succ y)$ where z is the analyzed alternative. This is not

Figure 2. Classical gambling approach

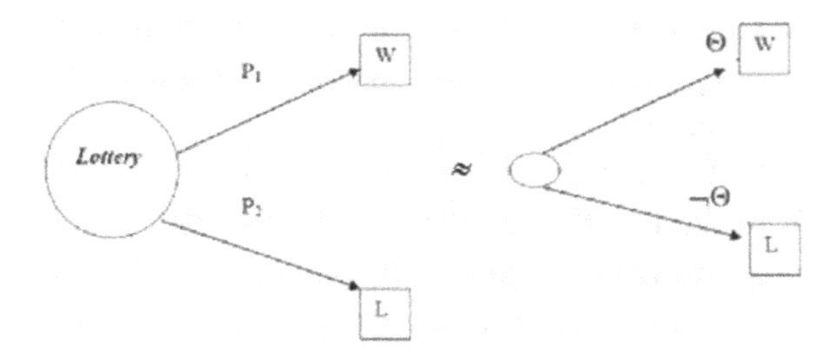

easy in the complex situations and multiattribute problems in the practice. All these difficulties explain the DM behavior observed in the famous Allais Paradox and in thousand examples discussed in the scientific literature (Cohen, 1988; Mengov, 2010).

Starting from the properties of the preference relation ($\succ$) and indifference relation ($\approx$) and from the weak points of the "gambling, lottery approach," we propose some changes in the process of evaluation. We accept all forms of preferences (Figure 3). We assume that the DM in the process of expressing preferences makes mistakes or has uncertainties in the choice. We suppose that this uncertainty or errors have random nature and may be represented as random variable with mathematical expectation zero and bounded variance, different in the different practical problems. We assume that this uncertainty is expressed in each comparison independently of the other comparisons (Aizerman, 1970; Pavlov, 1999, 2011).

In our approach, we define two sets $\mathbf{A}_u$ and $\mathbf{B}_u$ of positive preferences and negative preferences over the set of alternative $\mathbf{X}$. Let are the sets:

$$A_u=\{(\alpha,x,y,z)/(\alpha u(x)+(1-\alpha)u(y))>u(z)\},$$

$$B_u=\{(\alpha,x,y,z)/(\alpha u(x)+(1-\alpha)u(y))<u(z)\}.$$

These sets are defined by comparisons of lotteries of the simplest type

$$\left(<x,y,\alpha> \begin{smallmatrix}\succ\\ \approx\\ \prec\end{smallmatrix} <z,t,1> \right).$$

In our approach, the approximation of the utility function is constructed by pattern recognition of the set A_u and B_u. The proposed assessment process is machine learning based on the DM's preferences. The machine learning is a probabilistic pattern recognition because ($A_u \cap B_u \neq \varnothing$) and the utility evaluation is a stochastic approximation with noise (uncertainty) elimination.

2. PATTERN RECOGNITION OF POSITIVE AND NEGATIVE DM ANSWERS

Starting from the properties of the preference relation ($\succ$) and indifference relation ($\approx$) and from the weak points of the "lottery approach" we propose the next stochastic approximation procedure for evaluation of the utility function. In correspondence with the Proposition 4.1 in Chapter 3 it is assumed that ($X \subseteq P$), $((q,p) \in P^2 \Rightarrow (\alpha q+(1-\alpha)p) \in P$, for $\forall \alpha \in [0,1]$) and that the utility function $u(.)$ exists. We examine the following sets in agreement with the interval scale of utility measurement:

$$A_{u^*}=\{(\alpha,x,y,z)/(\alpha u^*(x)+(1-\alpha)u^*(y))>u^*(z)\},$$

$$B_{u^*}=\{(\alpha,x,y,z)/(\alpha u^*(x)+(1-\alpha)u^*(y))<u^*(z)\}.$$

The notation $u^*(.)$ describes the DM's empirical utility assessment. The following proposition is in the foundation of the used stochastic approximation approach (Pavlov, 2003, 2004):

Figure 3. Preferences expression

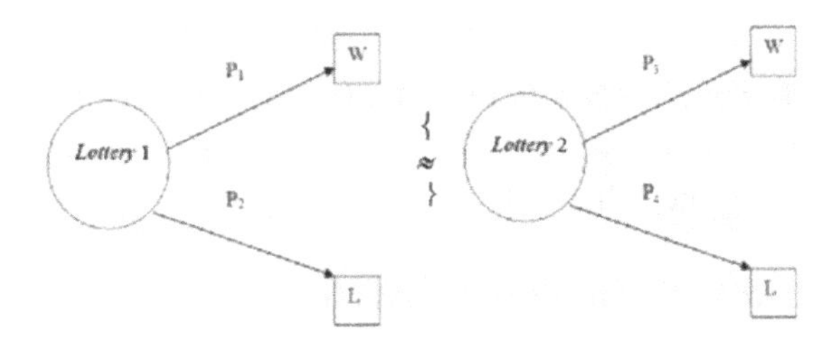

2.1. Proposition

We denote

$A_u=\{(\alpha,x,y,z)/(\alpha u(x)+(1-\alpha)u(y))>u(z)\}$.

If $A_{u1}=A_{u2}$ and $u_1(.)$ and $u_2(.)$ are continuous functions than is true $(u_1(.)=au_2(.)+b,\ a>0)$.

Proof

If $A_{u1}=A_{u2}=\varnothing$ than $u_1(.)$ and $u_2(.)$ are constants. If $A_{u1}\neq\varnothing$, than it exists $(x,y)\in X^2$, for which $u_1(x)=u_2(x)=1$ and $u_1(y)=u_2(y)=0$. This construction could be seen in theorem 2.1 in Chapter 4. Let for some $z\in\mathbf{X}$, is fulfilled $u_1(z)>u_2(z)$. The following cases are possible:

1/ $u_1(z)=u_1(x)$.

than

$u_2(z)<u_1(z)=\alpha u_1(x)+(1-\alpha)u_1(z)=\alpha.u_2(x)+(1-\alpha)u_2(x)$

Consequently $(\alpha,x,x,z)\in A_{u2}$ and $(\alpha,x,x,z)\notin A_{u1}$ and this is in contradiction with $A_{u1}=A_{u2}$.

2/It is accepted that $u_1(z)\neq u_1(x)$ and $u_1(z)\neq u_1(y)$. The following three cases are possible:

$u_1(z)<u_1(y)<u_1(x)$

$u_1(y)<u_1(z)<u_1(x)$

$u_1(y)<u_1(x)<u_1(z)$

Let us consider the first case. It does exist $\alpha'\in[0,1]$, for which is fulfilled the following for $\alpha>\alpha'$,

$((1-\alpha')u_1(z)+\alpha'u_1(x)-u_1(y))=0,\ (\alpha,x,z,y)\in A_{u1}$.

On the other side

$((1-\alpha')u_2(z)+\alpha'u_2(x)-u_1(y))<0$.

Consequently it exists $\alpha>\alpha'$, for what $(\alpha,x,z,y)\notin A_{u2}$ and this is in contradiction with $A_{u1}=A_{u2}$. The other two cases are considered in a similar way.

3/It is accepted that $u_1(z)=u_1(y)=0$. In this case from the fact that $u_2(z)<u_1(z)$ exists $\alpha\in(0,1)$ for what $(\alpha u_2(z)+(1-\alpha)u_2(x))<u_2(y)=0$. On the other side $(\alpha u_1(z)+(1-\alpha)u_1(x))>u_1(y)$ for $\forall\alpha\in[0,1)$. Consequently $A_{u1}\neq A_{u2}$, which is contrary to the presumption.

The proof is completed.

Proposition 2.1 in Chapter 6 shows one of the possible approaches for constructing the utility functions starting from the "preference" relation ($\succ$), namely in seeking an approximation of the set A_{u*}. The utility function is constructed by the means of stochastic recurrent procedures as a recognition of the set A_{u*} by learning the computer in the same preferences, as these of the expert (Aizerman, 1964, 1970). The transitivity breaches of the ($\succ$) and ($\approx$) are taken in like probability noise. This is in agreement with the prescriptive point of view, concerning the decision-making. The *utility evaluation*, the *utility function approximation* is constructed by pattern recognition of the set A_{u*} and B_{u*} and the proposed evaluation process is machine learning based on the DM's preferences. The machine learning is a probabilistic pattern recognition because $(A_{u*}\cap B_{u*}\neq\varnothing)$ and the evaluation is a stochastic approximation with noise (uncertainty) elimination. Key element in this solution is the Proposition 6.2.1.

The following presents the utility evaluation procedure (Pavlov, 1999, 2003, 2005):

The DM compares the "lottery" $<x,y,\alpha>$ with the simple alternative z, $z\in Z$ ***("better-*** $\succ$***, $f(x,y,z,\alpha)=1$," "worse-*** $\prec$***, $f(x,y,z,\alpha)=(-1)$" or "can't answer or equivalent- ~, $f(x,y,z,\alpha)=0$," $f(.)$ denotes the qualitative DM's answer).*** *This determine a learning point* ***((x,y,z,α), $f(x,y,z,\alpha)$).*** *The following recurrent stochastic algorithm constructs the utility polynomial approximation*

$$u(x) = \sum_i c_i \Phi_i(x):$$

$$c_i^{n+1} = c_i^n + \gamma_n \left[f(t^{n+1}) - \overline{(c^n, \Psi(t^{n+1}))} \right] \Psi_i(t^{n+1}) \quad (6.2.1)$$

$$\sum_n \gamma_n = +\infty, \sum_n \gamma_n^2 < +\infty, \forall n, \gamma_n > 0.$$

In the formula are used the following notations (based on A_u): $t=(x,y,z,\alpha)$, $\psi_i(t)=\psi_i(x,y,z,\alpha) = \alpha\Phi_i(x)+(1-\alpha)\Phi_i(y)-\Phi_i(z)$, where $(\Phi_i(x))$ is a family of polynomials. The line above the scalar product $\bar{v} = (c^n, \Psi(t))$ means: ($\bar{v} = 1$), if (v>1), ($\bar{v} = -1$) if (v<-1) and ($\bar{v} = v$) if (-1<v<1). The coefficients c_i^n take part in the polynomial presentation

$$g^n(x) = \sum_{i=1}^{n} c_i^n \Phi_i(x)$$

and

$$(c^n, \Psi(t)) = \alpha g^n(x) + (1-\alpha) g^n(y) - g^n(z) = G^n(x,y,z,\alpha)$$

is a scalar product. The learning points are set with a pseudo random sequence.

The mathematical procedure describes the following assessment process:

The expert relates intuitively the "learning point" $(x,y,z,\alpha))$ to the set A_{u*} with probability $D_1(x,y,z,\alpha)$ or to the set B_{u*} with probability $D_2(x,y,z,\alpha)$. The probabilities $D_1(x,y,z,\alpha)$ and $D_2(x,y,z,\alpha)$ are mathematical expectation of $f(.)$ over A_{u*} and B_{u*} respectively,

$(D_1(x,y,z,\alpha)=M(f/x,y,z,\alpha))$ if $(M(f/x,y,z,\alpha)>0)$,
$(D_2(x,y,z,\alpha)=(-)M(f/x,y,z,\alpha))$ if $(M(f/x,y,z,\alpha)<0)$.
Let $D'(x,y,z,\alpha)$ is the random value:
$D'(x,y,z,\alpha)=D_1(x,y,z,\alpha)$ if $(M(f/x,y,z,\alpha)>0)$;
$D'(x,y,z,\alpha)=(-D_2(x,y,z,\alpha))$ if $(M(f/x,y,z,\alpha)<0)$;
$D'(x,y,z,\alpha)=0$ if $(M(f/x,y,z,\alpha)=0)$.

We approximate $D'(x,y,z,\alpha)$ by a function of the type:

$$G(x,y,z,\alpha)=(\alpha g(x)+(1-\alpha)g(y)-g(z))$$

where

$$g(x) = \sum_i c_i \Phi_i(x). \quad (6.2.2)$$

The coefficients c_i^n take part in the polynomial approximation of $G(x,y,z,\alpha)$:

$$G^n(x,y,z,\alpha) = (c^n, \Psi(t)) = \alpha g^n(x) + (1-\alpha) g^n(y) - g^n(z),$$

$$g^n(x) = \sum_{i=1}^{N} c_i^n \Phi_i(x).$$

The function $G^n(x,y,z,\alpha)$ is positive over A_{u*} and negative over B_{u*} depending on the degree of approximation of $D'(x,y,z,\alpha)$. *The function* $g^n(x)$ *is the* utility polynomial *approximation of the utility function* u(.). The stochastic convergence of the evaluation (6.2.1) is analyzed in Aizerman (1970) and Pavlov (1999, 2005) and is described by the following stochastic procedure:

$$c_i^{n+1} = c_i^n + \gamma_n \left[D'(t^{n+1}) + \xi^{n+1} - \overline{(c^n, \Psi(t^{n+1}))} \right] \Psi_i(t^{n+1}), \quad (6.2.3)$$

$$\sum_n \gamma_n = +\infty, \sum_n \gamma_n^2 < +\infty, \forall n, \gamma_n > 0.$$

It is used the following decomposition:

$$\mathbf{f}(\mathbf{t}^{n+1}) = \left[D'(\mathbf{t}^{n+1}) + \xi^{n+1} \right]. \quad (6.2.4)$$

Following the approach in Raiffa (1968), Keeney (1993), and Pavlov (1999, 2005), DM compares the "lottery" <x,y,α>=(αx+(1−α)y) with the separate elements (alternatives) z∈**X**. DM expresses his preferences in a qualitative aspect - "better," "worse" or a refusal to choose—"equivalent," i.e.

$$(\alpha x+(1-\alpha)y) \underset{\prec}{\overset{\succ}{\approx}} z \tag{6.2.5}$$

This lottery is of the simplest possible type and is sufficient for the utility evaluation after Proposition 6.2.1. The expressed preferences, the answers of DM and comparisons are of cardinal (qualitative) nature and contain the inherent DM's uncertainty and errors.

3. UTILITY POLYNOMIAL APPROXIMATION AND CONVERGENCE OF THE PROCEDURE

The stochastic convergence of the Potential function method is analyzed in Aizerman (1970), Muller (2001), Kivinen (2004), and Pavlov (2011). The convergence of the evaluation (6.2.1) is based on the theorem 5.3.7. In this theorem the "Potential functions method" can be regarded as stochastic gradient procedure or more precisely as a realization of the extremal approach in the stochastic programming. Under the denotations in the previous paragraph, the stochastic recurrent utility evaluation procedure looks as follows:

$$\mathbf{c}_i^{n+1} = \mathbf{c}_i^{n} + \gamma_n \left[\mathbf{f}(\mathbf{t}^{n+1}) - \overline{(\mathbf{c}^{n}, \Psi(\mathbf{t}^{n+1}))}\right] \Psi_i(\mathbf{t}^{n+1}) \tag{6.3.1}$$

$$\sum_n \gamma_n = +\infty, \sum_n \gamma_n^2 < +\infty, \forall \mathbf{n}, \gamma_n \geq 0$$

In (6.3.1) $(\mathbf{c}^n, \Psi(\mathbf{t}))$ denotes the scalar product and it has the following expanded form:

$$(\mathbf{c}^n, \Psi(\mathbf{t})) = \alpha(\mathbf{c}^n, \Phi(\mathbf{x})) + (1-\alpha)(\mathbf{c}^n, \Phi(\mathbf{y})) - (\mathbf{c}^n, \Phi(\mathbf{z})) =$$
$$= \alpha g^n(x) + (1-\alpha) g^n(y) - g^n(z) = G^n(x, y, z, \alpha),$$

$$g^n(x) = \sum_{i=1}^{N} c_i^n \Phi_i(x). \tag{6.3.2}$$

We know that the line above $\overline{y} = \overline{(c^n, \Psi(t))}$ means that $\overline{y} = 1$, if y>1, $\overline{y} = -1$ if y<-1 and $\overline{y} = y$ if -1<y<1 . We set by definition

$$K(t,t) = (\Psi(t), \Psi(t)) =$$

$$= \sum_{i=1}^{N} (\alpha\Phi_i(\mathbf{x}) + (1-\alpha)\Phi_i(\mathbf{y}) - \Phi_i(\mathbf{z}), \alpha\Phi_i(\mathbf{x}) + (1-\alpha)\Phi_i(\mathbf{y}) - \Phi_i(\mathbf{z})) \tag{6.3.3}$$

Multiplying both sides of (6.3.1) by $\psi_i(t)$ and summing by i gives:

$$\mathbf{G}^{n+1}(\mathbf{t}) = \mathbf{G}^n(\mathbf{t}) + \gamma_n(\mathbf{f}(\mathbf{t}^{n+1}) - \overline{\mathbf{G}^n(\mathbf{t}^{n+1})})\mathbf{K}(\mathbf{t}, \mathbf{t}^{n+1}) \tag{6.3.4}$$

$$\sum_n \gamma_n = +\infty, \sum_n \gamma_n^2 < +\infty, \forall \mathbf{n}, \gamma_n \geq 0$$

From the formulae (6.3.4), (6.3.3), and (6.3.2) the following expression for $\alpha g^{n+1}(x)$ is derived:

$$\alpha g^{n+1}(x) = \alpha g^n(x) + \gamma_n(f(t^{n+1}) - \overline{G^n(t^{n+1})})(\alpha\alpha_{n+1}K'(x, x^{n+1}) + \tag{6.3.5}$$

$$+\alpha(1-\alpha_{n+1})K'(x, y^{n+1}) - \alpha K'(x, z^{n+1}))$$

$$\sum_n \gamma_n = +\infty, \sum_n \gamma_n^2 < +\infty, \forall \mathbf{n}, \gamma_n \geq 0$$

Box 1.

$$J_{D'}(G^n(x,y,z,a)) = M(\int_{D'(t)}^{G^n(t)} (\bar{y} - D'(t))dy) = \int (\int_{D'(t)}^{G^n(t)} (\bar{y} - D'(t))dy)dF \xrightarrow[n]{e.w.} \inf_{s(t)} (\int (\int_{D'(t)}^{s(t)} (\bar{y} - D'(t))dy)dF), \quad (6.3.6)$$

$$K'(x,y) = \sum_{i=1}^{N} \Phi_i(x)\Phi_i(y).$$

Similar presentations for the functions (1-α) $g^{n+1}(y)$ and $g^{n+1}(z)$ can be determined.

One of the basic results of the potential functions method shall be presented here in shortened form which is appropriate for the utility evaluation exposition. The convergence is based on the theorem 5.3.7 (Pavlov, 2005, 2011):

3.1. Theorem

Let in the recurrent formulas (6.2.1, 6.2.3, 6.3.1, 6.3.5 $t^1,\ldots,t^n,\ldots$ be a sequence of independent random values with one and the same distribution F, the sequence of random values $\xi^1,\xi^2,\ldots,\xi^n,\ldots$ satisfies the conditions:

$M(\xi^n/(x,y,z,\alpha),c^{n-1})=0$, $M((\xi^n)^2/(x,y,z,\alpha),c^{n-1})<d$,

$t^n=(x,y,z,\alpha)$, $d\in\mathbf{R}$.

Let $\Psi(t)$ be limited by a constant independent from t:

$$\|\Psi(\mathbf{t})\| < \mathbf{cte}.$$

Then it follows from the recurrent procedures (6.2.1, 6.2.3, 6.3.1) and from the theorem 3.7 in Chapter 5 convergence shown in Box 1 where e.w. denotes almost everywhere while M denotes the mathematical expectation. The function S(t) is expandable in the form of (6.3.2) and it fulfils

$$\int S(t)^2 dF < +\infty$$

and belongs to L_2 defined by the probability measure F.

The proof is based on the principal results of Aizerman (1970), chapter 4 concerning the extreme approach for convergence conditions of the potential functions method and on the theorem 3.7 in Chapter 5. The following statement is valid:

3.2. Proposition

Under the denotations above the following is fulfilled:

$$\int \int_{\mathbf{D}'(\mathbf{t})}^{\mathbf{G}^{\mathbf{n}}(\mathbf{t})} (\bar{\mathbf{y}} - \mathbf{D}'(\mathbf{t}))\mathbf{dydF} \geq \frac{1}{2} \int \overline{(\mathbf{G}^{\mathbf{n}}(\mathbf{t}) - \mathbf{D}'(\mathbf{t}))^2}\mathbf{dF}.$$

Proof

Let us consider the integral $\int_{D'}^{f} \bar{u}du$, where $-1 \leq D' \leq 1$.

$$= \frac{\mathbf{f}^2 - \mathbf{D}'^2}{2}, -1 \leq \mathbf{f} \leq 1$$

$$\int_{\mathbf{D}'}^{\mathbf{f}} \bar{\mathbf{u}}\mathbf{du} = \frac{1 - \mathbf{D}'^2}{2} + (\mathbf{f} - 1), \mathbf{f} > 1$$

$$= \frac{1 - D'^2}{2} + (|f| - 1), f < -1.$$

Taking account of the formula above, the three cases in Box 2 are possible:

For example the proof of (3) is:

$$\int_{D'(t)}^{G^n(t)} (\bar{u} - D'(t))du = \int_{D'(t)}^{-1} (\bar{u} - D'(t))du + \int_{-1}^{G^n(t)} (\bar{u} - D'(t))du =$$

$$= \frac{1}{2}(-1 - D'(t))^2 - (G^n(t) + 1)(1 + D'(t)) =$$

$$= \frac{1}{2}\overline{(G^n(t) - D'(t))^2} - (G^n(t) + 1)(1 + D'(t)), G^n(t) < -1$$

therefore

$$\int_{D'(t)}^{G^n(t)} (\bar{u} - D'(t))du = \frac{1}{2}\overline{(G^n(t) - D'(t))^2} + \nabla(G^n(t), D'(t))$$

in all three cases $\nabla(G^n(t), D'(t)) \geq 0$, thus the proof being completed.

Therefore, the following is fulfilled when taking account of Theorem 3.1 and Proposition 3.2 in Chapter 6.

3.3. Consequence

When account has been taken of (6.3.2), (6.3.6), and (6.3.7, shown in Box 3) it can be assumed that $g^n(x)$ is a polynomial approximation of the empirical utility u(.) if *n* is sufficiently great. The proposed algorithm allow two different forms (6.3.1, 6.3.5), in contrast to the algorithms shown in Chapter 5.

4. EXAMPLES OF EVALUATION AND NUMERICAL VERIFICATIONS

The dialogue between the expert and the computer was simulated, the expert being replaced by a model function in the numeric presentation (the solid line in Figure 4). In the modeling an additive noise determines the uncertainty in

Box 2.

$$\int_{D'(t)}^{G^n(t)} (\bar{u} - D'(t))du = \frac{1}{2}\overline{(G^n(t) - D'(t))^2}, -1 \leq G^n(t) \leq 1 \quad (1)$$

$$\int_{D'(t)}^{G^n(t)} (\bar{u} - D'(t))du = \frac{1}{2}\overline{(G^n(t) - D'(t))^2} + (1 - D'(t))(G^n(t) - 1), G^n(t) > 1 \quad (2)$$

$$\int_{D'(t)}^{G^n(t)} (\bar{u} - D'(t))du = \frac{1}{2}\overline{(G^n(t) - D'(t))^2} - (G^n(t) + 1)(1 + D'(t)), G^n(t) < -1 \quad (3)$$

Box 3.

$$\mathit{inf}_{s(t)} \int \int_{D'(t)}^{s(t)} (\bar{y} - D'(t))dydF \overset{e.w..}{\geq} \overline{\mathit{lim}}_{n}(\frac{1}{2} \int \overline{(G^n(t) - D'(t))^2}dF) \geq 0 \quad (6.3.7)$$

Figure 4. Evaluation with 512 "learning points"

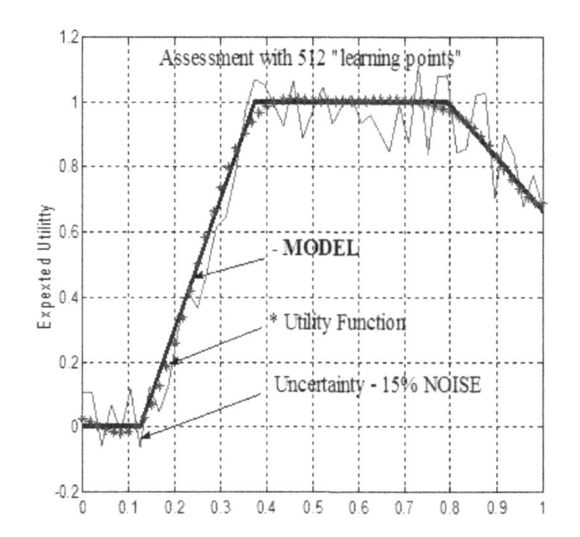

DM's answers. The expert was modeled being replaced by the seesaw lines in Figure 4 (Pavlov, 2005):

On the basis of the seesaw line in Figure 4 the answers are formed about the appurtenance of the "learning points" to $\mathbf{A}_u$ or $\mathbf{B}_u$ or a refusal to answer. The last case is modeled through setting the indistinguishability threshold. The learning points (x,y,z,α) are set with probability distribution F, for which x, y, z, and α are evenly distributed independent random values. The utility approximation is shown on the Figure 4 with stars line (*).

We can assess of the empirical risk in regards to Vapnik approach (Vapnik, 2006).

The empirical risk regarding Vapnik can be assessed with the function

$$G^n(x,y,z,\alpha) = \alpha g^n(x) + (1-\alpha)g^n(y) - g^n(z)$$

(Pavlov, 2003, 2005). The approximation function

$$g^n(x) = \sum_{i=1}^{n} c_i^n \Phi_i(x)$$

is randomized between 0 and 1 (Proposition 3. 4.1) and this causes the value ***abs***$(G^n(x,y,z,\alpha))$ to be limited by 1. The degree of the polynomial approximation

$$g(x) = \sum_{i=1}^{n} c_i \Phi_i(x)$$

is noted by n. The number of the "learning points" (x,y,z,α) in the learning sequence is noted by l. The "empirical" probability $P_e(c_i)$ is assessed as (n_w/l), where n_w is the number of wrong answers in the process of machine learning. The probability $P(c_i)$ is the probability of possible "wrong recognition" by the computer on the base of the constructed utility function. According to Vapnik the following is true:

$$P(c_i) \le P_e(c_i) + 2\frac{\ln(\frac{1,5l^n}{n!}) - \ln(\frac{\eta}{8})}{l}\left(1 + \sqrt{1 + \frac{P_e(c_i)l}{\ln(\frac{1,51^n}{n!}) - \ln(\frac{\eta}{8})}}\right) = P_m(c_i).$$

The above formula means that the probability $P(c_i)$ is smaller than the number $P_m(c_i)$ with probability bigger than (1-η) (in the case that (l > n)). Simple calculations show that this formula is useful in the case of (l > 1024). The specific presentation of the function $G^n(x,y,z,\alpha) = \alpha g^n(x) + (1-\alpha)g^n(y) - g^n(z)$ gives a hint that this valuation could be improved, because we obtained good results with 64 or 128 learning points (recognition of 97% of the learning points).

Our approach permits direct assessment of the dependence of the utility function on probability. Several non-expected utility theories have been developed in response of the transitivity violations (Shmeidler, 1989; Tversky, 1992; Mengov, 2010). Among these theories the rank dependent utility model and its derivative cumulative prospect theory are currently the most popular. In the rank Dependence Utility (RDU) the decision weight of an outcome is not just probability associated with this outcome. It is a function of both probability and the rank (the alternative) x:

$$RDU = \sum_{i=1}^{n} W(p_i)u(x_i)$$

Based on empirical researches several authors have argued that the probability weighting function W(.) has the inverse S-shaped form, which starts on concave and then becomes convex. Such a function is shown on Figure 5.

It is supposed that $W(p_i)= p_i+\Delta W(p_i)$. We search for an polynomial approximation with the use of procedure (6.2.1) of the kind $f_2(y,\alpha)$, $\alpha\in[0,1]$, $y\in \boldsymbol{X}$ following Nobel prizewinners Kahneman and Tversky (1992). Example of such an evaluated utility function $f_2(y,\alpha)$ is shown on Figure 6.

The explicit formula of the cumulative utility function $f_2(y)$ is:

$$f_2(y) = \int_0^1 f_2(y,\alpha)d\alpha\,.$$

Figure 5. Probability weighting function W(.)

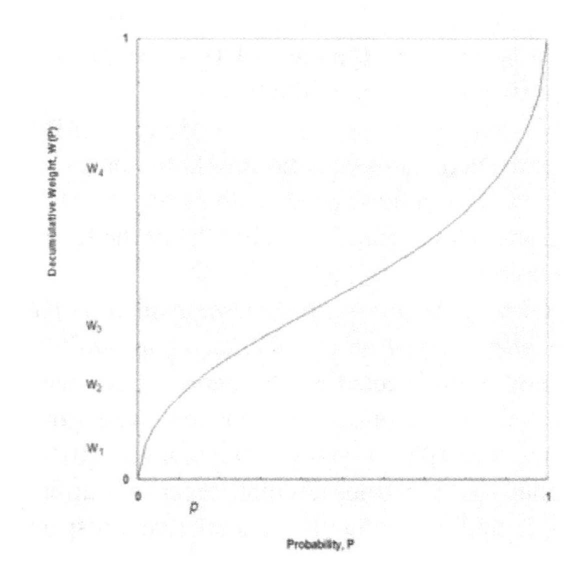

Figure 6. Utility function $f_2(y,\alpha)$

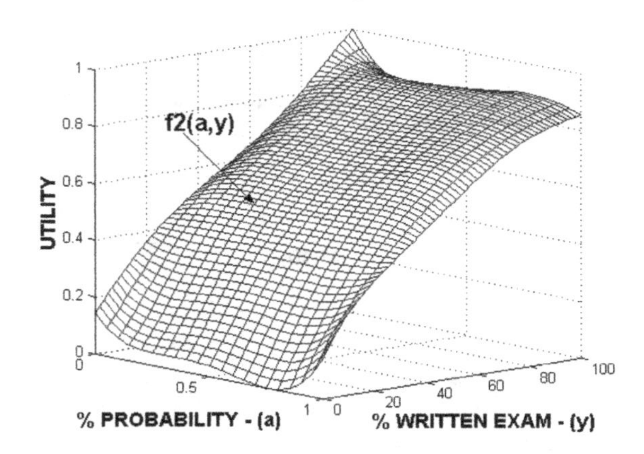

The utilities functions $f_2(y,\alpha)$ and $f_2(y)$ in Figure 6 and Figure 7 are constructed by 1024 DM's "learning points" $((x,s,z,\alpha), f(x,s,z,\alpha))$, $f(x,y,z,\alpha)\in\{1,-1,0\}$ with the use of a information system based on the mathematical results of the Paragraphs 6.3.2 and 6.3.3.

The seesaw line in Figure 7 is pattern recognition of the sets $\boldsymbol{A}_{u^*}$ and $\boldsymbol{B}_{u^*}$ and recognize correctly more than 95% of the DM's answers (Pavlov, 2011). This seesaw line is evaluated additionally and independently from the evaluation of the utility $f_2(y,\alpha)$ with 64 "learning points" (DM's answers). The cumulative utility function $f_2(y)$ is the solid line on Figure 7. This example shows that the procedure permits direct assessment of the utility function $f_2(y,\alpha)$ as a function of both probability and the rank (alternative) y, following the findings of Nobel prizewinners Kahneman and Tversky. If the cumulative function W(.) has a symmetric form (Figure 5) then is true:

$$\int_0^1 \Delta W(p)f_2(y)dp = 0\,. \qquad (6.4.1)$$

In this case the expected von Neuman and Morgenstern's utility function $f_2(.)$ is exactly the integral because p is evenly distributed in the pseudo-random sequence:

Figure 7. Utility function $f_2(y)$

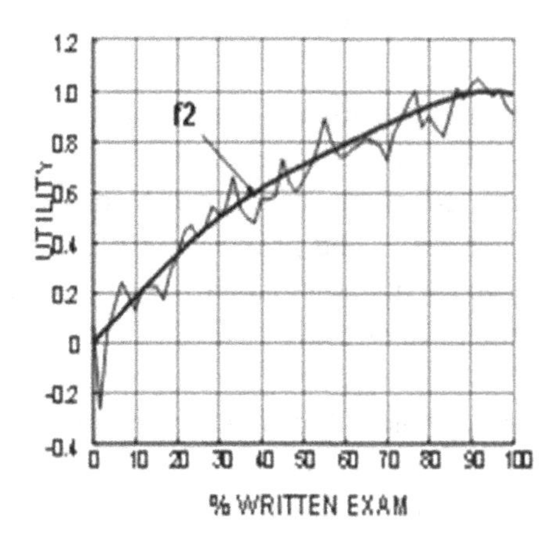

Figure 8. Pattern recognition of $f_2(y,\alpha)$

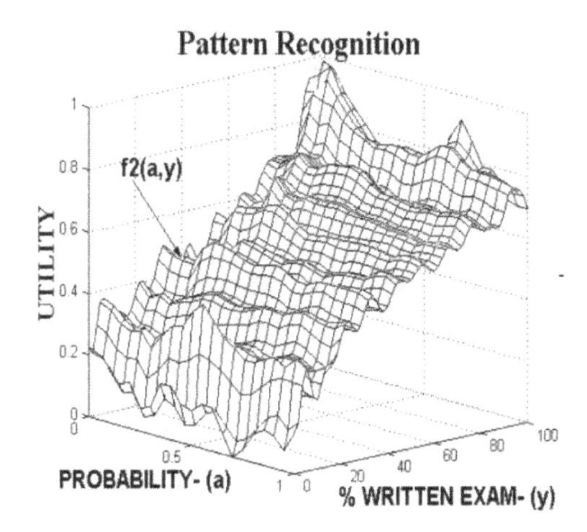

$$\int_0^1 W(p)f_2(y)dp = \frac{1}{2}u(y). \qquad (6.4.2)$$

Thus, we could first evaluate $f_2(y,\alpha)$, following Kaneman and Tversky, and after that we could apply formula (6.4.2). By this whey the "certainty effect" and "probability distortion" identified by Kahneman and Tversky could be reduced. For the application of these calculations the symmetry of the cumulative function W(.) is not obligatory but in this case are needed more complex lotteries. The determination as pattern recognition of the DM's function $f_2(y,\alpha)$ is shown on Figure 8 (Pavlov, 2010). The seesaw surface $f_2(y,\alpha)$ in Figure 8 is constructed by 1024 DM's "learning points" $((x,s,z,\alpha), f(x,s,z,\alpha)), f(x,y,z,\alpha) \in \{1, -1, 0\}$ (probabilistic pattern recognition of the sets A_{u*} and B_{u*}, $A_{u*} \cap B_{u*} \neq \varnothing$).

The last example concerns expert evaluation of the "best" specific growth rate of a fed-batch cultivation process. The specific growth rate of fed-batch cultivation processes determines the nominal technological condition. The complexity and the incomplete parametric information of the biotechnological fermentation process make difficult the determination of the "*best*" process parameter. A possible solution is the utilization of human estimations expressed by preferences. Our experience is that the human estimation of the process parameters of cultivation processes contains uncertainty in the range from 10% to 30%. Let $\boldsymbol{X}$ be the closed interval [0, 0.6] ($\boldsymbol{X}$={specific growth rates of the cultivation process-μ}=[0, 0.6]h^{-1}) and $\boldsymbol{P}$ is the convex subset of discrete probability distributions over $\boldsymbol{X}$. The seesaw line in Figure 8 is pattern recognition of the set $\boldsymbol{A}_{u*}$ and the set $\boldsymbol{B}_{u*}$. This seesaw line recognizes correctly more than 97% of the expert answers (technologist).

The polynomial approximation of the DM's utility $u(\mu)$ is the smooth line on the Figure 9 (the mathematical expectation). The expert utility function recognizes correctly more than 81% of the expert answers (used learning points). The maximum of the utility function determines the "best" parameter of the fed-batch cultivation process after the technologist.

The developed stochastic procedures could be used in complex control and management processes for exact mathematical description of the complex system "technologist or DM - mathematical model of the process." Such procedures permit iterative design of the control or management solutions in agreement with the DM's preferences.

Figure 9. Specific growth rates

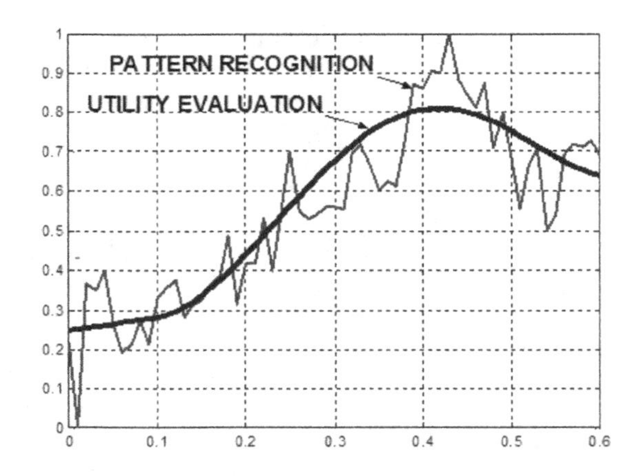

The experiments confirm the applicability of this approach. The suggested method has the following advantages:

- If the subjective and probability uncertainty of DM's preferences is interpreted as some stochastic noise, the achievements of stochastic approximation can be used for the recurrent evaluation of utility function with noise elimination as typical for stochastic approximation;
- After explaining the term "lottery," DM is relatively fast in learning the suggested methods according his/her qualification level (128 learning answers for about 45 minutes);
- The proposed assessment is machine learning based on DM's preference.
- The approach permits iterative and precise evaluation of the "best" parameters and iterative control design in agreement with the DM's preferences as maximum of the utility function.
- The stochastic programming approach permits evaluation of the dependence of the utility function on probability. For this purpose, we search for a polynomial approximation of the kind $u(x,\alpha)$, $\alpha\in[0, 1]$, following the findings of the Nobel prizewinners Kahneman and Tversky.

The utility function is a mathematical abstraction presented in the limits of the normative approach, the axiomatic systems of Von Neumann and Morgenstern. The proposition 6.2.1 reveals the existence of a mathematical expectation measured in the interval scale on the base of the DM's preferences expressed within the borders of the gambling approach. This mathematical expectation evaluated with the stochastic procedures could be interpreted as an approximation of the expected utility function.

REFERENCES

Aizerman, M., Braverman, E., & Rozonoer, L. (1964). Theoretical foundations of the potential function method in pattern recognition learning. *Automation and Remote Control*, *25*, 821–837.

Aizerman, M., Braverman, E., & Rozonoer, L. (1970). *Potential function method in the theory of machine learning*. Moscow, Russia: Nauka.

Cohen, M., & Jaffray, J.-Y. (1988). Certainty effect versus probability distortion: An experimental analysis of decision-making under risk. *Journal of Experimental Psychology. Human Perception and Performance*, *14*(4), 554–560. doi:10.1037/0096-1523.14.4.554

Farquhar, P. (1984). Utility assessment methods. *Management Science*, *30*, 1283–1300. doi:10.1287/mnsc.30.11.1283

Fishburn, P. (1970). *Utility theory for decision-making*. New York, NY: Wiley.

Keeney, R., & Raiffa, H. (1993). *Decision with multiple objectives: Preferences and value trade-offs* (2nd ed.). Cambridge, UK: Cambridge University Press. doi:10.1109/TSMC.1979.4310245

Kivinen, J., Smola, A., & Williamson, R. (2004). Online learning with kernels. *IEEE Transactions on Signal Processing*, *52*(8), 2165–2176. doi:10.1109/TSP.2004.830991

Larichev, O., & Olson, D. (2010). *Multiple criteria analysis in strategic sitting problems* (2nd ed.). London, UK: Springer.

Mengov, G. (2010). *Decision-making under risk and uncertainty*. Sofia, Bulgaria: Plovdiv–Janet 45.

Muller, K., Mika, S., Ratsch, G., Tsuda, K., & Scholkopf, B. (2001). An introduction to kernel-based learning algorithms. *IEEE Transactions on Neural Networks*, *12*(2), 181–205. doi:10.1109/72.914517

Pavlov, Y. (2004). Value based decisions and correction of ambiguous expert preferences: An expected utility approach. In *Proceedings of the International Conference BioPS 2004*. Sofia, Bulgaria: BioPS.

Pavlov, Y. (2005). Subjective preferences, values and decisions: Stochastic approximation approach. *Proceedings of the Bulgarian Academy of Sciences*, *58*(4), 367–372.

Pavlov, Y. (2010). Normative utility and prescriptive analytical presentation: A stochastic approximation approach. In *Social Welfare Enhancement in EU, 13th International Conference*. Sofia, Bulgaria: Government Publication. Retrieved from http://www.uni-sofia.bg/index.../SofiaConferenceProgramme2010.pdf

Pavlov, Y. (2011). Preferences based stochastic value and utility function evaluation. In *Proceedings of Informing Science & IT Education Conference (InSITE) 2011*, (pp. 404-411). Novi Sad, Serbia: InSITE. Retrieved from http://proceedings.informingscience.org/InSITE2011/index.htm

Pavlov, Y., & Ljakova, K. (2003). Machine learning and expert utility assessment. In *Proceedings of the International Conference – CompSysTech 2003*. Sofia, Bulgaria: CompSysTech. Retrieved from http://ecet.ecs.ru.acad.bg/cst/Docs/proceedings/S3A/IIIA-5.pdf

Pavlov, Y., & Tzonkov, St. (1999). An algorithm for constructing of utility functions. *Proceedings of the Bulgarian Academy of Sciences*, *52*(1-2), 21–24.

Pavlov, Y., & Vassilev, K. (1992). Recurrent construction of utility functions. *Comptes Rendus de l'Academie Bulgare de Science*, *3*, 5–8.

Raiffa, H. (1968). *Decision analysis*. Reading, MA: Addison-Wesley.

Shmeidler, D. (1989). Subjective probability and expected utility without additivity. *Econometrica*, *57*(3), 571–587. doi:10.2307/1911053

Tversky, A., & Kahneman, D. (1992). Advances in prospect theory: Cumulative representation of uncertainty. *Journal of Risk and Uncertainty*, *5*(4), 297–323. Retrieved from http://www.springerlink.com/content/lwr6176230786245/ doi:10.1007/BF00122574

Vapnik, V. (2006). *Estimation of dependences based on empirical data* (2nd ed.). London, UK: Springer.

Chapter 7
A Preferences-Based Approach to Subjective Probability Estimation

ABSTRACT

The idea that the subjective statistics and knowledge deserve confidence has age-long life. It is well known that humans have great capabilities to express precisely the gathered experience and to skillfully predict the events in the everyday life. The problem is that these empirical knowledge and adjustments are expressed mainly verbally and a difficult to measure level. The mathematical measurement of the subjective knowledge in the corresponding scales and their inclusion in mathematical models permit new management and control solutions in different areas of human activities.

It is well known that if a man judges for the probability of occurrence of an uncertainty event in the future based on his/her current experience and statistics, he\she is closed to optimal Bayesian statistics. In such a case, the utility theory and its prescription to make decisions based on the optimal mathematical expectation of the utility has scientific validity. The Bayesian approach is a natural basis on which human beings form their decisions, using their previous empirical experience.

Following the ideas of professor Raiffa, we can have the same attitude toward the subjective probabilities as with the objective probabilities, and we can use them freely in the theoretical constructions of the von Newman Utility theory. This is the subject of the chapter, evaluation of the subjective probability with the use of the stochastic programming. The probability is measured in an absolute scale in the context of the probability and measurement theory. Because of this, we can use the gambling approach to estimate

DOI: 10.4018/978-1-4666-2967-7.ch007

the DM's subjective probability as in the utility evaluations. Once again the authors solve the problem of best separation by using stochastic methods of the sets $\mathbf{A}_{u*}$ *and* $\mathbf{B}_{u*}$, $(\mathbf{A}_{u*} \cap \mathbf{B}_{u*}) \neq \varnothing$)). *The difference with the previous chapter is that now they seek the existence of number (p), and not of function. This makes the problem easier to solve. However, the question remains the same, elimination of errors and uncertainty, and the way this is achieved in the stochastic programming.*

1. PREFERENCES RELATIONS: GAMBLING APPROACH

In our everyday activity, we are constantly faced with the necessity of assessing uncertainty events and their effect on our life. For example, what will be the weather tomorrow, whether we should take a raincoat, whether to take a holiday and go on a trip, etc. On analyzing complex systems and processes, we face different forms of uncertainty while assessing various situations, parameters and others. On the one end of the spectrum, it is an uncertainty of a probability character just like the result after tossing a coin. The economic situation in 10 or 30 year's time is the uncertainty lying on the other end of the spectrum. Similar situations appear in complex biochemical processes as well, such as for example a subjective expert assessment of a biotechnological process of cultivation based on oblique factors or assessment of secondary metabolic product formation (See Chapter 9). So here raises the question whether we can rely on our subjective feelings and expectations in uncertainty while analyzing and decision-making in complex situations (so called uncertainty of second order) (Raiffa, 1986; Pfanzagl, 1971; Shmeidler, 1989; Larichev, 2010).

There are different opinions on that question. The most prevailing opinion is that if it is necessary to make a decision in a certain problem, than the subjective feelings, related to an existing uncertainty event have to be expressed by the language of the theory of probabilities and included in the decision-making process (Raiffa, 1986; Keeney, 1993; Pfanzagl, 1971). The idea that the subjective statistics and knowledge deserve trust has been around for centuries. It is known that a human being has great capacity to express precisely the nuances of the accumulated experience and to skillfully foresee the events in the everyday life. The psychologists, engineers, medical doctors, economists have everyday contact with such events and they freely use their subjective knowledge. The problem is that such empiric knowledge and predispositions are expressed mainly at verbal and difficult to measure level. Then the errors and contradictions with the aim of application of such knowledge and skill may increase in power. At such level, the human is lost in the multiattribute structure of the complex worldly and professional problems and has difficulties accounting for the nuances in his/hers empiric and long-term professional knowledge and the inter-relations in the structures of the sub-objectives of the main objective of the investigation. Without careful analysis, errors are often made and the influence of psychological factors as anger, fear, fatigue becomes a serious obstacle and source of errors. The mathematical measurement of the subjective knowledge in the respective scales and the description of the problem with models, reflecting mathematically precise these knowledge and the measurement scales, allow for the elimination of the mentioned above negative factors as a major influence in the decision process. Then a person, even if new in the professional field, makes decisions, if nothing else, in complete and non-contradictory agreement with his/hers personal knowledge and preferences.

The opinions of multitude scientists throughout the years is that a physiological mechanism in person's brain accounts for the frequency probability expressed in his/hers life and transforms it into force, which has as a result the rationality of the behavior and in the formation of ideas. Such scientists are Piaget, Inhelder, Ramsey, Savage, Fishburn, Raiffa (Fishburn, 1970; Savage, 1954; Mengov, 2010). In Bulgaria research in the field is conducted by Mengov, Popchev, Tenekidjiev, and others (Mengov, 2010; Popchev, 1986; Tenekidjiev, 2004). New knowledge and mathematical techniques are sought, which in dialogue mode between the man and computer can extract the empirical skills of the researchers and realize them synchronously with the other theoretical skills.

When we have objective statistics regarding the phenomenon of interest or/and under investigation we can evaluate the possibilities of appearance of the uncertainty events with the known probability distribution. These are the so-called objective probabilities. Our judgments in our profession and everyday life for complex situations and decisions in the conditions of risk are usually for a discrete number of variants, alternatives. If they are, for instance, (n) in number, then the probability distribution describes the uncertainty, which is evaluated in our decision:

$$\mathrm{p} = (\mathrm{p}_1, \mathrm{p}_2, .., \mathrm{p}_i, ..\mathrm{p}_n), \sum_{i=1}^{n} \mathrm{p}_i = 1.$$

The effect of the occurrence of one or another outcome variant in the particular problem is assessed by the chosen or evaluated by us utility function. This function is a measure for the effect of each outcome variant. Then using our utility function $\mathrm{u}(.)$ and following the ideology of the Bayesian approach, the effect of our participation in this situation with its inherent uncertainty is measured with the formula:

$$\mathrm{u}(\mathrm{p}) = \sum_{i} \mathrm{p}_i \mathrm{u}(\mathrm{x}_i), \ \ \mathrm{p} = (\mathrm{p}_1, \mathrm{p}_2, .., \mathrm{p}_i, ..\mathrm{p}_n), \sum_{i} \mathrm{p}_i = 1.$$

In other words, to each decision of ours and its respective action corresponds its inherent probability distribution. In our professional and everyday activities discussed at verbal level, we evaluate usually discrete probability distributions defined over the set of outcomes. De facto, our decision choice preferences are over the set of probability distributions defined over the set of possible outcomes or alternatives X. The utility function $\mathrm{u}(\mathrm{x}), \mathrm{x} \in \mathrm{X}$, is defined over the set of alternatives. Its mathematical expectation by the individual probability distributions is in fact the measure, according to the utility theory, by which we evaluate these probability distributions. In other words, these mathematical expectations are the way in which we evaluate our actions in conditions of risk and uncertainty.

These questions were discussed in the previous chapters. Here, in brief, we will remind some mathematical definitions (Fishburn, 1970). Let $\mathbf{Z}$ is a set of alternatives and $\mathbf{P}$ is a convex subset of discrete probability distributions over $\mathbf{Z}$. A utility function is any function $\mathbf{u}(.)$ which fulfils:

$$(p \succ q,\ (p,q) \in \mathbf{P}^2) \Leftrightarrow (\int \mathbf{u}(.)dp > \int \mathbf{u}(.)dq),\ (p,q) \in \mathbf{P}^2.$$

The DM's preference relation over $\mathbf{P}$ ($\mathbf{Z} \subseteq \mathbf{P}$) is expressed by($\succ$). Its induced indifference relation ($\approx$) is defined thus: $((x \approx y) \Leftrightarrow \neg((x \succ y) \vee (x \prec y)),\ (x,y) \in \mathbf{Z}^2)$. We denote with ($\int \mathbf{u}(.)dp$) integration based on the probability measure $p \in \mathbf{P}$.

The existence of an utility function $\mathbf{u}(.)$ over $\mathbf{Z}$ determines the preference relation ($\succ$) as a negatively transitive and asymmetric one and the indifference relation ($\approx$) as an equivalence (transitivity, symmetry, associativity). We mark the lottery "appearance of the alternative x with probability α and appearance of the alternative y with probability ($1-\alpha$)" as $<x,y,\alpha>$. It is assumed that an utility function $\mathbf{u}(.)$ exists and that is fulfilled $((q,p) \in \mathbf{P}^2 \Rightarrow (\alpha q + (1-\alpha)p) \in \mathbf{P}$, for $\forall \alpha \in [0,1])$. These conditions determine the utility function with precision up to an affine scale (interval scale), $\mathbf{u}_1(.) \approx \mathbf{u}_2(.) \Leftrightarrow \mathbf{u}_1(.) = \mathbf{a}\mathbf{u}_2(.) + \mathbf{b},\ \mathbf{a} > \mathbf{0}$ (Theorem 2.1 in Chapter 4).

The probability is measured in the absolute scale in the context of the probability theory and measure theory, if we assume that this is a positive Radon measure (Lebesgue measure) over a metric, locally compact, separable space, determined by its values over the compact subsets and equal to one over the whole space. This guarantees the uniqueness of this measure. Then we can use the gambling approach to evaluate this probability in the manner used in applying this approach to evaluate the utility (Raiffa, 1968; Farquhar, 1984; Keeney, 1993). This is the topic of the current chapter in the context of stochastic approximation as we use the expressed preferences of the decision maker without artificially seeking the equivalence of lotteries as is done it the classical sources on this topic (Farquhar, 1984). In our approach, we assume free expression of preferences, allowing for contradictions in their expression.

In chapter 3 of the present book, subjective probability was considered from two viewpoints. One is purely mathematical. On axiomatic basis is derived the existence utility function and probability measure over a finite discrete set of alternatives, which define the mathematical expectation as measure for the evaluation of the decision maker's opinion for events with probabilistic nature (Theorem 3.5.2). According to this theorem if we accept the axioms lying in the foundation of the theorem as sensible, then our attitude with respect to the uncertain events may be measured exactly as subjective probability over Boolean algebra (Parthasarathy, 1978; Shiryaev, 1989; Pfanzagl, 1971).

The second gives ground for the use of the subjective probabilities in the term of lotteries. It is given in detail in the book of Prof. Raiffa. It is more of a worldly empirical explanation of how our subjective attitudes to uncertainty in decision-making are described probabilistically. This ground is based to great extent on experience and informal reasoning founded on commonly accepted principles for logical soundness. Two main principles are used. The first is interchangeability of equivalent or indiscernible by utility events by the decision maker. If a lottery with win 1 or loss 0 is indiscernible in our concepts with respect to the uncertainty in the occurrence of event Θ (indifference with regard to the participation in the lottery or participation in the uncertain event with win 1 or loss 0) then the probability equivalent to this lottery is subjective probability with which we measure our attitude towards this uncertainty (Figure 1).

The second principle in Prof. Raiffa's approach is the assumed transitivity of preferences. We will give a short clarification of the above definition. With probability p_1 we win 1 or with probability p_2 we lose 0. In the right hand side of the figure is the occurrence or not of the uncertain event θ. If our participation in the lottery or the occurrence or not of the event θ, to us is with equal uncertainty with respect to the outcome, then we accept that our attitude towards the uncertainty of θ may be described with subjective probability equal to p_1:

$$P_{\text{subjectve probability}}(\theta) = p_1 \times 1 + p_2 \times 0 = p_1, \quad (p_1 + p_2 = 1).$$

Following the ideas of Prof. Raiffa we can treat the subjective probabilities as objective probabilities and use them freely in the definitions and formulations of von Neumann's utility theory. If the decision maker is agreeing with the principles of non-contradictory behavior: transitivity in preferences between lotteries (discrete probability

Figure 1. W=win, L=lose

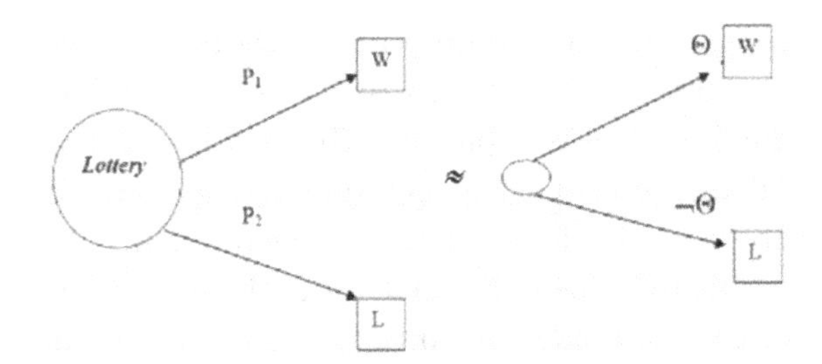

distributions over the set of alternatives), transitivity by relation indiscernibility or indifference (≈) and interchangeability of indiscernible lotteries, then we can treat our subjective uncertainty and subjective probabilities as objective when we use them in the process of decision formation and the mathematical formulations and proofs.

Our approach to decision-making and using mathematical definitions and constructions is more constructive in nature, descriptively prescriptive. The description of ideal, rational human in his/hers evaluations and preferences is not sought. What is ought is expression of internal conviction of the Decision maker in this choice, even with his/hers inherent contradiction and uncertainty. What is sought is consistent and logically sound solution, based on his/hers internal conviction and knowledge and the subsequent use of strict mathematical inference. The subject consciously and objectively expresses his/hers preferences. Our task in this chapter is to describe them strictly in mathematical definitions and conclusions. In this manner, the decision-making is cleared from internal inconsistency and contradiction. The final conclusion is mathematically founded and is based on the internal conviction and knowledge of the decision maker, which expresses the best (mathematically optimal) action in the solved problem.

Where is the difference between the presented here approach from the known from scientific literature application of the gambling approach for these problems? In the classical approach equivalence of lotteries is always sought. We know that the relation "equivalence" is the most often violated for the transitivity. Because of this in our approach, we assume that the subject (DM) in the process of expressing preferences makes errors or has uncertainty in the choice. We assume that this uncertainty or errors have random nature and may be represented by random variable with mathematical expectation zero and bounded variance, differing for the particular problems. We assume that this uncertainty at each comparison is expressed independently from the other comparisons. We define two sets $\mathbf{A}_{u*}$ and $\mathbf{B}_{u*}$ of the positive and negative preferences for comparison of lotteries of the type shown on Figure 1. The sets have the form (Figure 1):

$$\mathbf{A}_{u*} = \{(p_1) \in [0,1] / (<\mathbf{W},\mathbf{L},p_1>) \succ (<\mathbf{W},\mathbf{L},\theta>)\},$$

$$\mathbf{B}_{u*} = \{(p_1) \in [0,1] / (<\mathbf{W},\mathbf{L},p_1>) \prec (<\mathbf{W},\mathbf{L},\theta>)\}.$$

In the above notation ($<\mathbf{W},\mathbf{L},\theta>$) is a lottery in the right hand side of Figure 1. In the conditions of strict preferences it is fulfilled ($\mathbf{A}_{u*} \cap \mathbf{B}_{u*}$)=∅, which is in agreement with the normative approach. But in practice, due to the errors, uncertainty in the expert preferences and thresholds of indiscernibility the two sets intersect ($\mathbf{A}_{u*} \cap \mathbf{B}_{u*}$)≠∅. Then we once again solve the problem of best separation by using stochastic methods of the sets $\mathbf{A}_{u*}$ and $\mathbf{B}_{u*}$, ($\mathbf{A}_{u*} \cap \mathbf{B}_{u*}$)≠∅). The difference with the previous chapter is that now we seek the existence of number (p), and not of function. In addition, this number is the probability in our prescription in the particular problem. This makes the problem much easier to solve than the previous problem for constructing utility function. However, the question remains the same. Can the noise (the errors and the uncertainty in the preferences) be eliminated the way this is achieved in the stochastic programming? To solve this problem we will use the classic stochastic procedure of Robbins-Monro, described in chapter 5 (Aizerman, 1970; Ermolev, 1976; Robbins, 1951).

2. PATTERN RECOGNITION OF POSITIVE AND NEGATIVE DM ANSWERS: ROBBINS-MONRO STOCHASTIC APPROXIMATION

We will start this section with information about a contemporary scientific research confirming that humans are good Bayesian statisticians (Mengov, 2010). The American psychologists Griffiths and Tenenbaum during the last decade (2006), by analyzing intuitive evaluations in the conditions of repetitive life situations, have proved the statistical optimality of human assessment (Griffiths, 2006). The major idea of this study is that humans process the new data about the surrounding world by interpreting them in the framework of a built in their consciousness probability model (Mengov, 2010). That means that the Bayesian approach was a natural basis on which human beings form their decisions, using their previous empirical experience. In Mengov's review of Griffiths and Tenenbaum research it is said that if a man judges for the probability of occurrence of a given event in the future based on his/hers current experience and statistics, he\she is closed to optimal Bayesian statistician. Then the presence of Bayesian approach found in the research of scientists like von Neumann, Raiffa, Fishburn, Keeney in the last seven decades proves once more the validity of their conclusions and theoretical frameworks. In such case the utility theory and its prescription to make decision based on the optimal mathematical expectation of the utility has another scientific validation.

The people surveyed Griffiths and Tenenbaum have made forecasts with high precision the median, which is a sufficient condition for recognizing the type of the distribution (Mengov, 2010). In such way, their investigation has affirmed the capabilities of people to recognize probability distributions in their everyday lives and to form their evaluations close to the optimal with regard to Bayesian approach. In their research, they have analyzed distributions for the duration of the career of the American congressmen and the reign of Egyptian pharaohs. There were analyzed also hard to describe analytically distributions such as time needed for baking food in the oven. The answers and conclusions are that the human is precise in his estimates for contemporary events and more imprecise for remote in time events like the reign of the Egyptian pharaohs before 2000 years. The conclusion is that the more knowledge and experience we have for the everyday event, the more precisely we forecast its occurrence, and vice versa, the less knowledge we have, like the history of Egypt, the worse the forecasts we make. The general conclusion is to work with good experts and people having practical experience in the field and use more rationally mathematics and its theoretical and application results! Now let us look into the mathematics itself.

If the uncertainty is of the second type (cardinal subjective information and subjective feelings about the uncertainty [Larichev, 2010]) then P_i ($i=1\div n$) are subjective probabilities which reflect our subjective feelings and expectations about the process (Raiffa, 1968):

$$(P_1, P_2, .., P_i, ..P_n), \sum_{i=1}^{n} P_i = 1.$$

The question is how to assess the probabilities P_i. Scientific literature suggests the following algorithm (Raiffa, 1968). Two potential outcomes (**W**) and (**L**) are chosen (**W**$\succ$**L**). They can be defined as the best and the worst. We mark the lottery "appearance of the alternative (**W**) with probability P (uncertainty event θ) and appearance of the alternative (**L**) with probability (1 - P)" as < **W**, **L**, P> (Figure 2).

In the classical approach we seek for a situation (event) S (W $\succ$ S $\succ$ L) equivalent to the "lottery" (Farquhar, 1984):

Figure 2. Lottery <**W**, **L**, *P*>

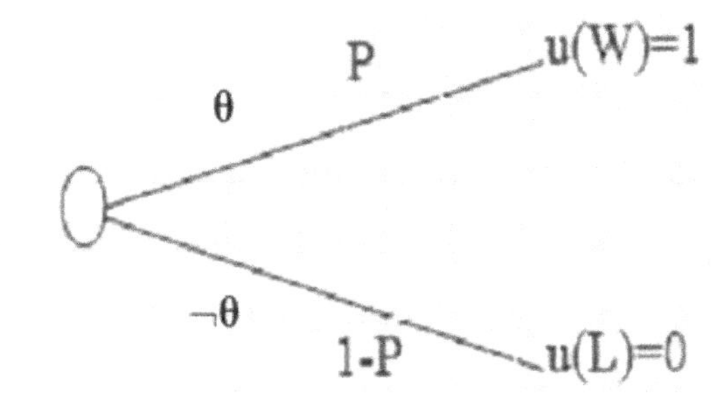

Figure 3. Set [0, 1] separated into three areas

0 X1 P X2 1

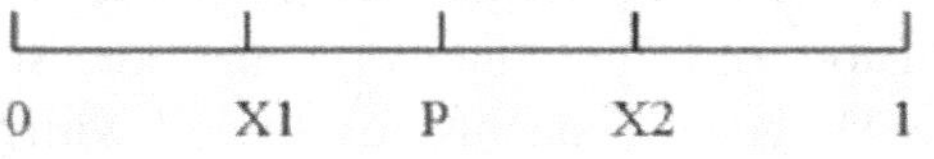

$$((\mathbf{u(W)P} + \mathbf{u(L)}(1 - \mathbf{P})) \approx \mathbf{u(S)})$$
$$\Rightarrow \mathbf{P} = \frac{(\mathbf{u(S)} - \mathbf{u(L)})}{(\mathbf{u(W)} - \mathbf{u(L)})}.$$

This procedure can be criticized because for every expert there is a real-defined limit of indiscernibility—the natural "indistinguishability" threshold of every expert and necessity of estimation of the utility u(.). Besides the expert information includes probability and subjective uncertainty and mistakes (Mengov, 2010).

There is another solution to the problem above. The expert is directly asked whether the tested uncertainty event θ (with hypothetical probability P) (Lottery $(<\mathbf{W}, \mathbf{L}, \theta>)$, Figure 2) is more probable than the lottery $(<\mathbf{W}, \mathbf{L}, \alpha>)$ which has the probability α, $\alpha \in [0,1]$ (Figure 1). The set [0, 1] is separated into three areas (Figure 3).

The values (probabilities) of $[0, X_1]$ interval are assessed as smaller (less probable) than P. In the interval $[X_2, 1]$ each of the values (probabilities) are assessed as bigger (more probable) than P. The values inside the interval $[X_1, X_2]$ are indiscernible for the expert in regard to P (uncertainty event θ). If X_1 and X_2 are estimated we can use the following formulae for *assessement of the subjective probability* P:

$$\mathbf{P} = \mathbf{X}_1 + \frac{\mathbf{X}_2 - \mathbf{X}_1}{2}.$$

For this purpose we put

$(\lambda_1 = X_1)$, $(\lambda_2 = X_2 - X_1)$, $(\lambda_3 = 1 - X_2)$.

It is obvious that $\lambda_{1+} \lambda_2 + \lambda_3 = 1$. Let X be the set:

$X = \{(\lambda_1, \lambda_2, \lambda_3) / \lambda_1 + \lambda_2 + \lambda_3 = 1\}$,

and η is a uniformly distributed random value in [0, 1], $\eta \in [0, 1]$. We define the following random vector $\chi = (\eta_1, \eta_2, \eta_3)$, where:

1. If $(<\mathbf{W}, \mathbf{L}, \eta> \prec <\mathbf{W}, \mathbf{L}, \mathbf{P}>) \Rightarrow \chi = (\eta_1 = 1, \eta_2 = 0, \eta_3 = 0)$, $\eta \in \mathbf{B}_{\mathbf{u}*}$;
2. If $(<\mathbf{W}, \mathbf{L}, \eta> \succ <\mathbf{W}, \mathbf{L}, \mathbf{P}>) \Rightarrow \chi = (\eta_1 = 0, \eta_2 = 0, \eta_3 = 1)$, $\eta \in \mathbf{A}_{\mathbf{u}*}$;
3. If indiscernibility $(<\mathbf{W}, \mathbf{L}, \eta> \approx <\mathbf{W}, \mathbf{L}, \mathbf{P}>) \Rightarrow \chi = (\eta_1 = 0, \eta_2 = 1, \eta_3 = 0)$.

Let χ^n is a sequence of independent random values with equal to χ distribution. We will use the recurrent stochastic procedure in Box 1 where Pr_X is the projection over the set X. (Aizerman, 1970; Ermolev, 1976; Robbins, 1951).

It is well known that the recurrent stochastic procedure minimizes the function (Chapter 5; Gikhman, 1988):

$$\mathbf{F}(\lambda_1, \lambda_2, \lambda_3) = \mathbf{M}(\eta_1 - \lambda_1)^2 + \mathbf{M}(\eta_2 - \lambda_2)^2 + \mathbf{M}(\eta_3 - \lambda_3)^2,$$

Box 1.

$$(\lambda_1^{n+1}, \lambda_2^{n+1}, \lambda_3^{n+1}) = \mathbf{Pr}_X[(\lambda_1^n, \lambda_2^n, \lambda_3^n) - \gamma_n((\eta_1^n, \eta_2^n, \eta_3^n) - (\lambda_1^n, \lambda_2^n, \lambda_3^n))],$$
$$\sum_1^\infty \gamma_n = \infty, \sum_1^\infty \gamma_n^2 < \infty, \gamma_n \geq 0, \forall n \in N, \qquad (7.1.1)$$

where M is the mathematical expectation.

Function $\mathbf{F}(\lambda_1, \lambda_2, \lambda_3)$ is convex in regard to $(\lambda_1, \lambda_2, \lambda_3)$ as well the set X, which allows including Pr_X in the recurrent stochastic procedure (formula 7.1.1), where X is the set where we seek the solution.

The minimum of $\mathbf{F}(\lambda_1, \lambda_2, \lambda_3)$ is at $\lambda_i = M(\eta_i), i = 1 \div 3$. The proof is very easy. Let χ is a random value and $\theta = M(\chi)$. Then

$$M(\chi-\beta)^2 = M(\chi-\theta+\theta-\beta)^2$$

$$= M(\chi-\theta)^2 + (\theta-\beta)^2 + 2.\ M(\chi-\theta).\ (\theta-\beta)$$

$$= M(\chi-\theta)^2 + (\theta-\beta)^2 \geq M(\chi-\theta)^2,\ \beta \in R\ .$$

The projection $(\lambda_1', \lambda_2' \lambda_3') = Pr_X(\lambda_1, \lambda_2, \lambda_3)$ is defined as follows:

- We put $\Delta = 1 - \lambda_1 - \lambda_2 - \lambda_3$;
- For $i = 1 \div 3$, $\lambda'_i = \lambda_i + \Delta/3$.

The construction of the projection is shown in the two-dimensional space (Figure 4).

It is obvious that the vector (a→b) is parallel to the line x=y. Point (b) is the projection of (a) over segment x+y=1 (x>0, y>0) (set X).

Figure 4. Pr_X over X

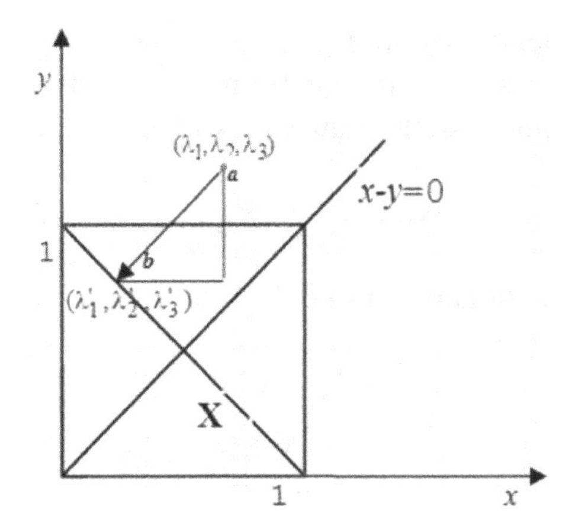

Figure 5. W=win, L=lose

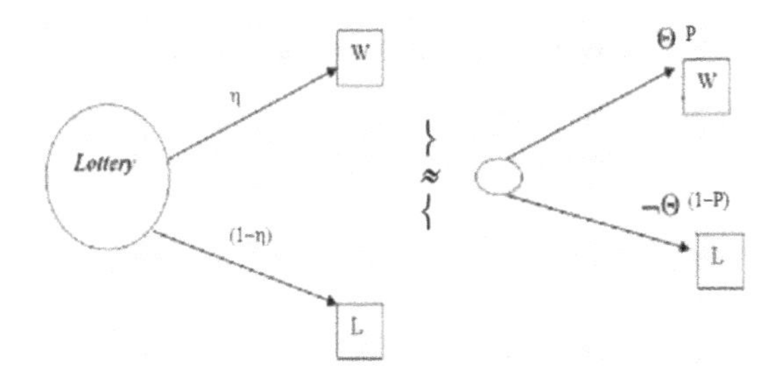

The stochastic recurrent procedure is the following. We establish apriori N, the number of questions and expert answers gambling with figures as Figure 5. In this assessment we use the pseudo-random sequence Lp_τ instead of the evenly distributed random value η (Sobol, 1973). This defines number of questions $N = 2^p$ (p is integer: 8, 16, 32, 64…). At the n^{-th} questions we proceed as following. We compare the uniformly distributed random value η, $\eta \in [0, 1]$ with the subjective feelings of the decision maker (Figure 5).

Then we define for every question an expert answer shown in Box 2.

Then the N expert answers χ are processes in cycle or randomly by the procedure (7.1.1), for example in T steps (T >> N). Then we estimate P by the formula:

$$\mathbf{P} = \lambda_1^T + \frac{\lambda_2^T}{2}.$$

We made numerical experiments with the recurrent procedure (formula 7.1.1). After setting apriori the assessed probability P and the indistinguishability threshold respectively $\Delta/2 = 0.05$, $\Delta/2 = 0.1$, $\Delta/2 = 0.2$, and testing the model with T = 1000, are obtain results respectively shown in Tables 1, 2, and 3.

We use Lp_τ the pseudo-random sequence because the random values generator not always gives a good approximation for the evenly distributed random value. It is also important the quick convergence of procedure (7.1.1) towards

Box 2.

If $(\eta < P)$, $(<\mathbf{W}, \mathbf{L}, \eta> \prec <\mathbf{W}, \mathbf{L}, P>) \Rightarrow \chi = (\eta_1 = 1, \eta_2 = 0, \eta_3 = 0)$, $\eta \in \mathbf{B}_{u*}$;	(1)
If $(\eta > P)$, $(<\mathbf{W}, \mathbf{L}, \eta> \succ <\mathbf{W}, \mathbf{L}, P>) \Rightarrow \chi = (\eta_1 = 0, \eta_2 = 0, \eta_3 = 1)$, $\eta \in \mathbf{A}_{u*}$;	(2)
If indiscernibility $(\eta \approx P) \Rightarrow \chi = (\eta_1 = 0, \eta_2 = 1, \eta_3 = 0)$.	(3)

Table 1. Results

P / questions N=32	Δ/2=0.05	Δ/2=0.1	Δ/2=0.2
0.321	0.328	0.359	0.344
0.505	0.516	0.516	0.516
0.731	0.734	0.749	0.734

Table 2. Results

P /questions N=64	Δ/2=0.05	Δ/2=0.1	Δ/2=0.2
0.321	0.328	0.336	0.336
0.505	0.515	0.508	0.515
0.731	0.734	0.742	0.734

Table 3. Results

P / questions N=128	Δ/2=0.05	Δ/2=0.1	Δ/2=0.2
0.321	0.324	0.328	0.328
0.505	0.512	0.508	0.511
0.731	0.734	0.734	0.734

the mathematical expectation of the random vector η, as well as the opportunity of recurrence of questions and correction of the answers.

From this modeling the following conclusions can be derived:

- The precision of the results increases along with the number of the questions.
- The precision weakly depends on the expert's indistinguishability threshold.
- There are good results when the number of questions is low (8, 16, 32, or 64) (they demand about 15-20 minutes).

Experiments and the suggested modeling show that the expert can answer 64 questions for about 30 minutes, which proves the suggested procedure to be practically applicable.

When we have discrete distributions with more probabilities, we can use the following *algorithm for finding the projection* $\boldsymbol{Pr}_x$ at each intermediate step of the stochastic recurrent procedure (formula 7.1.1). This algorithm ends in finite number of steps (Ermolev, 1976). Let the set X be defined with the following formula:

$$X = \{x = (x_1, x_2, .., x_n) / (\sum_{i=1}^{n} a_i x_i = b), (x_i \geq 0), i = 1, 2, 3, ..., n\}.$$

The projection of an arbitrary point $(y_1, y_2, .., y_n)$ over the set X is reduced to minimization of the formula:

$$\sum_{i=1}^{n} (y_i - x_i)^2 .$$

Let $x^0 = (0, 0, .., 0)$ and on step (k) is obtained the approximation $x^k = (x^k_1, x^k_2, .., x^k_n)$, in which n_k values are different from zero. Ordered by magnitude, let the values look like:

$$x^k_1 > 0, x^k_2 > 0, ..., x^k_{n_k} > 0, x^k_{n_k+1} = 0, .., x^k_n = 0.$$

Additionally on step (k) is fulfilled:

$$\frac{(x^k_1 - y_1)}{a_1} = ... = \frac{(x^k_{n_k} - y_{n_k})}{a_{n_k}} < \frac{(x^k_{n_k+1} - y_{n_k+1})}{a_{n_k+1}} \leq .. \leq \frac{(x^k_n - y_n)}{a_n} .$$

We put:

$$\pi_1^k = \frac{(x_{n_k+1}^k - y_{n_k+1})}{a_{n_k+1}} - \frac{(x_{n_k}^k - y_{n_k})}{a_{n_k}},$$
$$\pi_2^k = \frac{b - (\sum_{i=1}^{n_k} a_i x_i^k)}{(\sum_{i=1}^{n_k} a_i^2)},$$
$$x_j^{k+1} = \begin{cases} x_j^k + \pi^k a_j,\ j = 1,2,....,n_k \\ x_j^k,\ j = n_k + 1,............,n \end{cases},$$
$$\pi^k = \min(\pi_1^k, \pi_2^k).$$

If on this step we obtain $(\sum_{i=1}^{n} a_i x_i^{k+1} < b)$, then the described placements are applied for x^{k+1}. If it is fulfilled,

$$(\sum_{i=1}^{n} a_i x_i^{k+1} = b),$$

then $\Pr_X(y_1, y_2, .., y_n) = x^{k+1}$. In this last case, the solution x^{k+1} has the property, there exists a number μ such that:

$$x_j^{k+1} - y_j = \mu a_j, \quad x_j^{k+1} > 0,$$
$$x_j^{k+1} - y_j \geq \mu a_j, \quad x_j^{k+1} = 0.$$

These are necessary and sufficient conditions for the optimality of the solution x^{k+1}.

We can combine the stochastic procedure (7.1.1) with the just described algorithm as follows. On the current step we determine a random vector

$\chi = (\chi^1, ..., \chi^i, ..., \chi^n) = ((\eta^1_1, \eta^1_2, \eta^1_3), ..., (\eta^i_1, \eta^i_2, \eta^i_3), ..., (\eta^n_1, \eta^n_2, \eta^n_3))$.

We use the formula 7.1.1, and for determining the projection

$\Pr_X((\lambda_1^1, \lambda_2^1, \lambda_3^1)^k, .., (\lambda_1^i, \lambda_2^i, \lambda_3^i)^k, .., (\lambda_1^n, \lambda_2^n, \lambda_3^n)^k)$

on the kth step we use described above algorithm. After the end of the iterative process, we can use the formula:

$$\mathbf{P}_1 = \lambda_1^{1T} + \frac{\lambda_2^{1T}}{2}, .., \mathbf{P}_n = \lambda_1^{nT} + \frac{\lambda_2^{nT}}{2}$$

In this manner, we can evaluate every subjective discrete distribution.

3. USE CASES

We will demonstrate the approach proposed in the previous section to a real-world example. We will evaluate our concept for the weather on the next day. We assume that the most preferred in our concepts for such forecast is (**W**-sunny) and the least preferred at the moment of the study is (**L**-rainy). We will use comparisons and answers with our preferences with respect to lotteries of the type shown on Figure 5. In evaluating our subjective notions was used an information system prototype developed on the basis of the algorithms proposed in the previous sections. The result constructed with 32 answers is shown on Figure 6.

When posing the question through the gambling approach it is important what is the semantics of the lotteries. For instance, the lotteries in the considered example may be of the type: "tomorrow the weather will be ((α100)%) sunny and ((1- α)100)% rainy ($\alpha \in [0,1]$)" and this concept of ours is compared against our expectation for the next day. The answers may be "I expect better day," "I cannot decide under these values," or "I expect that the day will be more rainy or cloudier." In the solution shown on Figure 6 there is clearly seen some contradiction in the answers belonging to the set $\mathbf{A}_{u*}$.

In each particular problem, the semantics is different and, obviously, the answers and errors in the preferences depend on it. Due to this, for

Figure 6. Probability evaluation of the weather tomorrow

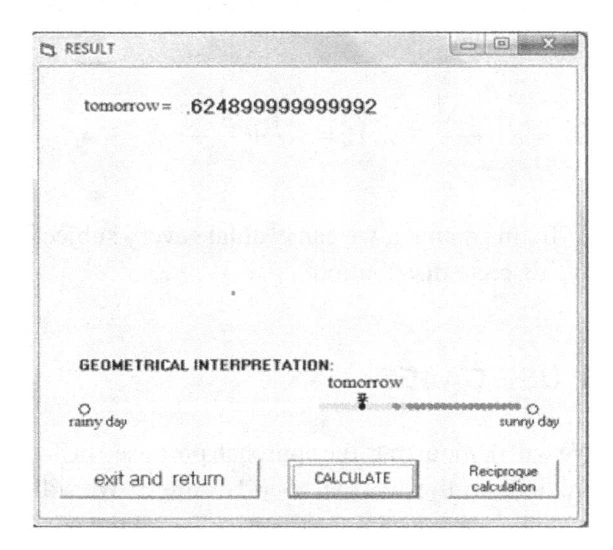

each problem is necessary that the semantics of the lotteries be well placed as questions.

A very interesting application is the use of the proposed approach for evaluation of the coefficients, obtained for the decomposition of utility, multiattribute function into functions of fewer variables. Such decomposition was discussed in chapter 4, section 4.3.

As an example, we will consider the calculation of coefficients K_1 and K_2, participating in the representation of the utility function, discussed at the end of chapter 4. We will remind its form (Pavlov, 1996):

$$U(Y) = K_2f_2(Y2) + K_1f_1(Y1, Y3, Y4, Y5) =$$

$$= K_2f_2(Y2) + K_1[\{K_3f_3(Y3) + K_4f_4(Y4) +$$

$$K_{34}f_3(Y3)f_4(Y4)\}K_{134} + K_{115}f_{15}(Y1, Y5) +$$

$$+ K_{11345}f_{15}(Y1, Y5)\ [K_3\ f_3(Y3) + K_4f_4(Y4) +$$

$$K_{34}f_3(Y3)f_4(Y4)].$$

The variables are as follows:

- Yl: Digestion speed [10-30 g/l].
- Y 2: Ambient temperature influence/fermentor volume [10-10^4 m^3].
- Y3: Substrate concentration [20-90 g/l].
- Y4: Substrate type - % of dry substance [2.5-6.0%].
- Y5: Process temperature [25-42°C].

We already know that the utility functions are measured through the gambling approach in the interval scale. Because of this, each partial utility function of one or two variable in the above formula may be normalized between 0 and 1 or between any two values we want (Keeney, 1993). This is achieved easily. After the construction of the utility functions with the algorithms given in chapter 6, we subtract from the function its minimal value and we divide the result to the difference between its maximal and minimal value. For such transformation, the maximum becomes 1 and the minimum 0. This is made for the function U(Y), as well as for its composing functions. As an example for the complete function U(Y), we can make the following normalizations (Pavlov, 1996):

$U(Y^m1, Y^m2, Y^m3, Y^m4, Y^m5)=U(10, 10, 90, 2.5, 25)=0$ and approximately

$U(Y^*1, Y^*2, Y^*3, Y^*4, Y^*5)=U(25, 10^4, 20, 5.3, 38)=1$.

This is true if all partial functions are normalized between 1 and 0.

The same can be said for the function

$K_2f_2(Y2)=U(Y^m1, Y2, Y^m3, Y^m4, Y^m5)= U(10, Y2, 90, 2.5, 25)$.

For the utility function $f_2(.)$ is fulfilled:

$f_2(10)=U(Y^m1, 10, Y^m3, Y^m4, Y^m5)/K_2=0$ and $f_2(10^4)=U(Y^m1, 10^4, Y^m3, Y^m4, Y^m5)/K_2=1$.

The similar is valid for the function $f_1(.)$

Box 3.

$$\left\{\begin{matrix} \mathrm{U}(25,\ 10^4,\ 20,\ 5.3,\ 38) \times \alpha \\ + \\ \mathrm{U}(10,\ 10,\ 90,\ 2.5,\ 25) \times (1-\alpha) \end{matrix}\right\} \begin{matrix} \succ \\ \approx \\ \prec \end{matrix} \mathrm{U}(10,\ 10^4,\ 90,\ 2.5,\ 25).$$

$K_1 f_1$(Y1, Y3, Y4, Y5)= U(Y1, Y^o2, Y3, Y4, Y5)=U(Y1, 10, Y3, Y4, Y5).

If all partial functions are normalized between 0 and 1 the following is valid for the utility function $f_1(.)$ (...):

f_1(10, 90, 2.5, 25)= U(10, 10, 90, 2.5, 25)/K_1=0 and approximately

f_1(25, 20, 5.3, 38)= U(25, 10, 20, 5.3, 38)/ K_1=1.

Taking into account that U(25, 10^4, 20, 5.3, 38)=1, U(10, 10, 90, 2.5, 25)=0 and $f_2(10^4)$=U(10, 10^4, 90, 2.5, 25)/K_2=1, we can form lotteries of the kind shown in Box 3.

The question to the decision maker are like lotteries in which we vary only the values by $\alpha \in [0,1]$ and this is shown on Figure 7.

This evaluation is relatively easy. We have constant values for the three five dimensional vectors and only three events related to them. We change only the weight coefficient α. We only evaluate the coefficient K_2, since from the normalization it is clear that K_1 is equal to (K_1=1- K_2). Obviously from the general formula it follows K_2=U(10, 10^4, 90, 2.5, 25). If the utility function is constructed the coefficient K_2 is known. However, multiattribute utility function of 5 variables is difficult to evaluate. We evaluate the coefficient K_2 as a demonstration, only with 8 questions and the result is shown on Figure 8.

The obtained result in this particular case is 0.250. When the objective functions are measured between zero and one we always know where is the zero point. It is easy to determine after the utility functions participating in the decomposition of the multiattribute utility are estimated and approximated by polynomials. Then instead of using

Figure 7. Lotteries in which is varying only α

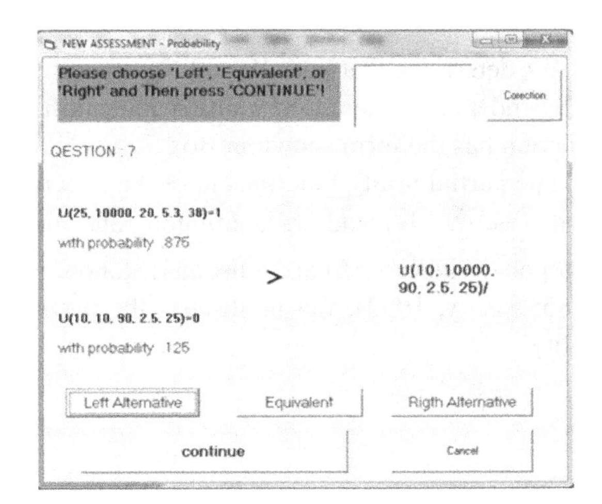

Figure 8. Evaluation of K_2=U(10, 10^4, 90, 2.5, 25) with 8 questions

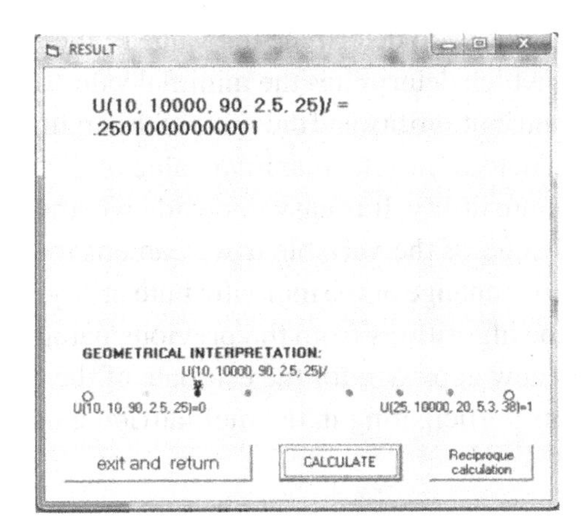

Box 4.

$$\left\{\begin{array}{l} U(10+\alpha(25\text{-}10),\ 10+\alpha(10^4\text{-}10),\ 90+ \\ +\alpha(20\text{-}90),\ 2.5+\alpha(5.3\text{-}2.5),\ 25+\alpha(38\text{-}25)) \end{array}\right\} \begin{array}{c}\succ\\ \approx\\ \prec\end{array} \left\{ U(10,\ 10^4,\ 90,\ 2.5,\ 25) \right\}.$$

the lottery in the shown above form and include the probabilities in an obvious form we can include these probabilities as a percentage shift of the value of the argument which determines the minimal value of the multiattribute utility towards the value of the argument, which determines the maximal value of the multiattribute utility. If between the two values of the arguments of the multiattribute utility function the function is monotonically increasing and continuous then there exists only one value of the new variable (α) for which we obtain equality between the left hand side and the right hand side of the lottery.

In this case, the comparison is only between the two values, similarly to the comparisons for the value function. Then the previous lottery and comparisons are shown in Box 4. Such comparisons are already easy for people with education different from mathematics.

This approach is admissible because we can construct analytically in advance all particular sub-objective utility functions participating in the decomposition of the multiattribute utility. After that, it is easy to determine the value of the argument, which determines the minimal value of the multiattribute utility and the value of the argument, which determines the maximal value of the multiattribute utility. It is easy to decide whether via the change of the variable α we can ensure monotonous change of the multiattribute utility. This way the algorithms from the previous paragraph allow new approach for the estimate of the coefficients participating in the multiattribute utility. The DM's preferences are expressed as in the process of evaluation of a value function.

We shall demonstrate this approach using the utility function in the example of optimal route selection in city street network, discussed in chapter 4. We remember the sub-objectives and the appropriate criteria (Stanoulov, 1985):

- X_{11}-length of the road with deteriorated surface in per cent [0, 100]%;
- X_{12}-length of the route in kilometers [0.1, 10] km;
- X_{13}-number of compulsory stops per 1 km (presence of intersections, traffic lights, etc.) [0, 10];
- X_2-overload of the traffic, number of vehicles beyond the road segment's capacity in per cent [0, 100] %.
- The utility independence is the following:
- X_1={ X_{11}, X_{12}, X_{13}} from X_2;
- X_2 from X_1;
- {X_{13}, X_2} from{X_{11}, X_{12}}.

We determine that both X_2 and X_{13} are utility independent from here. The multiattribute utility function has the form shown in Box 5.

The partial utility functions have the general form $\left(a_i e^{c_i x} - b_i\right)$ and are continuous and monotonous, where a_i, c_i, b_i are constants (Stanoulov, 1985; Keeny, 1993). Let us discuss the partial utility

Box 5.

$$\mathbf{k'u(X)} + 1 = \mathbf{U(X_{13},X_2)} \times \{\mathbf{k'[u(0;X_{12};0;0)} \times \times \left(1\text{-}\mathbf{u(X_{11};0.1;0;0)}\right) + \mathbf{u(100;X_{12};0;0)u(X_{11};0.1;0;0)]} + 1\} = = \left[\mathbf{k'u(0;0.1;X_{13};0)} + 1\right] \times \left[\mathbf{k'u(0;0.1;0;X_2)} + 1\right] \times \times \{\mathbf{k'[u(0;X_{12};0;0)} \times \left(1\text{-}\mathbf{u(X_{11};0.1;0;0)}\right) + + \mathbf{u(100;X_{12};0;0)u(X_{11};0.1;0;0)]} + 1\}.$$

$$\mathbf{U(X_{13},X_2)} = \left[\mathbf{k'u(0;0.1;X_{13};0)} + 1\right] \times\left[\mathbf{k'u(0;0.1;0;X_2)} + 1\right] = \{\mathbf{k_1U_1(X_{13})} + \mathbf{k_2U_2(X_2)} + \mathbf{K_{12}U_1(X_{13})U_2(X_2)}.$$

We measure of the functions $\mathbf{U(X_{13},X_2)}$, , $\mathbf{U_1(X_{13})}$, and $\mathbf{U_2(X_2)}$ between 0 and 1. The maximum of the partial utility function $\mathbf{U(X_{13},X_2)}$ is in (0, 0) and the minimum is in (10, 100). Then coefficient k_1 could be determined by comparisons of the following type:

$$\left\{\mathbf{U}(\alpha 10;\ \alpha 100)\right\} \begin{matrix} \succ \\ \approx \\ \prec \end{matrix} \left\{\mathbf{U}(0;\ 100)\right\}, \alpha \in [0,1].$$

In the preferences, there is not uncertainty and probabilities, only professional consideration are rendered on account of. The coefficient k_2 could be determined by the same approach using the algorithms described in the previous paragraph. The coefficient k_{12}=(1- k_1- k_2) and so on.

In conclusion of this chapter, we would like to point out again that it is very important the way of the semantics is given in every lottery and in every particular problem. This can improve or decrease the preciseness of the evaluations and to increase or decrease the errors of the evaluations of the solution taker.

REFERENCES

Aizerman, M., Braverman, E., & Rozonoer, L. (1970). *Potential function method in the theory of machine learning*. Moscow, Russia: Nauka.

Ermolev, Y. M. (1976). *Methods of stochastic programming*. Moscow, Russia: Nauka.

Farquhar, P. (1984). Utility assessment methods. *Management Science*, *30*, 1283–1300. doi:10.1287/mnsc.30.11.1283

Fishburn, P. (1970). *Utility theory for decision-making*. New York, NY: Wiley.

Gikhman, A., Skorokhod, V., & Yadrenko, M. (1988). *Probability theory and mathematical statistics*. Kiev, Russia: Vyshcha Shkola.

Grifits, T. L., & Tenenbaum, J. B. (2006). Optimal prediction in everyday cognition. *Psychological Science*, *17*(9), 767–773. doi:10.1111/j.1467-9280.2006.01780.x

Keeney, R., & Raiffa, H. (1993). *Decision with multiple objectives: Preferences and value trade-offs* (2nd ed.). Cambridge, UK: Cambridge University Press. doi:10.1109/TSMC.1979.4310245

Larichev, O., & Olson, D. (2010). *Multiple criteria analysis in strategic sitting problems* (2nd ed.). London, UK: Springer.

Mengov, G. (2010). *Decision-making under risk and uncertainty*. Sofia, Bulgaria: Publishing house JANET 45.

Parthasarathy, K. R. (1978). *Introduction to probability and measure*. Berlin, Germany: Springer-Verlag.

Pavlov, Y., Grancharov, D., & Momchev, V. (1996). Economical and ecological utility oriented analysis of the process of anaerobic digestion of waste waters. *European Journal of Operational Research*, *88*(2), 251–256. doi:10.1016/0377-2217(94)00193-6

Pfanzagl, J. (1971). *Theory of measurement*. Wurzburg, Germany: Physical-Verlag.

Popchev, I. (1986). *Technology for multi-criteria evaluation and decision-making*. Sofia, Bulgaria: BISA.

Raiffa, H. (1968). *Decision analysis*. Reading, MA: Addison-Wesley.

Robbins, H., & Monro, S. (1951). A stochastic approximation method. *Annals of Mathematical Statistics*, *22*(3), 400–407. doi:10.1214/aoms/1177729586

Savage, J. (1954). *The foundations of statistics*. New York, NY: Wiley.

Shiryaev, A. N. (1980). *Probability*. New York, NY: Springer.

Shmeidler, D. (1989). Subjective probability and expected utility without additivity. *Econometrica*, *57*(3), 571–587. doi:10.2307/1911053

Sobol, I. (1973). *Monte-Carlo numerical methods*. Moscow, Russia: Nauka.

Stanoulov, N., Pavlov, Y., & Tonev, M. (1985). Multi-criteria routes selection in urban conditions. In I. Popchev & P. Petrov (Eds.), *Optimization, Decision-Making, Microprocessors systems, Sofia 1983- Eight Bulgarian-Polish Symposium*. Sofia, Bulgaria: Publishing House of Bulgarian Academy of Sciences.

Tenekedjiev, K. (2004). *Quantitative decision analyze: Utility theory and subjective statistics*. Sofia, Bulgaria: Marin Drinov.

Chapter 8
Extrapolation Methods in Control and Adaptive System

ABSTRACT

The problem for the design and prognosis can be described in the following way. There is a finite number of points in the space of the situations or alternatives, where some solutions exist (in the general case these are vectors or matrices). Then the problem for the estimation and prognosis as an extrapolation estimate is to estimate the new solution in the aspects of the chosen vector or matrix description in situations that are not in the initial set of the alternatives. The choice of the methodology depends on whether the process is deterministic or random in nature. When the process is deterministic, the problem of prognosis and construction can be described using interpolation or extrapolation. The interpolation means determination of the value of the function or model in the points of interest using the values of the function or the solutions of the model in other points. The authors look for solution inside the interval or the segment of the observation. In the problem of extrapolation, they look for the function values outside the interval of the observation. Moreover, for complex processes and new products, the data is often scarce.

In the initial stages of the choice of approaches and methods, the heuristic of the investigator is very important, because in most of the cases there is a lack of measurements or even clear scales under which to implement these measurements and computations. This stage is often outside of the strict logic and mathematics and is close to the art, in the widest sense of the word. For complex systems and practical problems, the explicit description of such an algorithm is a difficult problem; often it is not solvable, but the existence of beforehand solutions and realizations allow one to make meaningful prognosis estimates. Such an approach is the method of "Multidimensional Linear Extrapolation" (MLE) described in this chapter. The main idea is that close situations give close solutions. The method is very efficient and gives many good solutions for difficult modeling problems.

DOI: 10.4018/978-1-4666-2967-7.ch008

1. THEORETICAL BASIS OF EXTRAPOLATION IN CONTROL DESIGN

The indeterminacy of the initial data is an objective condition in which decisions are made for the creation of new samples of technical or other products in the diverse human activity. This indeterminacy is especially extensive in the initial stages and the preliminary planning, as it can be in the structure and/or the parameters of the product (Gig, 1981). From the mentioned here, it is clear that in such tasks it is necessary to use an approach of consecutive stages, phases in which programs are applied with the aim of achieving one or several intermediate objectives. The aspiration for quantity measurements, estimates, and prognosis at all phases of the investigations and the creation of the new sample products is natural. However, this task is solved with very scarce initial information, especially in the initial development phase. Precisely this objective condition has to be accounted for when choosing the mathematical methods for design and prognosis and construction of new and non-standard products or new control solutions (Pfanzagl, 1971; Raiffa, 1968).

In the initial stages of the choice of approaches and methods, the heuristic of the investigator is very important, because in most of the cases there is a lack of measurements or even clear scales under which to implement these measurements and computations.

This stage is often outside of the strict logic and mathematics and is close to the art, in the widest sense of the word, to choose the right decision among great number of circumstances and often without associative examples of similar activity (Gig, 1981; Rastrigin, 1981). Often intuitively insignificant circumstances have to be ignored and the main solution has to be chosen as a process of recognition. We call this the art to orient the scientist in this initial undetermined situation.

In these initial stages mathematical methods are used which in their concept allow for broad semantic interpretations, flexibility and iterativity in the mathematical constructs in the models as well as in the control solutions. The models which are built at this initial stage are based on preliminary experience with other, often quite different products or solutions, or are formed by training.

For this reason, the methods for design and prognosis here have to facilitate solutions in multidimensional spaces, to allow for easy correction and easy tweaking of the objectives to adequately handle the accumulating basic information in the process of the investigation and to admit iterative development of the modeling process. Such an approach is the method of "Multidimensional Linear Extrapolation" (MLE), designed by the Russian scientist Rastrigin (Rastrigin, 1981, 1986).

The solution of the problem of the identification in the broad sense of the word or the prognosis is determined by the nature of the process or the nature of the basic information for the process. The information in most general terms can be of quantitative or qualitative type. In the beginning, it is obligatory to have a stage of classification, determination of the basic scales and the way of measurements taking into account the type of basic information (Pfanzagl, 1971; Gig, 1981).

The choice of the methodology depends on whether the process is deterministic or random in nature. When the process is deterministic, the problem of prognosis and construction can be described using interpolation or extrapolation. The interpolation means determination of the value of the function or model in the points of interest using the values of the function or the solutions of the model in other points. We are looking for solution inside the interval or the segment of the observation.

In the problem of extrapolation, we are looking for the function values outside the interval of the observation. Moreover, for complex processes and new products the data is often scarce.

It is possible that the function modeling the product is known in advance or in the harder case these functions are unknown and only their approximation is sought based on scarce initial information. If we add the presence of a noise or internal initial indetermination in the data and the multidimensionality of the process then the choice of the mathematical methods and models becomes severely limited.

In the case when the processes are stochastic we are looking for the prognosis of mathematical expectation and other statistical quantities let us call them averaged prognosis estimates (Jonson, 1977). This added to the internal initial indetermination in the data, imposes even stricter requirements for the choice of the mathematical tools for estimation and prognosis (Braverman, 1983). We can say that the method of the multidimensional linear extrapolation is one of a very small number of methods, which can be used in these conditions.

The problem for the design and prognosis can be described in the following way (Rastrigin, 1981, 1986). There is a finite number of points in the space of the situations or alternatives, where some solutions exist (in the general case these are vectors or matrices).

Then the problem for the estimation and prognosis as an extrapolation estimate is to estimate the new solution in the aspects of the chosen vector or matrix description in situations, which are not in the initial set of the alternatives.

We start with the mathematical description of what we said above (Rastrigin, 1981, 1986). It is supposed that we have (n) design projects X_i and developed projects Y_i, $i=1 \div n$, as an a priori given data in a form of the table of precedents:

$$M = \begin{Vmatrix} X_1; Y_1 \\ \dots\dots \\ X_n; Y_n \end{Vmatrix} = \begin{Vmatrix} x_{11}, x_{21}, \dots, x_{k1}; y_{11}, y_{21}, \dots, y_{r1}; \\ \dots\dots\dots\dots\dots\dots\dots\dots\dots\dots\dots \\ x_{1n}, x_{2n}, \dots, x_{kn}; y_{1n}, y_{2n}, \dots, y_{rn}; \end{Vmatrix},$$

$$Xi = (x_{1i}, x_{2i}, \dots, x_{ki}), i = 1 \div n,$$
$$Yi = (y_{1i}, y_{2i}, \dots, y_{ri}), i = 1 \div n.$$

In other words from the beforehand investigations or experiments we recognize the dependence:

$$Yi = F^0(X_i), i = 1 \div n.$$

We are looking for the solution of the following problem. Let us be given new task X_{n+i} and from it we would like to prognose new project Y_{n+i}, based on the known matrix of precedents M. We are looking for formal solution, irrespective of the algorithm F^0, to prognose the final results Y dependent on the input data X.

For complex systems and practical problems the explicit description of such an algorithm is difficult problem, often it is not solvable, but the existence of beforehand solutions and realizations allow us to make meaningful prognosis estimates.

This task can be extrapolation or interpolation problem depending on where is the position of the new task in the preliminary known data. The aim is to create convenient and simple enough method, which permits solution of a wide range of extrapolation problems, going beyond the borders of the classical design and control problem.

Towards the so formulated extrapolation problem, we put the following requirements:

The developed method must restore exactly all the a priori data shown in the a priori matrix of the precedents, i.e. if in the input of the algorithm we put X_i we must receive Y_i at the output:

$$Yi = \Psi(Xi, M), i = 1 \div n.$$

The extrapolation algorithm has to be vector algorithm, i.e. X has to be a vector and the output Y can be a scalar quantity. Then the solution has the form:

$$y_i = \Psi_i(Xj, M), i = 1 \div r.$$

Such a solution will transform the more complicated vector problem to r-times simpler scalar solution.

3. The complexity of the problem must grow with respect to (n) and (κ) no faster than linearly. This is pointed out in the scientific literature as a serious requirement and often it is not satisfied.
4. The method for prognosis and extrapolation has to be feasible and to work well with small values of the quantity (n). That means that the extrapolation algorithm has to admit arbitrary values of the quantity (n), including one.
5. Independently of the difficulty and complexity of the solved problem, the structure of the dependence between the input $\boldsymbol{X}$ and the output $\boldsymbol{Y}$ of the extrapolation problem has to be simple and easy enough to allow programming implementation and practical use.

In details, these requirements are discussed in the literature (Rastrigin). Every extrapolation problem is based on some formal representation of the algorithm $\boldsymbol{F^0}$ or in other words the dependence between the end prognosis results $\boldsymbol{Y}$ and the input data or, in general terms, the initial causes $\boldsymbol{X}$.

This formalization is necessary mainly in order to receive definitive final estimate under a given input of the algorithm and also with the aim of optimization by a given criterion. This formalization can be represented symbolically as:

$$F = \varphi(M, \Theta).$$

In this case $\boldsymbol{M}$ is a table or protocol for different known in advance realizations of the causal connection $\boldsymbol{F^0}$ between the input and output quantities and Θ is an a priori representation of the form of the synthesized inter-relation $\boldsymbol{F}$, which may be structure or degree of polynomial, minimum residual variance and others.

In this book, the protocol is the matrix of precedents $\boldsymbol{M}$ of the real causal relation or the operator $\boldsymbol{F^0}$. The model $\boldsymbol{F}$ of the real connection is supposed to be in the general case vector function, which can be described using the following formalization:

$$F = (f_1, ..., f_n).$$

Or the main task is synthesizing (n) in number scalar functions of the type:

$$y_i = f_i(x), i = 1 \div n.$$

In order to solve this problem, different extrapolation methods have been designed, like for example the least squares method, different optimization problems, the use of spline functions and others. The different methods of synthesizing of the model $\boldsymbol{F}$ differ in the way of the description of the a priori information. For example if Θ gives us some structure $\boldsymbol{F}_{structure}$ and criterion for the best description of the model $\boldsymbol{F}$, then the task can be reduced to the best parametrical description of the problem.

$$y = F_{structure}(X, C), C = (c_1, c_2, ..., c_r),$$

$$Q(F_{structure}(X, C)M) \xrightarrow[\min C \in S]{} C^* = (c^*_1, c^*_2, ..., c^*_r).$$

In the above formulas by $\boldsymbol{Q}$ is denoted the criterion, used for the parametrical optimization with the aim of the best structural representation. By $\boldsymbol{S}$ we denote the domain, in which parameters can take value and by $\boldsymbol{C^*}$ we denote the optimal solution. As efficiency criterion $\boldsymbol{Q}$ in a lot of problems the residual variance is used:

$$Q = \sum_{i=1}^{n} \left| F_{structure}(X, C) - y_i \right|^2.$$

The most difficult part of these problems is the description of the corresponding structure. For the adequate description of the structure $\boldsymbol{F}_{structure}(.)$ in more complicated problems scientific theory concerning the real causal link F^0 is necessary. In the general case, this condition is not satisfied and then we cannot say anything definitive about $F_{structure}$. In this case, point 5 from the conditions simply cannot be satisfied.

In a lot of cases this structural dependence is a functional decomposition to a family of orthogonal functions $\{ \Phi(X) \}$:

$$y = \sum_{i=1}^{s} c_i \times \varphi_i(X),$$

$$\Phi(X) = \{\varphi_1(X), \varphi_2(X), \ldots\ldots, \varphi_s(X)\}.$$

For the least squares method the parameters C^* are obtained as a solution of the following well-known problem:

$$Q(C) = \sum_{j=1}^{n} \left| y = \sum_{i=1}^{s} c_i \times \phi_i(X) - y_j \right|^2 \xrightarrow[\min C]{} .$$

If the input quantity X is a vector, then for (s) greater than 2, the complexity of the problem grows nonlinearly (Rastrigin, 1986). The method which satisfies the five conditions discussed above is MLE. In its essence this methods can be fully described in terms of the linear algebra.

2. LINEAR SPACES AND LINEAR TRANSFORMATIONS

For the needs of our further work, we need definitions, facts, and theorems from the linear algebra. We start with one definition (Lang, 1965; Dixmier, 1967; Strang, 1976; MacLane, 1974).

2.1. Definition

A vector space over the field of the real numbers R is a set V together with two binary operations that satisfy the eight axioms listed below. In the list below, let $\boldsymbol{u}$, $\boldsymbol{v}$, $\boldsymbol{w}$ be arbitrary vectors in V, and a, b be scalars in F. Elements of V are called *vectors*. Elements of R are called *scalars*:

- Associativity of addition: $v_1 + (v_2 + v_3) = (v_1 + v_2) + v_3$;
- Commutativity of addition: $v_1 + v_2 = v_2 + v_1$;
- Identity element of addition: There exists an element $\mathbf{0} \in V$, called the *zero vector*, such that $v + \mathbf{0} = v$ for all $v \in V$;
- Inverse elements of addition: For all $v \in$V, there exists an element $-v \in V$, called the *additive inverse* of v, such that $v + (-v) = \mathbf{0}$;
- Distributivity of scalar multiplication with respect to vector addition: $s(v_1 + v_2) = sv_1 + sv_2$;
- Distributivity of scalar multiplication with respect to field addition: $(n_1 + n_2)v = n_1 v + n_2 v$;
- Respect of scalar multiplication over field's multiplication: $n_1 (n_2 v) = (n_1 n_2) v$;
- Identity element of scalar multiplication: $1v = v$, where 1 denotes the multiplicative identity in $\boldsymbol{R}$.

A basis exists in any finite vector space. Basis reveal the structure of vector spaces in a concise way. A basis is defined as a finite set $B = \{v_i\}_{i \in I}$ of vectors v_i indexed by some index set I that spans the whole space, and is minimal with this property. The former means that any vector v can be expressed as a finite sum, called linear combination of the basis elements:

$$v = \sum_{j=1}^{s} a_j \times v_j .$$

In the formula the a_k are scalars and v_k ($k = 1, \ldots, s$) elements of the basis B. the elements of the bases B are linearly independent. A set of vectors is said to be linearly independent if none of its elements can be expressed as a linear combination of the remaining ones. Equivalently, an equation can only hold if all scalars a_1, ..., a_n equal zero:

$$(\sum_{j=1}^{s} a_j \times v_j = 0) \Rightarrow (a_j = 0,\ j = 1 \div s).$$

Linear independence ensures that the representation of any vector in terms of basis vectors, the existence of which is guaranteed by the requirement that the basis span V, is unique. This is referred to as the coordinatized viewpoint of vector spaces, by viewing basis vectors as generalizations of coordinate vectors x, y, z in $\boldsymbol{R}^3$ and similarly in higher dimensional cases. For example the coordinate vectors $\mathbf{e}_1 = (1, 0, ..., 0)$, $\mathbf{e}_2 = (0, 1, 0, ..., 0)$, to $\mathbf{e}_s = (0, 0, ..., 0, 1)$, form a basis of $\boldsymbol{R}^s$, called the standard basis, since any vector $(x_1, x_2, ..., x_n)$ can be uniquely expressed as a linear combination of these vectors:

$$(x_1, x_2, .., x_s) = x_1(1, 0, .., 0) + x_2(0, 1, 0, .., 0) + ... + x_s(0, .., 0, 1) = x_1\mathbf{e}_1 + x_2\mathbf{e}_2 + ... + x_s\mathbf{e}_s.$$

Every vector space has a basis. All bases of a given vector space have the same number of elements. It is called the *dimension* of the vector space, denoted dim V.

The relation of two vector spaces can be expressed by *linear map* or *linear transformation*. They are functions that reflect the vector space structure. They preserve sums and scalar multiplication:

$f(\mathbf{x} + \mathbf{y}) = f(\mathbf{x}) + f(\mathbf{y})$ and $f(a \cdot \mathbf{x}) = a \cdot f(\mathbf{x})$ for all $\mathbf{x}$ and $\mathbf{y}$ in V, all a in F.

An *isomorphism* is a linear map $f{:}V \to W$ such that there exists an inverse map $g{:}W \to V$, which is a map such that the two possible compositions $f\,_o\,g{:}W \to W$ and $g\,_o\,f{:}V \to V$ are identity maps. Equivalently, f is both one-to-one (injective) and onto (surjective). If there exists an isomorphism between V and W, the two spaces are said to be *isomorphic*; they are then essentially identical as vector spaces, since all identities holding in V are, via f, transported to similar ones in W, and vice versa via g. Linear maps $V \to W$ between two fixed vector spaces form a vector space $\mathrm{Hom}_F(V, W)$, also denoted $L(V, W)$.

Once a basis of V is chosen, linear maps $f{:}V \to W$ are completely determined by specifying the images of the basis vectors, because any element of V is expressed uniquely as a linear combination of them. If dim V = dim W, a 1-to-1 correspondence between fixed bases of V and W gives rise to a linear map that maps any basis element of V to the corresponding basis element of W. It is an isomorphism, by its very definition. Therefore, two vector spaces are isomorphic if their dimensions agree and vice versa. Another way to express this is that any vector space is completely classified (up to isomorphism) by its dimension, a single number. In particular, any n-dimensional R-vector space V is isomorphic to R^n.

Matrices are a useful notion to encode linear maps. A typical n-by-s matrix is:

$$\boldsymbol{A} = \begin{bmatrix} a_{11} & .... & a_{1k} \\ ... & ... & ... \\ a_{n1} & ... & a_{nk} \end{bmatrix}.$$

They are written as a rectangular array of scalars as in the image at the right. Any n-by-s matrix A gives rise to a linear map from R^n to R^m, by the following:

$$x = (x_1, ..., x_k) \to$$

$$(y = (\sum_{j=1}^{k} a_{1j} \times x_j, \sum_{j=1}^{k} a_{2j} \times x_j, .., \sum_{j=1}^{k} a_{nj} \times x_j)'),$$

where $(.)'$ denotes transposition, $y = \boldsymbol{A}x$.

Moreover, after choosing bases of V and W, any linear map $f: V \rightarrow W$ is uniquely represented by a matrix via this assignment. An n-by-n (square) matrix **A** is called invertible if there exists an n-by-n matrix **B** such that $AB = BA = I_n$, where $\mathbf{I}_n$ denotes the n-by-n identity matrix and the multiplication used is ordinary matrix multiplication. If this is the case, then the matrix **B** is uniquely determined by **A** and is called the inverse of **A**, denoted by $\mathbf{A}^{-1}$. It follows from the linear space theory that if ($AB = I$) for finite square matrices **A** and **B**, then also ($BA = I$). Let **A** be a square n by n matrix over the field $\boldsymbol{R}$ of real numbers. The following statements are equivalent: **A** is invertible, det(**A**) $\neq 0$ and the (rank $\mathbf{A} = n$). The equation $\mathbf{Ax} = \mathbf{0}$ has only the trivial solution $\mathbf{x} = \mathbf{0}$. The equation $\mathbf{Ax} = \mathbf{b}$ has exactly one solution for each **b** in $\boldsymbol{R}^n$, ($\mathbf{x} \neq \mathbf{0}$). The columns of **A** are linearly independent and form a basis of $\boldsymbol{R}^n$. The linear transformation mapping **x** to **Ax** is a bijection from $\boldsymbol{R}^n$ to $\boldsymbol{R}^n$. The transpose $\mathbf{A}^T$ is an invertible matrix. Furthermore, the following properties hold for an invertible matrix **A**:

- $(\mathbf{A}^{-1})^{-1} = \mathbf{A}$;
- $(k\mathbf{A})^{-1} = k^{-1}\mathbf{A}^{-1}$ for nonzero scalar k;
- $(\mathbf{A}^T)^{-1} = (\mathbf{A}^{-1})^T$;
- For any invertible n-by-n matrices **A** and **B**, $(\mathbf{AB})^{-1} = \mathbf{B}^{-1}\mathbf{A}^{-1}$. More generally, if $\mathbf{A}_1,\ldots,\mathbf{A}_k$ are invertible n-by-n matrices, then $(\mathbf{A}_1\mathbf{A}_2\cdots\mathbf{A}_{k-1}\mathbf{A}_k)^{-1} = \mathbf{A}_k^{-1}\mathbf{A}_{k-1}^{-1}\cdots\mathbf{A}_2^{-1}\mathbf{A}_1^{-1}$;
- $\det(\mathbf{A}^{-1}) = \det(\mathbf{A})^{-1}$.

For the needs of the definition of the MLE method and the description of its capabilities we need the definition and the properties of the pseudo-inverse matrices.

In mathematics, and in particular linear algebra, a *pseudo-inverse* A^+ of a matrix A is a generalization of the inverse matrix (Arthur, 1972). The most widely known type of matrix pseudo-inverse is the *Moore–Penrose pseudo-inverse*, which was independently described by E. H. Moore in 1920, Arne Bjerhammar in 1951 and Roger Penrose in 1955. A common use of the Moore–Penrose pseudo-inverse is to compute a 'best fit' (least squares) solution to a system of linear equations that lacks a unique solution. Another use is to find the minimum Euclidean norm solution to a system of linear equations with multiple solutions. The algebraical and someway obscure definition of pseudo-inverse matrix is the following (Arthur, 1972):

2.2. Definition

For a matrix A, a Moore–Penrose pseudo-inverse (hereafter, just pseudo-inverse) of A is defined as a matrix A^+ satisfying all of the following four criteria:

- ($AA^+A=A$);
- ($A^+AA^+= A^+$);
- (AA^+ is Hermitian);
- (A^+A is also Hermitian).

Described in such way the pseudo-inverse matrix does not explicitly show its true features. We are going to state three theorems via which the properties and the capabilities of the pseudo inverse matrix become obvious (Arthur, 1972).

2.3. Theorem

Let z be n-dimensional vector and A be a matrix with dimension (nxm). Then there exist a unique vector $\hat{u}$ with minimal Euclidean norm, such that it minimizes:

$$\left\| z - A\,\hat{u} \right\|^2 .$$

The vector $\hat{u}$ is the unique vector which lies in the subspace generated by the columns of the transpose matrix A^T and satisfies the equation $\hat{z} = A\hat{u}$.

The vector $\hat{z}$ is the projection of the vector z in the subspace generated by the columns of the matrix A.

2.4. Theorem

From all vectors minimizing $\left\| z - A\hat{u} \right\|^2$ the vector $\hat{u}$ with minimal Euclidean norm is the unique vector of the type $\hat{u} = A^T y$, satisfying the equation:

$$A^T A u = A^T z.$$

2.5. Lemma

For every symmetric matrix A with elements real numbers the limit

$$P_A = \lim_{\delta \to 0}(A + \delta I)^{-1} A = \lim_{\delta \to 0} A(A + \delta I)^{-1}$$

exist. For every vector u the vector $\hat{u} = P_A u$ is the projection of u in the subspace generated by the columns of the matrix A.

Then the following fundamental theorem shows the basic properties of the pseudo-inverse matrices.

2.6. Theorem

For every matrix A with elements real numbers and dimension (nxm) the limit

$$A^+ = \lim_{\delta \to 0}(A^T A + \delta I)^{-1} A^T = \lim_{\delta \to 0} A^T (A A^T + \delta I)^{-1}$$

exists. For every vector z the vector $\hat{u} = A^+ z$ is a projection of the vector z in the subspace, generated by the columns of the matrix A^T and it is with minimal Euclidean norm and minimizes:

$$\left\| z - A\hat{u} \right\|^2 .$$

When the matrix is invertible the pseudo inverse matrix is the inverse matrix. Now we can formally describe the method of the Multiattribute Linear Extrapolation.

3. MULTIATTRIBUTE LINEAR EXTRAPOLATION WITH EUCLIDEAN NORM

The method of the Multiattribute (Multidimensional) Linear Extrapolation (MLE) requires relatively small number of measurements and it has shown good results when modeling complex systems in practice (Rastrigin, 1981, 1986). By T we denote the "situations"and by S the values of the measurements of the "object"—the final results. We suppose that the functional dependency between T and S is denoted as S=F(T). The mathematical essence of the MLE is starting from a not very large number of situations T_i and the known decisions for them S_i (in general vectors) to estimate F (Rastrigin, 1981, 1986):

$$(T_i \Rightarrow S_i,\ i=1 \div n) \Rightarrow F,$$

$$Q(T_i, S_i)=\text{extremum}_{A_i \in \Omega(T_i)},$$

where Q is an optimization criterion, using which we choose A_i or in fact the description of the experimental data for the concrete estimated object for the moment and $\Omega(T_i)$ is the set of potential solution decisions for the moment T_i. The aim is to prognosticate S_{i+1}, S_{i+2}, S_{i+3} for chosen T_{i+1}, T_{i+2}, T_{i+3}. The MLE methodology is the following:

1. Under the hypothesis that the transformation S=F(T) is close to linear we can use the method of MLE.
2. We make the projection S* in the m-dimensional space with basis (S_j, $j=1\div n$), caused by the known training situations F_i, $i=1\div n$, where S* are the empirical data for the process and (S_j, $j=1\div n$) are created by F_i, $i=1\div n$.
3. We represent the sought solution F in the space generated of the training situations F_i, $i=1\div n$ with the same coefficients with which the projection Pr(S*) is represented.

In the general case MLE gives an aproximate solution, since the final answer is a projection in a finite dimensional linear space generated by the vectors of the known to the current moment precedents. In fact the main idea is that close situations give to us close solutions. The method is very efficient and gives good solutions in a great number difficult for modeling problems.

For some problems and prognoses it can give exact, in an analytical sense, solutions. For example, if we describe in polynomials the behaviour of the system till the moment as a function of the time, then the future solutions can be considered as an extrapolation of the already known polynomial.

Let us be given m moments in which we have done the measurements. We denote by t_1 the initial and by t_m the current moment. We build polynomial model in the interval [t_1, t_m] with one of the suitable for these aim approximation methods. Such methods can be regression analysis, stochastic approximation, etc. Let the approximation polynomial has the following form:

$$P(t) = \sum_{i=0}^{k} x_i t^i .$$

We choose n training polynomials in accordance with the nature of the problem or derived from general theoretical settings. One of the ways for finding training sample of polynomials is to use polynomial constructed from the empirical data by adding some noise at the coefficients.

By using the set of the new coefficients, we build a set of training polynomials. Then the future solutions are projected in the linear space generated by the training sample of polynomials and this projection is the prognosis polynomial. In other words we expect that close situations will generate close solutions.

After the mathematical description of the MLE method in polynomial form, we are going to give a real example of concrete implementation, where we are going to demonstrate the just mentioned construction. Let the values of the training polynomials at the moments t_i, $i=1\div m$ are a_{ij}, for $j=1\div n$:

$$A_j = (a_{1j}, a_{2j}, \dots, a_{ij}, \dots, a_{mj}),\ j = 1 \div n.$$

The coefficients of the polynomials form the following $(k+1)$ dimensional vectors:

$$X_j = (x_{0j}, x_{1j}, \dots, x_{kj}),\ \ j = 1 \div n.$$

According to point 3 the optimal solution X* has the form:

$$x_0^* = x_{01} + \sum_{j=1}^{n-1} \lambda_j \times (x_{0\,(j+1)} - x_{01})$$

....

$$x_k^* = x_{k1} + \sum_{j=1}^{n-1} \lambda_j \times (x_{k\,(j+1)} - x_{k1})$$

The above vector equation can be written in the following form:

$$(\vec{X}^* - \vec{X}_1) = X_{matrix} \vec{\lambda} .$$

If we multiply the corresponding coefficient of the concrete j^{th} polynomial with the corresponding degree of the chosen time t_i and sum them ($i=1\div m$, $j=1\div n$) we are going to obtain the following system of linear equations:

$$a_{ij} = x_{0j} + \sum_{s=1}^{k} x_{sj} \times t_i^s,\ i = 1 \div m,\ j = 1 \div n,$$

$$a_1^* = a_{11} + \sum_{j=1}^{n-1} \lambda_j \times (a_{1(j+1)} - a_{11})$$

....

$$a_m^* = a_{m1} + \sum_{j=1}^{n-1} \lambda_j \times (a_{m(j+1)} - a_{m1})$$

The linear equation written above can be written form the position of the MLE in the following short form

$$\left\| (\vec{a}^* - \vec{a}_1) - A\vec{\lambda} \right\| = \min,$$

$$\vec{\lambda} = A^+ (\vec{a}^* - \vec{a}_1).$$

In this case $A^*=(a_1^*,a_2^*,...,a_m^*)$ are the measured values till the moment t_m of the investigated process.

Solving the system of linear equations with respect to λ and applying the obtained solution, we find the solution X^* of the given problem. In the general case ($n\neq(m+1)$) and in practice often ($m < (k+1)$). Moreover, the measurements can be taken within different time intervals, which make the procedures very useful for the practice.

Now we are going to consider a real example in which we are going to demonstrate the capabilities of the MLE and some of the difficulties in its application in polynomial form. Let us consider a process modeled via stochastic approximation with polynomials of 25 degree shown on Figure 1 (Aizerman, 1970; Ermolev, 1976).

The polynomials of the training sample, obtained via stochastic approximation are from such a high degree, because of the features of the process. There is an area in which the process is almost straight line following an area in which the process has developed dynamically.

Let us try using the polynomials from the Figure 1 to construct a polynomial training sample and to make a prognosis for the process. It is known that if we put noise at the coefficients of the approximating the real data polynomial with the aim of generating a training sample and prognosis, the mathematical expectation is exactly the initial polynomial.

We are going to test this process on real data taken from technical documentation. The result is shown on Figure 2.

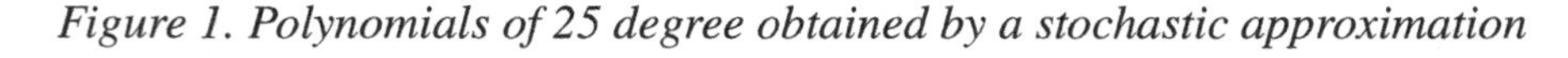

Figure 1. Polynomials of 25 degree obtained by a stochastic approximation

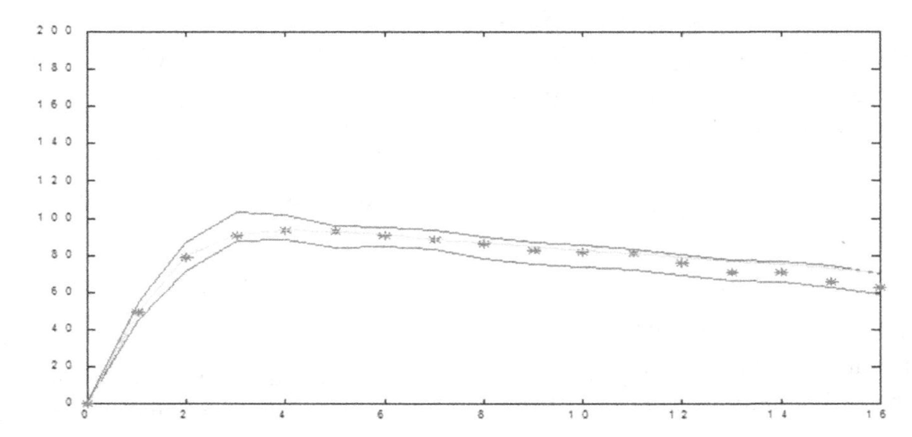

Figure 2. Prognosis after modeling with stochastic approximation and bad training polynomials

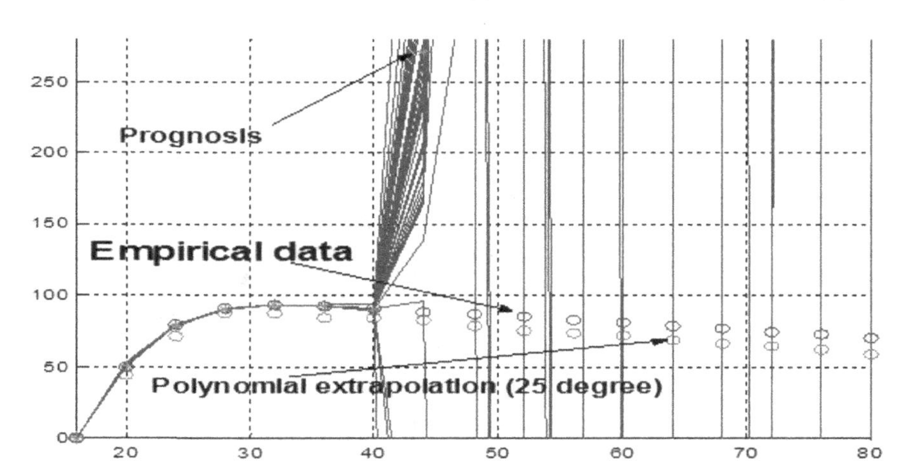

Figure 3. Prognosis with 6 measurements and 15 training polynomials

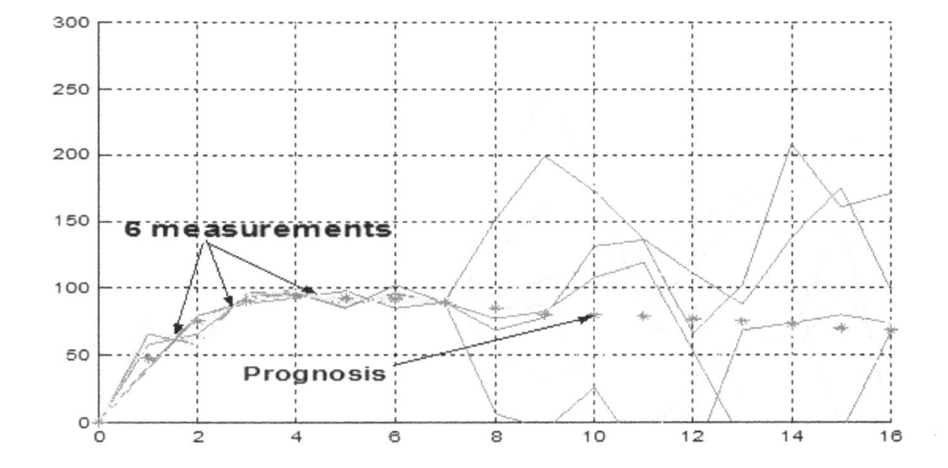

On Figure 2 the number of the measurements is 13, and the number of the training polynomials is insufficient. On Figure 3 the picture is better.

The result shown on Figure 1 is bad, because the approximation is not good and after putting noise at the coefficients of the polynomial from degree 25 the projection space does not reflect the process. This means that the space, generated from the training sample does not reflect the essence of the process in the a priori measurements. It is necessary to have an additional preliminary processing to achieve better approximation and to take advantage of the capabilities of the MLE method. Due to the insufficient number of measurements, the forecasting polynomials are "shaken" around the points of the measurements. The small number of the measurement points and the bad training sample allows great freedom of the forecasting polynomials. This fixes the polynomial very precisely in the observation interval, but outside this interval, the forecasts are not good enough. It is obvious that in the case the training sample is not good.

Figure 4. Prognosis with 13 polynomials and 7 measurements

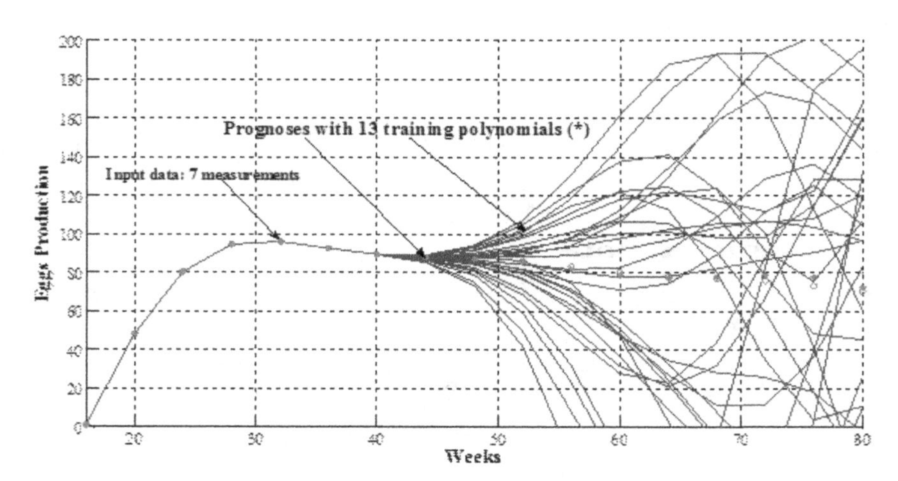

Figure 5. Prognosis using training sample obtained via regression

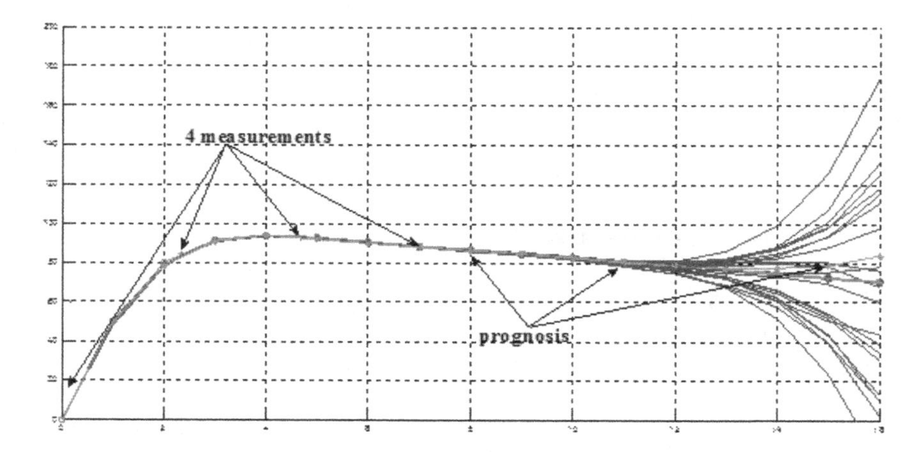

On Figure 4 we increase the number of the prognoses and after taking their average we get better results. This means that we project vector with dimension 13 in subspace by using a matrix with dimension (13, 15). The polynomial training space is with dimension 15, which is a subspace of the 25 dimensional Euclidean linear space. We can see the diversity of the projection subspaces can result from inaccurate initial measurements (the bad approximation polynomial of Figure 1). It is obvious that in this case the training sample is not good. We make another computation, as we use the classical regression analysis (Jonson, 1977; Braverman, 1983) in order to form approximation polynomial shown on Figure 5.

The good correspondence between the model and the approximating polynomial is obvious. The prognosis is a part of an investigation, described in the literature (Ljakova, 2002). In practice in order to obtain such a low error good initial data in sufficient quantity is necessary.

4. RECURRENT FORMULAS

For the computation, we can use the theorem of Greville and the Greville's recurrent formulas derived from it. If as a result of the new measurements the dimension of the matrix increases we can use the following recurrent relations (Arthur, 1972; Rastrigin, 1986):

$$A_{m+1} = (A_m, a_{m+1}),$$

$$A^+_{m+1} = (A^+_m \times (I - a_{m+1} \times K^T_{m+1})) \quad .$$

where the matrix K_{m+1} has the form:

$$K_{m+1} = \begin{cases} \dfrac{(I - A_m \times A^+_m) \times a_{m+1}}{\left\|(I - A_m \times A^+_m) \times a_{m+1}\right\|^2}, & if \quad \| \ \| \neq 0 \\ \dfrac{A^{+T}_m \times A_m + a_{m+1}}{1 + \left\|A^+_m \times a_{m+1}\right\|^2} \end{cases}$$

Based on this there are derived recurrent formulas for the computation of the unknown coefficients. This auxiliary mathematical instrument would be very useful for processes, which are being renovated continuously and require continuous recalculation of the pseudo-inverse matrices A^+.

REFERENCES

Aizerman, M., Braverman, E., & Rozonoer, L. (1970). *Potential function method in the theory of machine learning*. Moscow, Russia: Nauka.

Arthur, A. (1972). *Regression and the moor-penrose pseudo-inverse*. New York, NY: New Academic Press.

Braverman, E., & Muchnik, I. (1983). *Structure methods for the processing of empirical data*. Moscow, Russia: Nauka.

Dixmier, J. (1967). *Cours de mathematiques*. Paris, France: Gautier-Villars.

Ermolev, Y. (1976). *Stochastic programming methods*. Moscow, Russia: Nauka.

Gig, J. (1981). *Applied general systems theory*. Moscow, Russia: Mir.

Jonson, N., & Leone, F. (1977). *Statistics and experimental design*. New York, NY: John Wiley & Sons.

Lang, S. (1965). *Algebra*. Reading, MA: Addison-Wesley.

Ljakova, K., & Pavlov, Y. (2002). Modeling and prognosis in bird-production decision support. *Automatica & Informatics*, *1*, 42–48.

MacLane, S., & Birkhoff, G. (1974). *Algebra*. New York, NY: The Macmillan Company.

Pfanzagl, J. (1971). *Theory of measurement*. Wurzburg, Germany: Physical-Verlag.

Raiffa, H. (1968). *Decision analysis*. Reading, MA: Addison-Wesley.

Rastrigin, L. (1981). *Adaptation of complex systems*. Riga, Russia: Zinatne.

Rastrigin, L., & Ponomarev, Y. (1986). *Extrapolation methods for design and control*. Moscow, Russia: Mashinostroene.

Strang, G. (1976). *Linear algebra and its applications*. New York, NY: Elsevier.

Chapter 9
Preferences, Utility Function, and Control Design of Complex Cultivation Process

ABSTRACT

This chapter demonstrates the flexibility and the diversity of the potential functions method and its conjunction with the utility theory when it describes completely analytically the complex system "decision maker-dynamical process." The utility analytical descriptions have been built concerning the attitude of the technologist toward the dynamic process. Using these approach factors as ecology, financial perspective, social effect can be taken into account. They are included in the expert preferences via the expert attitude towards them.

The analytic construction of the utility function is an iterative "machine-learning" process. This interactivity allows a new strategy in the process of control design and in the control of the system with human participation in the final solution. The first and the most important effect of this strategy is the possibility for the analytical description of such complex systems. This has been achieved for the first time in scientific practice.

The second effect is the introduction of the iterativity in the process of forming the control as is used naturally and harmonically computer and analytical mathematical techniques.

The third effect is the fact that the process of training can be reversed towards the trainer technology expert with the aim of additional analysis and corrections.

In the control design are overcome restrictions connected with the observability of the Monod kinetics and with the singularities of the optimal control of Monod kinetic models.

DOI: 10.4018/978-1-4666-2967-7.ch009

1. DESCRIPTION OF THE PROBLEM

Fermentation processes are relatively difficult objects for control. Their features have been discussed repeatedly. This has led to search for solutions via approaches and methods related to a wide range of contemporary mathematical areas. Such areas are the classical control theory (linear systems, nonlinear systems, stochastic systems, adaptive systems), the theory of systems with distributed parameters (distributed systems), the theory of variable structure systems and their main direction sliding mode control, etc. In the last decade up to date methods and approaches in the areas of functional analysis, differential geometry and its modern applications in the areas of nonlinear control systems as reduction, equivalent diffeomorphic transformation to equivalent systems and optimal control have been used (Neeleman, 2002; Pavlov, 2005; Tzonkov, 2006).

For some problems and tasks these approaches allowed completely new interpretations and solutions. For example important biotechnological parameters of the process as the specific growth rate of the biomass can be determined only approximately by the Biotechnologist. The reason is that they not only define the quality of the yield product, but can depend on other difficult for mathematical modeling factors, such as market prices, economic and ecological considerations, social effect and others.

In our practice, such estimates have been made with precision 10-30% from maximal growth rate of the cultivation process. It is obvious that such one-step determination cannot be effective. Here the results obtained from the Decision-making theory, the Utility theory and methods for evaluations and estimations of the utility functions as stochastic programming can be used (Pavlov, 2010). One analytic description of the thinking of the biotechnologist with respect to the biotechnological process as a utility function will allow a new mathematical thinking with respect to the system Technologist-process and new mathematical modeling description as Technologist-dynamical model. This will allow far more flexible and exact description of the complex economic, ecological, social, and other factors as a part of the mathematical model and control solutions.

Among the most-widely used control models for Biotechnological Processes are the so-called unstructured models, based on mass balance. In these models, the biomass is accepted as homogeneous, without internal dynamic. Most widely used are models based on the description of the kinetic via the well-known equation of Monod or some of its modifications. As an example, a fed-batch biotechnological process is well described via well-known Monod nonlinear model (Neeleman, 2002; Tzonkov, 2006). The rates of cell growth, sugar consumption, concentration in a yeast fed-batch growth process are commonly described for all functional states according to the mass balance as follows:

$$\begin{aligned} \dot{X} &= \mu_m \frac{S}{K_S + S} X - \frac{F}{V} X \\ \dot{S} &= -k\mu_m \frac{S}{K_S + S} X + \frac{F}{V}(S_0 - S) \\ \dot{V} &= F \end{aligned} \qquad (9.1.1)$$

Here X is the concentration of the biomass, S is the substrate concentration, *V* is the volume of the bioreactor. The maximal growth rate is denoted by μ_m and K_S is the coefficient of Michaelis-Menten. With *k* we denote a constant typical for the corresponding process. The feeding rate is denoted by *F*. If the process is continuous (*F*/*V*) is substituted by the control *D*, the dilution rate of the Biotechnological Process (BTP). The third equation is dropped off. This differential equation is often part of more general and complex dynamic models.

One of the most important characteristics of biotechnological processes, which make the control design more difficult, is the change of cell population state. Here it is necessary to emphasize once more the strong nonlinearity of such models, which combined with the internal instability of the BTP leads to serious difficult to overcome control problems (Neeleman, 2002; Tzonkov, 2006).

An additional obstacle is the existence of noise of non-Gaussian type. This type of noise appears in the measurement process as well as in the process of the determination of the structure parameters of the model, however, may be the most serious obstacle provoked by the differences in the rate of changes of the elements of the state space vector of the control system. Combined with the strong nonlinearity of the control system of the Monod type this feature of the control system leads to numerical instability or to unsatisfactory performance of the control algorithms.

The use of the classical methods of the linear control theory is embarrassed, mainly due to the fact that the noise in the system is not of Gaussian or colored type. The changes of the values of the structural parameters of the Monod kinetics models also lead to bad estimates when using Kalman filtering. This is the reason for search of other new control approaches and methods. In addition, the Monod kinetics models are characterized by another feature of the optimal control solutions. The dynamic optimization based on the Pontryagin maximum leads to singular optimal control problems.

This type of models has another serious flaw. Using classical linearization and control solutions via the feeding rate, the linear system is not observable (Wang, 1987). This is very serious flaw of this class of models, because it restricts the capabilities for identification of the system and hence the possibilities for larger freedom in the development of the algorithms for control. We have to mention that the possibilities of choice of the control input are restricted mainly to the feeding rate of the fed-batch process or to dilution rate for continuous process.

For some types of BTP it is possible to choose the control via the temperature and/or via changing the acidity in the bioreactor. What we mean is indirect influence on the biomass maximum growth rate, which, according to the contemporary researches, is one of the main factors, determining the quality of the cultivation process. Here the problem about the observability of the control system arises again, because this directly concerns the possibility for satisfactory identification of the state space vector of the control system and more specifically for the determination of the specific growth rate of the BTP. This group of questions is still a topical branch in the theory of the biotechnological control systems (Neeleman, 2002; Tzonkov, 2006).

Good control solution results are shown in control systems based on the solutions of variable structure systems. Here a different problem arises. Good control via sliding mode control is possible when the system is led to the initial position in a point from the area of the "equivalent control solution." This is a specific task for fed-batch biotechnological control via information about the growth rate of the biomass.

The above problems have led to development of extended dynamical models in which the dynamic of the changes of the growth rate of the BTP is described by separate equation on the general differential equation. Such extended observable model based on Monod kinetics is offered by the American scientist Wang (Monod-Wang model) (Wang, 1987).

The usage of the control via growth rate of the biomass is up to date branch in the scientific research in the area of the bioprocess control systems. In addition to such a growth rate control solution in this chapter mathematically accurate approach is substantiated, a unified approach which allows the usage of expert information. The expert can be modeled and the expert information can be described as a utility function and this function can be included as a criterion for control of the complex system technologist-biotechnological process.

2. UNSTRUCTURED BIOTECHNOLOGICAL MODELS: MONOD KINETIC MODELS, MONOD-WANG MODELS

We have seen that the kinetic of Monod is described by model (formula 9.1.1). The above problems and the non-observability of the Monod model has led to the development of the widened dynamical models, in which the dynamics of the specific growth rate of the BTP is described via separate equation in the system of differential equations. The dynamics of the growth rate μ in the Monod-Wang model is modeled as a first order lag process with rate constant m, in response to the deviation in the growth rate. This model called also the model of Monod-Wang determines a linear observable system in the classical linearization (Pavlov, 2007; Wang, 1987).

$$\begin{aligned} \dot{X} &= \mu X - \frac{F}{x_4} X \\ \dot{S} &= -k\mu X + \frac{F}{V}(S_0 - S) \\ \dot{\mu} &= m(\mu_m \frac{S}{K_S + S} - \mu) \\ \dot{V} &= F \end{aligned} \qquad (9.2.1)$$

The dynamical model concerns a fed-batch biotechnological process. Obviously, model (formula 9.1.1) is a singular form of model (formula 9.2.1). The comparison of both models is shown in Figures 1, 2, 3, 4, 5, and 6.

One description of the fed-batch process looks like (Pavlov, 2007; Tzonkov, 2006).

$$\begin{aligned} \dot{X} &= \mu X - \frac{F}{V} X, \\ \dot{S} &= -k\mu X + (So - S)\frac{F}{V}, \\ \dot{\mu} &= m(\mu_m \frac{S}{(K_S + S)} - \mu), \\ \dot{V} &= F, \\ \dot{E} &== k_2 \mu E - \frac{F}{V} E, \\ \dot{A} &== k_3 \mu X - \frac{F}{V} A. \end{aligned} \qquad (9.2.2)$$

Here X denotes the concentration of biomass, [g/l]; S – the concentration of substrate (glucose), [g/l]; V - bioreactor volume, [l]; F – substrate feed rate, [h^{-1}]; S_0 – substrate concentration in the feed, [g/l]; μ_{max} – maximum specific growth rate, [h^{-1}]; K_S – saturation constant, [g/l]; k and k_2 – yield coefficients, [g/g], m – rate coefficient [-]; E – the concentration of ethanol, [g/l]. The dynamics of μ in the Monod-Wang model is modeled as a first order lag process with rate constant m, in response to the deviation in μ. The fifth equation describes the production of ethanol (*E*). This equation could be omitted in the case when ethanol production is not possible, as is in the *E. coli* bioprocess. The last equation describes the production of acetate (*A*). This equation is dynamically equivalent to the first one after the implementation of a simple transformation ($X = (1/k_3) A$).

The best description is given by the so called model of Wang-Yerusalimsky:

$$\begin{aligned} \dot{X} &= \mu X - \frac{F}{V} X, \\ \dot{S} &= -k\mu X + (So - S)\frac{F}{V}, \\ \dot{\mu} &= m(\mu_m \frac{S}{(K_S + S)} \frac{k_E}{(k_E + X)} - \mu), \\ \dot{V} &= F, \\ \dot{E} &== k_2 \mu E - \frac{F}{V} E, \\ \dot{A} &== k_3 \mu X - \frac{F}{V} A. \end{aligned} \qquad (9.2.3)$$

Figure 1. Growth of the biomass (Monod model)

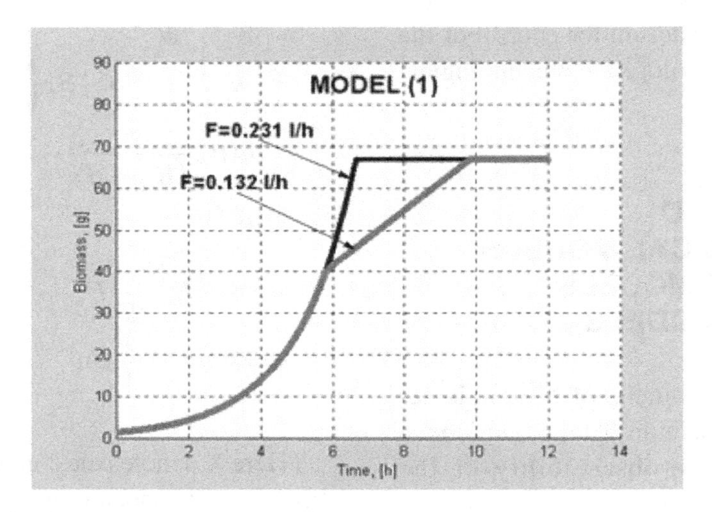

Figure 2. Growth of the biomass (Wang-Monod)

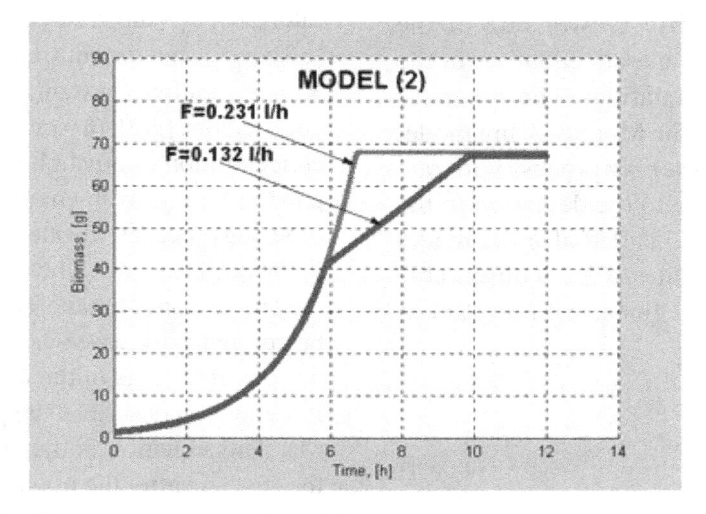

Figure 3. Specific growth rate using Monod model

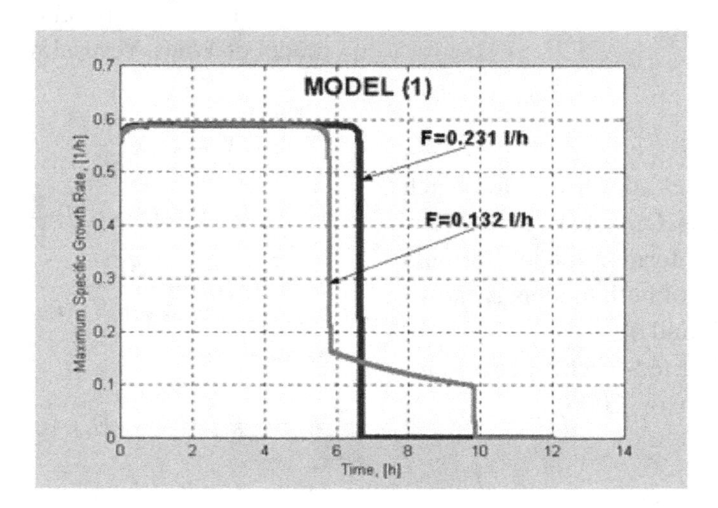

Figure 4. Specific growth rate using Wang-Monod model

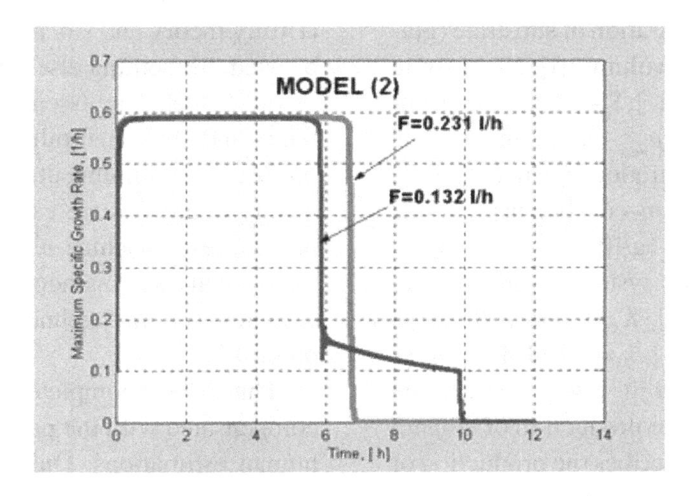

Figure 5. Feeding rate using Monod model

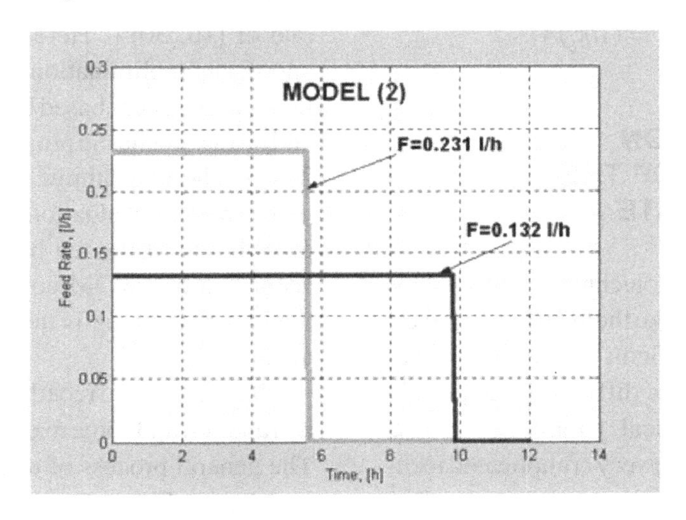

Figure 6. Feeding rate using Wang-Monod model

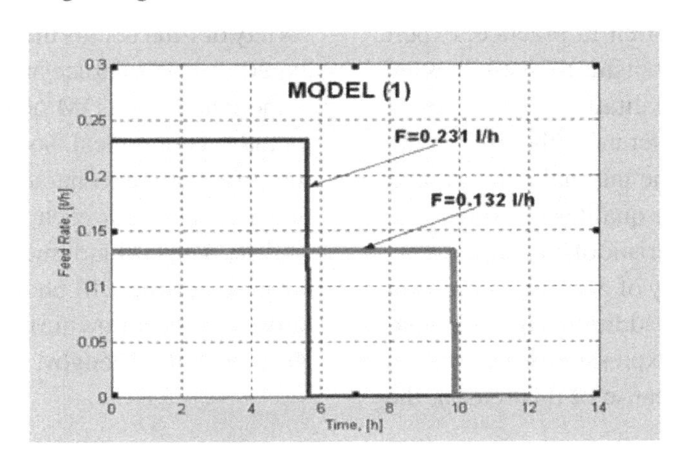

In the formula X is the concentration of biomass, [g/l]; S–the concentration of substrate (glucose), [g/l]; V-bioreactor volume, [l]; F–substrate feed rate (control input), [h^{-1}]; S_0–substrate concentration in the feed, [g/l]; μ_{max}-maximum specific growth rate, [h^{-1}]; K_S–saturation constant, [g/l]; k, k_2 and k_3–constants, [g/g]; m–coefficient [-]; E–the concentration of ethanol, [g/l]; A–the concentration of acetate [g/l]. The system parameters are as follows: μ_m=0.59 [h^{-1}], K_S=0.045 [g/l], m=3 [–], S_0=100 [g/l], k=2 [–], k_2=3.79 [–], k_3=1/71 [–], k_E=50 [–], F_{max}= 0.19 [h^{-1}], V_{max}=1.5 [ ]. The 5th equation describes the production of ethanol (E). The last equation describes the production of acetate (A). This equation is dynamically equivalent to the first one after the implementation of a simple transformation ($X = (1/k_3)A$).

3. UTILITY FUNCTION DETERMINATION OF THE BEST GROWTH RATE

The complexity of the biotechnological systems and their singularities make them difficult objects for control. They are difficult to control also because of the fact that it is difficult to determine their optimal technological parameters. These parameters can depend on very complicated technological, ecological, or economical market factors. Their taking into account in one mathematical model directly is impossible for the time being. Because of this reason often in practice, expert estimates are used. From outside the estimates are expressed only by the qualitative preferences of the technologist. The preferences themselves are in rank scale and bring the internal indetermination, the uncertainty of the qualitative expression, which is a general characteristic of human thinking. Separately the complexity of general economical or social factors brings an additional indetermination, even mistakes in the expression of the decision maker's preferences. Because of this reason, the mathematical models from the previous chapters (Utility theory and stochastic programming) can be used. Something else, these methods because of their stochastic essence eliminate the uncertainty and could neutralize the wrong answers if one uses the gambling utility evaluation approach when preferences are expressed combined with procedure for machine-learning. Thus, we achieve totally analytical mathematical description of the complex system "technologist-biotechnological process."

Thus, the incomplete information usually is compensated with the participation of imprecise human estimations. Our experience is that the human estimation of the process parameters of a cultivation process contains uncertainty at the rate of [10, 30]%. Here is used a mathematical approach for elimination of the uncertainty in the DM's preferences based both on the Utility theory and on the Stochastic programming (Pavlov, 2010, 2011). The algorithmic approach permits exact mathematical evaluation of the optimal specific growth rate of the fed-batch cultivation process according to the DM point of view even though the expert thinking is qualitative and pierced by uncertainty.

We are going to recall the definition of the utility function and some mathematical enunciations. The general process of using the utility function is shown in Figure 7.

Standard description of the utility function application is presented by Figure 7. There are a variety of final results that are consequence of the expert or DM's choice and activity. This activity is motivated by a DM objective, which possibly includes economical, social, ecological or other important process characteristics. A utility function $U(.)$ assesses each of this final results (μ_i, i=1÷n). The DM judgment of the process behavior based of the DM choice is measured quantitatively by the following formula (Fishburn, 1970; Keeney, 1993; Mengov, 2010).

Figure 7. Utility function application

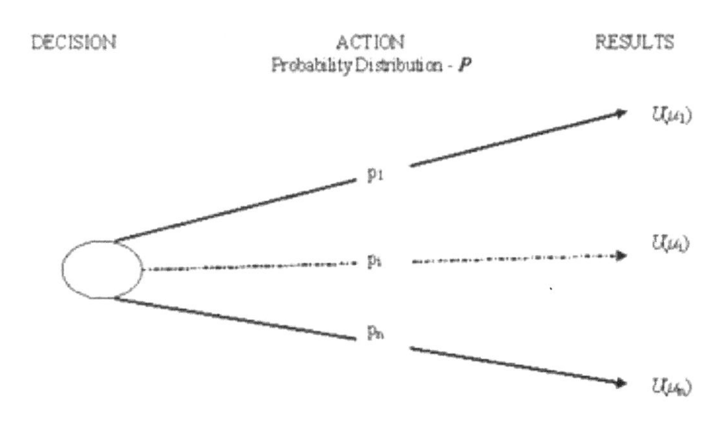

$$U(p) = \sum_i U(\mu_i) p_i, \qquad (9.3.1)$$

where $p = (p_1, p_2, .., p_i, ..p_n), \sum_i p_i = 1.$

We denote with p_i subjective or objective probabilities, which reflect the uncertainty of the final result. Let **Z** be the set of alternatives (**Z** ≡ {specific growth rates} ≡ [0, 0.6]) and **P** be a convex subset of discrete probability distributions over **Z**. The expert expresses his "preferences" over **P** ($\succ$) (**Z** ⊆ **P**). We know that the utility function is defined in the interval scale (in the proposed conditions) (Fishburn, 1970; Keeney, 1993). A decision support system for individual utility *U*(.) evaluation is used, Figure 8.

The utility function *U*(.) is approximated by a polynomial:

$$U(\mu) = \sum_{i=0}^{6} c_i \mu^i . \qquad (9.3.2)$$

The polynomial representation permits analytical determination of the derivative of the utility function and easy implementation in the optimization and control. We already know that we are looking for the approximate evaluation of the set $\boldsymbol{A}_{u*}$ of the positive preferences and to the set $\boldsymbol{B}_{u*}$ of the negative preferences:

$$\boldsymbol{A}_{u*} = \{(x,y,z,\alpha)/\ (\alpha u^*(x)+(1-\alpha)u^*(y))>u^*(z)\},$$

$$\boldsymbol{B}_{u*} = \{(x,y,z,\alpha)/\ (\alpha u^*(x)+(1-\alpha)u^*(y))<u^*(z)\}.$$

The star in the notations means an empirical estimate of the utility of the technologist. The utility function itself is built as a recurrent procedure for the recognition of the set A_{u*}, this procedure in its essence can be considered as a machine-learning computer training on the same preferences as these of the expert (Ayzerman, 1970; Pavlov, 2005, 2011). The disturbance in the transitivity in the relations ($\succ$) and (≈) are considered as a noise caused by the uncertainty in the expert answers according with the prescriptive decision making approach.

In the base of the chosen approach DM compares "lotteries" ($\alpha x+(1-\alpha)y$, $x,y\in$ **Z**, $\alpha\in$**[0,1]**) with simple alternatives $z\in$**Z** and the answer is determined from him ("better," "worse," or "indifference, equivalency, or impossibility for explicit delimitation") (Farquhar, 1984; Keeney, 1993). The Biotechnologist compares "lottery" <*x,y,α*> with z and for every point (x, y, z, α), for every comparison determines his answer: $f(x, y, z, \alpha)$=1 for ($\succ$), $f(x, y, z, \alpha)$=-1 for ($\prec$) and $f(x, y, z, \alpha)$=0 for (≈). The function f(x, y, z, α) is a probability function, subjective characteristic of the DM depicturing intuition and empirical knowledge and also including subjective and the probability

Figure 8. A decision support system for individual utility U(.) evaluation

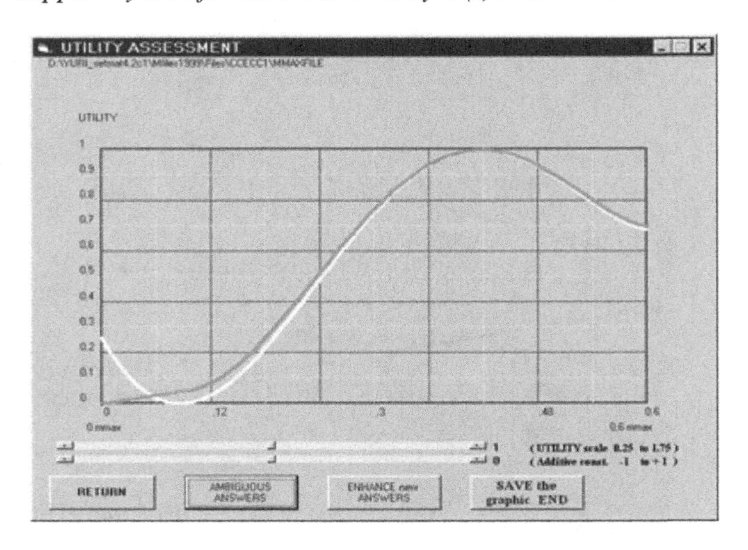

uncertainty of the answers. Using this approach the "training points" $(x, y, z, \alpha, f(x, y, z, \alpha))$ are formed from the DM (the Biotechnologist). In the machine learning recurrent procedure "the training point" $(x, y, z, \alpha, f(x, y, z, \alpha))$ is treated as point from the set A_u with probability $D_1(x, y, z, \alpha)$ or a point from B_u with probability $D_2(x, y, z, \alpha)$).

We suppose that (x, y, z, α) are given by the probability distribution $F(x,y,z,\alpha)$. In fact this is a pseudo random Lp_τ sequence (Sobol). Then the probabilities $D_1(x, y, z, \alpha)$ and $D_2(x, y, z, \alpha)$ are the conditional mathematical expectations of $f(.)$ over the sets $\boldsymbol{A}_u$ and $\boldsymbol{B}_u$, respectively. We denote by M the mathematical expectation, in this case the conditional mathematical expectation of $f(.)$ with fixed (x, y, z, α). With $D'(x, y, z, \alpha)$ we denote the conditional random value:

$$D'(x, y, z, \alpha) = \begin{cases} D_1(x, y, z, \alpha), & \text{when } M(f/x, y, z, \alpha)>0, \\ -D_2(x, y, z, \alpha), & \text{when } M(f/x, y, z, \alpha)<0, \\ 0, & \text{when } M(f/x, y, z, \alpha)=0. \end{cases}$$

The measurable function $D'(x, y, z, \alpha)$ is approximated by function of the type $G(x, y, z, \alpha)= (\alpha g(x)+(1-\alpha)g(y)-g(z))$. The function $G(x, y, z, \alpha)$ is positive on A_u and negative on B_u concordant with the degree of the approximation to $D'(x, y, z, \alpha)$. The function g(x) can be accepted as an approximation of the utility function $U(.)$ according to teorem 6.3.1 in chapter 6. The process of the recognition of the sets $\boldsymbol{A}_{u*}$ и $\boldsymbol{B}_{u*}$ is shown on Figure 9.

In fact total recognition is impossible, because $(\boldsymbol{A}_{u*} \cap \boldsymbol{B}_{u*} \neq \varnothing)$ of the wrong preferences of the technologist caused by the threshold of the indetermination in his preferences and the uncertainty of his preferences (Kahneman, 1979; Pavlov, 2011).

The next step is the determination of the polynomial approximation of the utility function based on the process of the recognition of $\boldsymbol{A}_{u*}$ и $\boldsymbol{B}_{u*}$. The evaluated expert utility function $U(.)$ is shown on Figure 10.

This utility function recognizes correctly more than 81% of the expert answers. It is possible that we use two approaches for approximation of the utility function. The first approach is direct stochastic approximation with a polynomial of lower degree. The second approach is a regression model based on broken (scissors) line of the recognition of the sets. The second approach is more effective and permits the utilization of powerful information systems as MATLAB (MathWorks Inc.).

Figure 9. Pattern recognition

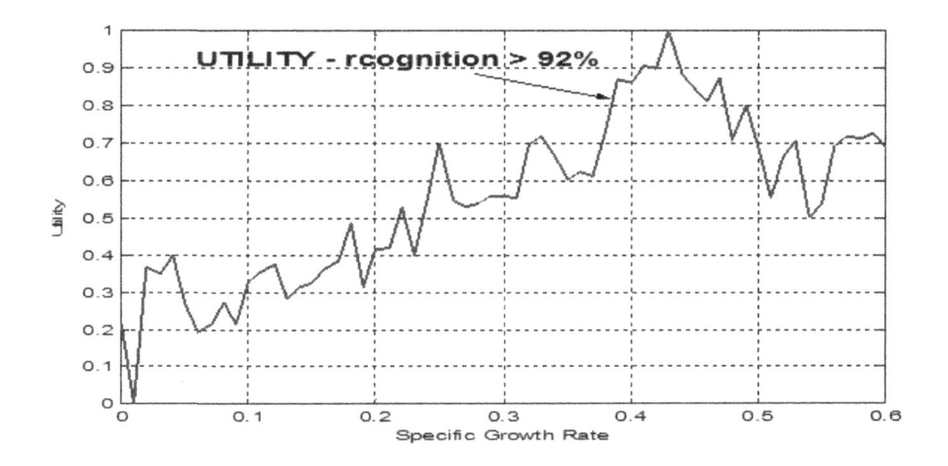

Figure 10. Expert utility versus growth rate

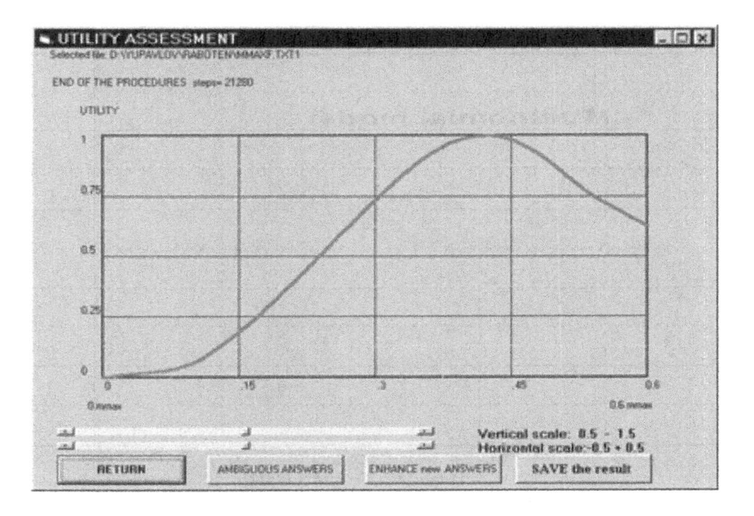

The two processes, the recognition of the sets A_{u*} и B_{u*} and the buildings of the utility function (regression or stochastic approximation) are shown on Figure 11 build one on top of the other.

It is possible that we look for modeling of the utility function with higher degree of the polynomial. For example, a multinomial model of degree 17 gave minimal residual sum of squares – 0.0044. This is shown on Figure 12.

The process of giving training questions and receiving DM answers in the frames of the gambling approach in the built information system for individual utility function evaluation is shown on Figure 13.

In the end, let us summarize. The utility function in this investigation is evaluated with 64 learning DM's answers, sufficient for a primary orientation in the problem. The proposed procedure and its modifications are machine learning (Ayzerman, 1970; Pavlov, 2005, 2011). The computer is taught to have the same preferences as DM. DM is comparatively quick in learning to

Figure 11. Utility evaluation of the specific growth rate

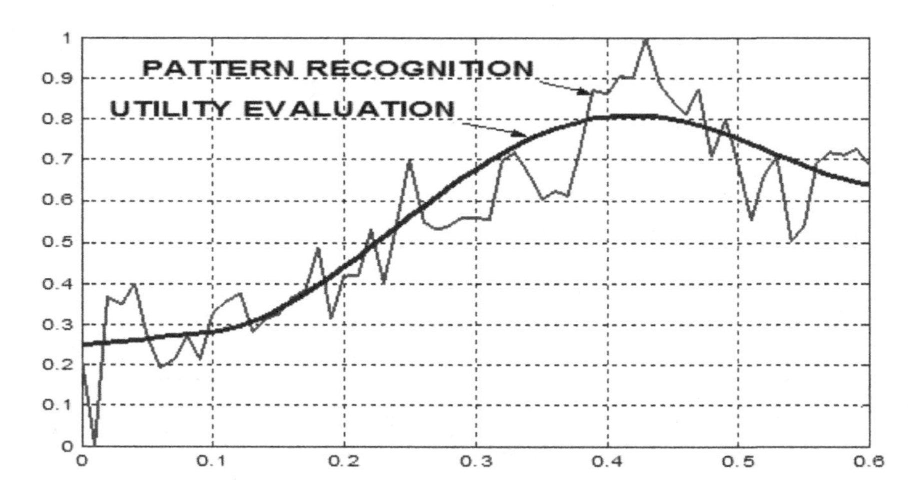

Figure 12. Model with minimal residual sum of squares

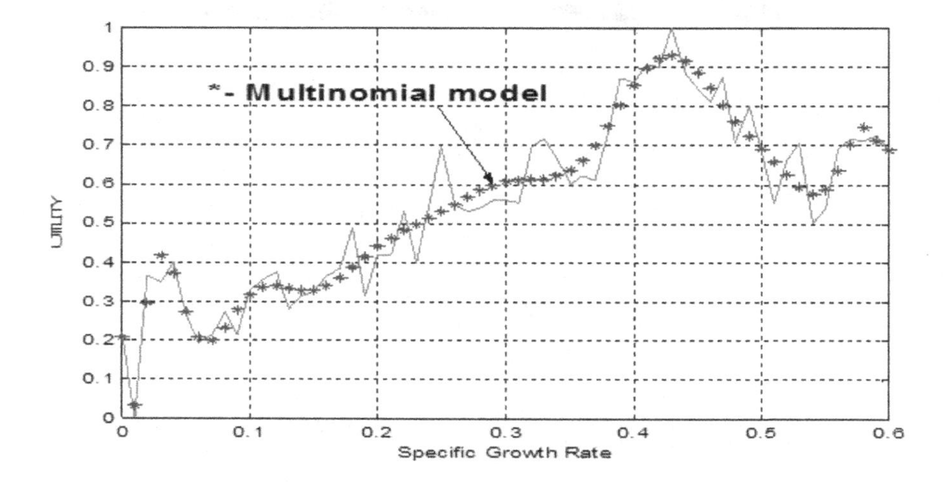

operate with the procedure. For example, a session with 128 questions (learning points and DM answers $(x,y,z,\alpha, f(x,y,z,\alpha))$ takes approximately no more than 45 minutes.

4. OPTIMAL CONTROL

We start with defining of optimal control with the help of the Brunovsky normal form of Monod-Wang model (Pavlov, 2007, 2008). After the analytical form of the utility function is known, we use it as an optimal control criterion. Both, the utility function and the differential equation of the process give to us mathematical description of the complex system "Biotechnologist – fed-batch process." We preserve the notation $U(.)$ for the DM utility. On the next step we are going to use the equation and the model of Mono-Wang-Yerusalimsky. The reason is that they have the same equivalent Brunovsky normal form. The control design of the fed-batch process is based on the next subsidiary optimal control problem:

Figure 13. Decision support system

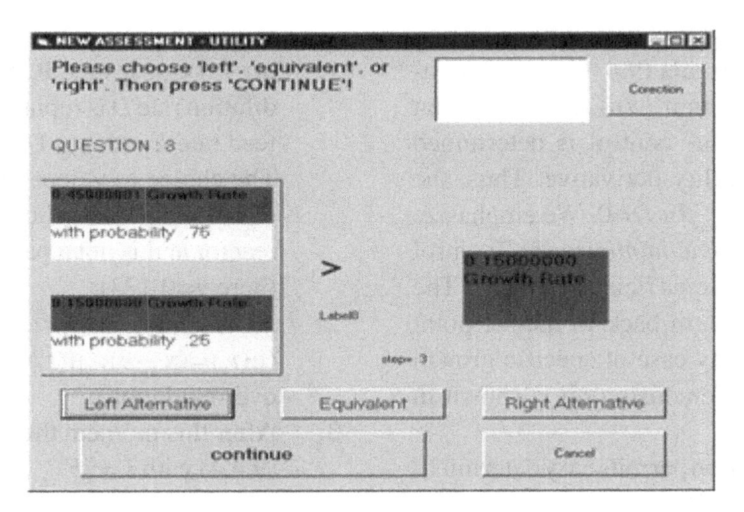

Max(U(μ(Tint))), where the variable μ is the specific growth rate, (μ∈[0, μmax], D∈[0, Dmax]). Here U(μ) is an aggregation objective function (the utility function – Figure 10) and D is the control input (the dilution rate):

$$\max(U(\mu)),\ \mu \in [0, \mu_{\max}],\ t \in [0, T_{\text{int}}],\ D \in [0, D_{\max}]$$

$$\dot{X} = \mu X - DX$$

$$\dot{S} = -k\mu X + (So - S)D$$

$$\dot{\mu} = m(\mu_m \frac{S}{(K_S + S)} - \mu) \qquad (9.4.1)$$

wh en T_{int} is sufficiently small, the optimization is in fact "time minimization." The differential equation in (9.4.1) describes *a continuous fermentation process*. The model permits exact linearization to the following Brunovsky normal form (Goursat, as regard to the differential forms) (Elkin, 1999; Pavlov, 2001):

$$\dot{Y}_1 = Y_2,$$

$$\dot{Y}_2 = Y_3, \qquad (9.4.2)$$

$$\dot{Y}_3 = W.$$

Here *W* denotes the control input of the model (9.4.2). The derivative of the function Y_3 determines the interconnection between *W*-model (9.4.2) and *D*-model (9.4.1). It can be seen that this model is linear and easily allows application of the Pontryagin maximum principle. The new state space vector (Y_1, Y_2, Y_3) is the following:

$$Y_1 = u_1,$$

$$Y_2 = u_3(u_1 - ku_1^2),$$

$$Y_3 = u_3^2(u_1 - 3ku_1^2 + 2k^2u_1^3) + m(\mu_m \frac{u_2}{(K_S + u_2)} - u_3)(u_1 - ku_1^2),$$

$$\begin{bmatrix} u_1(X,S,\mu) \\ u_2(X,S,\mu) \\ u_3(X,S,\mu) \end{bmatrix} = \begin{bmatrix} \frac{X}{S_o - S} \\ S \\ \mu \end{bmatrix}. \qquad (9.4.3)$$

The control design is a design based on the Brunovsky normal form and application of the Pontryagin's maximum principle step by step for sufficiently small time periods *T* (Aleksiev, 1979; Hsu, 1972). The optimal control law has the analytical form (Pavlov, 2007, 2010):

$$D_{opt} = sign\left(\left(\sum_{i=1}^{6} ic_i\mu^{(i-1)}\right)(T - t)\left[\frac{(T-t)\mu(1-2kY_1)}{2} - 1\right]\right)D_{\max},$$
$$where: sign(r) = 1, r > 0,\ sign(r) = 0, r \le 0. \qquad (9.4.4)$$

The time interval T can be of the order of the step of discretization of the differential equation solver. The sum in formula (9.4.4) is the derivative of the utility function $U(\mu)$. It is clear that the "time-minimization" control is determined by the ***sign*** of the utility derivative. Thus, the control input is $D=D_{max}$ or $D=0$. We emphasize that the solution is a *"time-minimization"* control (if the time period T is sufficiently small). The control brings the system back to the set point for minimal time in any case of specific growth rate deviations. The demonstration is shown in Pavlov (2007, 2008).

The previous solution permits easy determination of the control stabilization *of the fed-batch process*. The control law is based on the solution of the following optimization problem:

Max(U(μ(Tint))), where the variable μ is the specific growth rate, (μ∈[0, μmax], F∈[0, Fmax]). Here U(μ) is the utility function in Figure (10) and F is the control input (the substrate feed rate):

$$\begin{aligned} &\max(U(\mu(T_{\text{int}}))),\ \mu \in [0, \mu_{\max}],\ t \in [0, T_{\text{int}}],\ F \in [0, F_{\max}] \\ &\dot{X} = \mu X - \frac{F}{V}X \\ &\dot{S} = -k\mu X + (So - S)\frac{F}{V} \\ &\dot{\mu} = m(\mu_m \frac{S}{(K_S + S)} - \mu) \\ &\dot{V} = F \end{aligned} \tag{9.4.5}$$

The control law *of the fed-batch process* has the same form (Formula 9.4.4) because $D(t)$ is replaced with $F(t)/V(t)$ in the fed-batch model. Thus, the feeding rate $F(t)$ takes $F(t)=F_{max}$ or $F(t)=0$, depending on $D(t)$ which takes $D=D_{max}$ or $D=0$. *We conclude that the control (9.4.4) bring the system to the optimal point (optimal growth rate) with a" time minimization" control, starting from any deviation point of the specific growth rate (Figure 14).* Thus, we design the following control law (Pavlov, 2007, 2008):

1. At the interval $[0, t_1]$ the control is "time-minimization" control (9.4.4), where $\mu(t_1)=(\mathbf{x}_{30}-\varepsilon)$, $\varepsilon > 0$, $\mathbf{x}_{30}=max(U(\mu))$. The dilution rate D is replaced with the substrate feed rate $F=\gamma F_{max}$, $1\geq\gamma>0$, when $D=D_{max}$. The choice of γ depends on the step of the equation solver and of the type of the bioreactor and is not a part of the optimization (here γ=0.123);
2. At the interval $[t_1, t_2]$ the control is $F=0$ ($\mu(t_1)=(\mathbf{x}_{30}-\varepsilon)$, $\mu(t_2)=\mathbf{x}_{30}$ - to avoid an overregulation);
3. After this moment the control is the control (9.4.4) with $F=\gamma F_{max}$, when $D=D_{max}$ (chattering control with $1\geq\gamma>0$).

The deviation of the fed-batch process with this control is shown in Figures 14 and 15.

After the stabilization, the process can be maintained around the optimal parameters with control in sliding mode (Figure 14 and Figure 15). Possible solution in sliding mode is alternation of μ_m (as a function of the temperature and the acidity in the bioreactor) or alternation of F (Emelyanov, 1993).

An important moment in the leading of the control system in a point of the area of the optimal profile is the determination of moment t_1 and moment t_2. The solution described in the previous paragraph is a chattering control. Here we determine a smooth control solution for determination of the interval $[t_1, t_2]$ - determination of the moments t_1 and t_2. Here is supposed that $\mu_i(0)<\mu_e$, $S(0)<S_e$ where:

$$\begin{aligned} &x_{30} = \mu_e = \mu_m \frac{S_e}{K_s + S_e} \quad (\text{optimal pont}) \\ &\Rightarrow \quad S_e = \frac{K_s\mu_e}{\mu_m - \mu_e}, x_{30} = 0.31. \end{aligned} \tag{9.4.6}$$

The optimization problem is:

Figure 14. Stabilization of the fed-batch process

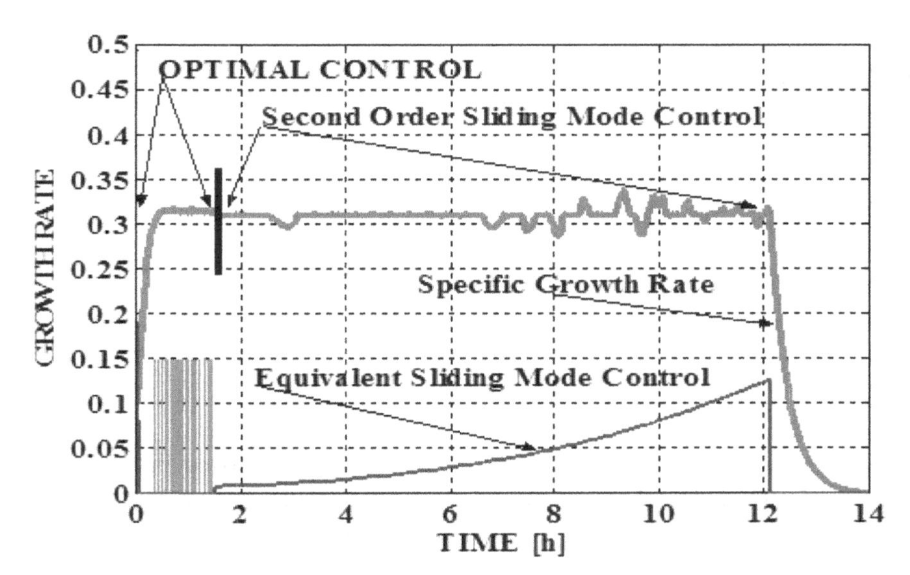

Figure 15. Substrate concentrations in the bioreactor

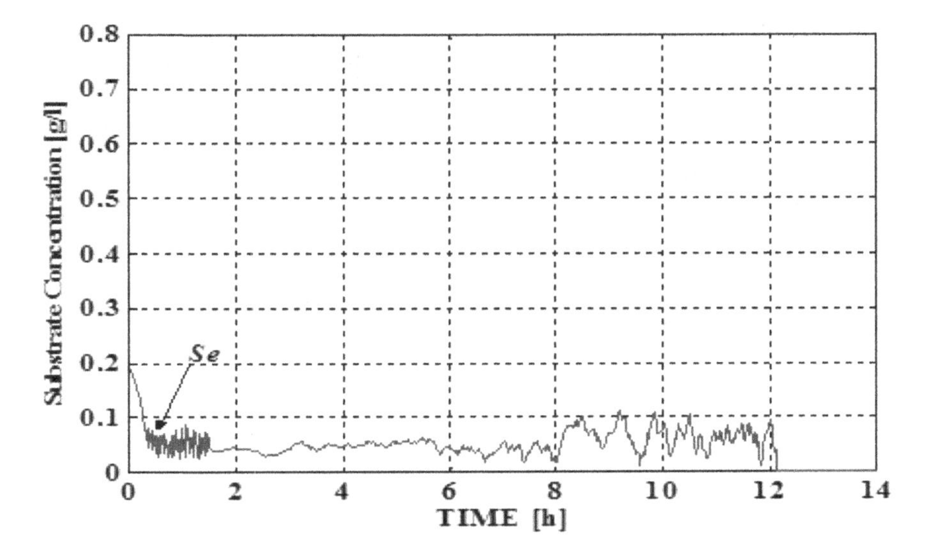

$$\min(\int_0^{t2}\left|\mu_e - \mu(t)\right|dt),\ \mu \in [0, \mu_{\max}],\ t \in [0, t_2],$$
$$F \in [0, F_{\max}], \mu_e = 0.31,$$
$$\dot{X} = \mu X - \frac{F}{V}X$$
$$\dot{S} = -k\mu X + (So - S)\frac{F}{V}$$
$$\dot{\mu} = m(\mu_m \frac{S}{(K_S + S)} - \mu)$$
$$\dot{V} = F \qquad (9.4.7)$$

A supplementary condition is "elimination of any overregulation" of μ. The optimal control law, according to the previous solution is:

1. At the time interval $[0, t_1]$ the control is $F(t)=\gamma F_{max}$. This is proved in Pavlov (2004), (γ=0.123, this parameter is not a part of the optimization);
2. At the interval $[t_1, t_2]$ the control is F=0 (to avoid any overregulation). The growth rate changes from μ_n to μ_e where $\mu(t_1)=\mu_n$, $(\mu(t_2)=\mu_e \wedge S(t_2)=S_e)$ and $\mu_n<\mu_e$; We effect this by determination of a manifold on the base of the problem (Formula 9.4.7) (see formula 9.4.9). When the state vector crosses over this manifold the control becomes $F(t_1)=0$. The moment t_2 is the moment of over crossing the manifold $(\mu=\mu_e)\cap(d\mu/dt=0)$;
3. After the moment t_2 the control is $F(t)=(kX(t)\mu(t)V(t)/(So-S_e)$, where $X(.)$ is the quantity of biomass in the bioreactor (if $F(t)>F_{\max}$ we pose $F=F_{\max}$).

The determination of moment t_1 and moment t_2 is based on the following optimal control problem:

$$\min(\int_{t_1}^{t_2}(\mu_e - \mu(t))dt),\ \mu \in [0, \mu_{\max}],\ t \in [t_1, t_2],$$
$$F \in [0, F_{\max}], \mu_e = 0.31,$$
$$\dot{X} = \mu X$$
$$\dot{S} = -k\mu X$$
$$\dot{\mu} = m(\mu_m \frac{S}{(K_S + S)} - \mu)$$
$$\mu(t_2) = \mu_e, \quad \dot{\mu}(t_2) = 0. \qquad (9.4.8)$$

Here is supposed that $\mu(0)<\mu_e$. We propose the next numerical solution for determination of an approximation of moment t_1:

- If $S(0)>S_n$ t_1 is equal to zero (t_1=0). In this case the initial conditions are equivalent to these of model (9.2.2). An overregulation is possible, because the process is uncontrollable from moment 0 to moment t_2;
- If $S(0)<S_n$ and $\mu(0)<\mu_e$; the moment t_1 is the moment at which the state vector of the differential Equation (9.4.8) crosses over the following manifold:

$$\mathit{Manifold} \quad (X, S, \mu, \mu_e, Ks) = -\mu_e$$
$$+\exp(-f(t_1))\mu(t_1) +$$
$$+\mu_m m\left[f(t_1) + \frac{Ksf(t_1)}{(S_e - S(t_1))}\ln\left|\frac{Ks + S(t_1)}{(Ks + S_e)}\right|\right]$$
$$-\mu_m m^2\left[\frac{mf(t_1) - 1}{m^2} + \frac{\exp(-f(t_1))}{m^2}\right] = 0,$$

$$\mathit{where}\ f(t_1) = \left[\frac{\left[\frac{(\mu_e - \mu(t_1))}{m} + \ln(\frac{X(t_1) + S(t_1) - S_e}{X(t_1)})\right](S_e - S(t_1))}{\mu_m\left[S_e - S(t_1) + Ks\ln\left|\frac{Ks + S(t_1)}{(Ks + S_e)}\right|\right]}\right]. \qquad (9.4.9)$$

The moment t_2 is determined approximately as follows $t_2=t_1+f(t_1)$. The solution is shown on Figure 16.

The optimal profile of the fed-batch cultivation, during the whole time period is shown on Figure 17. This control law determines the same optimal profile as solution as the chattering controls (9.4.4).

The manifold (9.4.9) is determined based on the optimization problems (Formula 9.4.7 and 9.4.8) with some simplifications and approximations of the integral. We underline that this is a numeric solution. The moment t_2 is determined approximately. If the process do not reaches the value μ_e at the calculated time t_2 (because this is a numerical solution) the step is repeated iteratively.

The next step is determination of the Brunovsky normal form of Wang-Yerusalimsky model and time minimization optimal control. In the beginning we shall apply the mathematical approach and diffeomorfic transformations to a Wang-*Yerusalimsky* kinetic model which describes a continuous cultivation process. After that we shall show that the time minimization optimal control of the fed-batch process has the same form as that of the continuous process. The *continuous process* is described dynamically by the following model (Pavlov, 2008):

$$\begin{aligned}
\dot{X} &= \mu X - DX, \\
\dot{S} &= -k\mu X + (So - S)D, \\
\dot{\mu} &= m(\mu_m \frac{S}{(K_S + S)} \frac{k_E}{(k_E + X)} - \mu), \\
\dot{E} &== k_2 \mu E - DE.
\end{aligned} \tag{9.4.10}$$

Here above D denotes the dilution rate. We apply the following transformation to Wang-Yerusalimski model:

$$\begin{aligned}
u_1 &= \frac{X}{(So - S)}, \\
u_2 &= S, \\
u_3 &= \mu, \\
u_4 &= \frac{E}{(So - S)}.
\end{aligned} \tag{9.4.11}$$

The dynamical model of the continuous process obtains the next equivalent form:

Figure 16. Optimal profile with the manifold

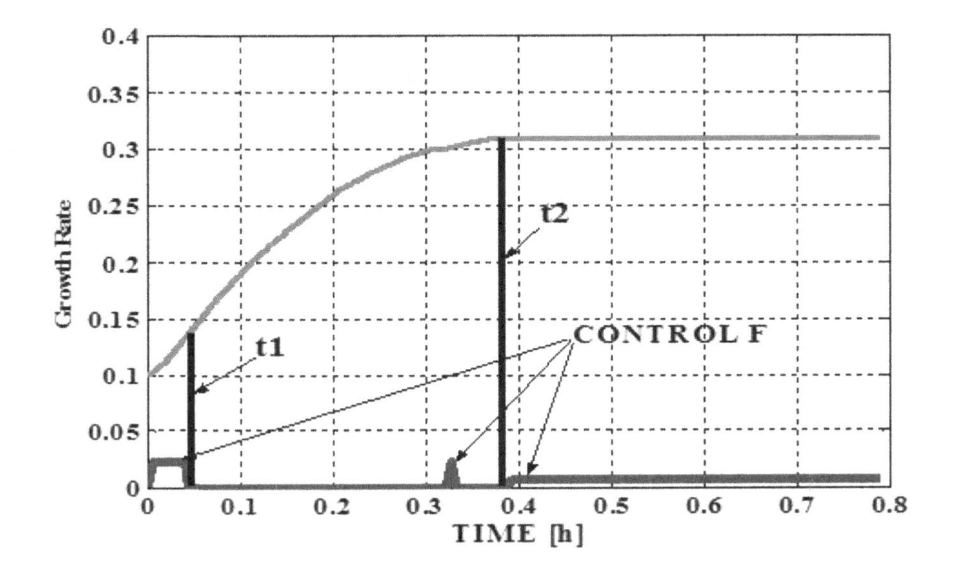

Figure 17. Optimal profile of the growth rate

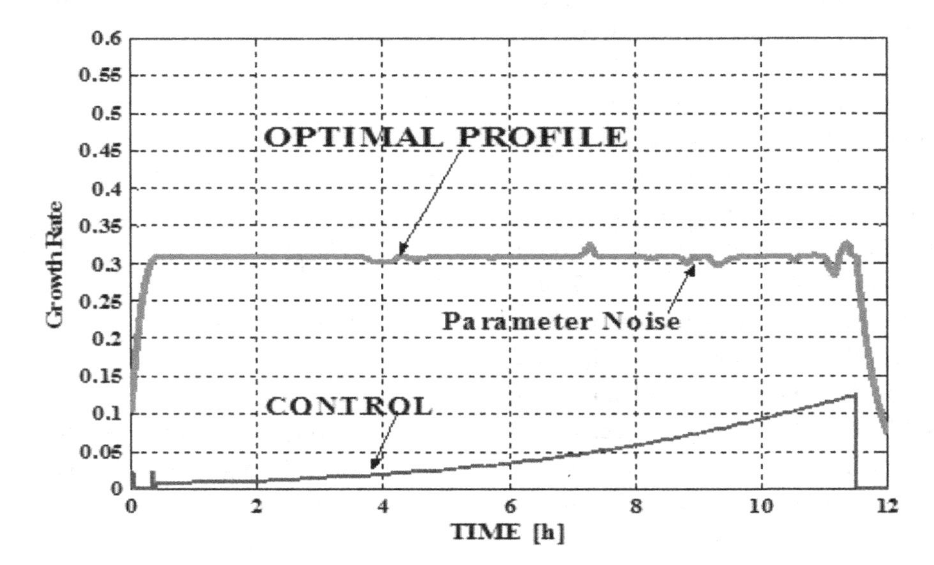

$$\dot{u}_1 = u_3u_1 - ku_1^2 u_3,$$
$$\dot{u}_2 = -ku_1(So - u_2)u_3 + D(So - u_2),$$
$$\dot{u}_3 = m(\mu_m \frac{u_2}{(K_S + u_2)} \frac{k_E}{(k_E - u_1(So - u_2))} - u_3),$$
$$\dot{u}_4 = u_4u_3(k_2 - ku_1). \tag{9.4.12}$$

The next step is application of the GS algorithm for exact linearization to Brunovsky normal form, published by Gardner and Shadvwick in 1992 (Elkin, 1999; Pavlov, 2001). The new equivalent model of model (9.4.10) and model (9.4.12) has the form:

$$\dot{Y}_1 = Y_2$$
$$\dot{Y}_2 = Y_3$$
$$\dot{Y}_3 = W \tag{9.4.13}$$
$$\dot{Y}_4 = Y_4Y_2\frac{(k_2 - kY_1)}{Y_1(1 - kY_1)}.$$

The state vector of model (9.18) has the following explicit extended form:

$$Y_1 = u_1$$
$$Y_2 = u_3(u_1 - ku_1^2)$$
$$Y_3 = u_3^2(u_1 - 3ku_1^2 + 2k^2u_1^3)$$
$$+m(u_1 - ku_1^2)(\mu_m \frac{u_2k_E}{(K_S + u_2)(k_E + u_1(So - u_2))} - u_3)$$
$$Y_4 = u_4. \tag{9.4.14}$$

The control input of the model (9.4.13) is W and it has the huge analytical form in Box 1.

The control input D of the continuous *Wang-Yerusalimsky* model takes part in the last mathematical expression of formula (9.4.15) and has the form shown in Box 2.

The last equation of *Wang-Yerusalimsky* model (9.4.13) can be solved by separation of the variables:

$$\dot{Y}_4 = Y_4Y_2\frac{(k_2 - kY_1)}{Y_1(1 - kY_1)} = Y_4\dot{Y}_1\frac{(k_2 - kY_1)}{Y_1(1 - kY_1)}, \tag{9.4.17}$$

Box 1.

$$W = 2u_3(u_1 - 3ku_1^2 + 2k^2u_1^3)m(\mu_m \frac{u_2}{(K_S + u_2)} \frac{k_E}{(k_E + u_1(So - u_2))} - u_3) + \\ + u_3^3(1 - 6ku_1 + 6k^2u_1^2)(u_1 - ku_1^2) + \\ + u_3 m(1 - 2ku_1)(u_1 - ku_1^2)(\mu_m \frac{u_2}{(K_S + u_2)} \frac{k_E}{(k_E + u_1(So - u_2))} - u_3) - \\ - u_3 m(u_1 - ku_1^2)^2 \mu_m \frac{u_2}{(K_S + u_2)^2} \frac{k_E}{(k_E + u_1(So - u_2))^2}(So - u_2) - \\ - m^2(u_1 - ku_1^2)(\mu_m \frac{u_2}{(K_S + u_2)} \frac{k_E}{(k_E + u_1(So - u_2))} - u_3) + \\ + m(u_1 - ku_1^2) \times \\ \times \frac{\mu_m k_E(K_S + u_2)(k_E + u_1(So - u_2)) - \mu_m u_2 k_E(k_E + u_1 So - K_S u_1 - 2u_1 u_2)}{(K_S + u_2)^2(k_E + u_1(So - u_2))^2} \times \\ \times[-ku_3u_1(So - u_2) + (So - u_2)D] \tag{9.4.15}$$

Box 2.

$$f_{input}(u_1, u_2, u_3, u_4, D) = \\ = m(u_1 - ku_1^2) \times \\ \times \frac{\mu_m k_E((K_S + u_2)(k_E + u_1(So - u_2)) - u_2(k_E + u_1 So - K_S u_1 - 2u_1u_2))}{(K_S + u_2)^2(k_E + u_1(So - u_2))^2} \times \\ \times(So - u_2)D = \\ = m(u_1 - ku_1^2)\frac{\mu_m k_E(K_S k_E + K_S u_1 So + u_1 u_2^2)}{(K_S + u_2)^2(k_E + u_1(So - u_2))^2}(So - u_2)D \tag{9.4.16}$$

$$\frac{\dot{Y}_4}{Y_4} = \dot{Y}_1 \frac{(k_2 - kY_1)}{Y_1(1 - kY_1)} \tag{9.4.18}$$

Consecutively the variable Y_4 depends only on Y_1 and can be described analytically by Y_1. The solution of equation (9.4.18) is:

$$Y_4 = k_4 Y_1^{k_2} \left|1 - kY_1\right|^{(1-k_2)}, \quad k_4 \in R. \tag{9.4.19}$$

That is why *Wang-Yerusalimsky* models (Formula 9.4.10) and (Formula 9.4.13) are equivalent dynamically to the following Brunovsky normal form (Elkin, 1999; Pavlov, 2008):

$$\dot{Y}_1 = Y_2 \\ \dot{Y}_2 = Y_3 \\ \dot{Y}_3 = W \tag{9.4.20}$$

It is clear that if k_2=0 or k_2=1 (formula 9.4.19) the variables X and E in model (Formula 9.4.13) are equivalent with precision up to an affine transformation and the model is over-regulated.

We continue the investigation with a mathematical technique described for Monod-Wang model. At least the *Wang-Yerusalimsky* model has an essential new restriction. We note with μ_e the growth rate and with X_e the biomass concentration in steady state in Wang-Monod kinetic model. The moment t_1 is determined when the state vectors of the Monod model across the manifold described by formula (9.4.9). The moment t_2 is the moment of intersection of another manifold $(\mu=\mu_e)\cap(d\mu/dt=0)$. The solution needs determination of the substrate concentration S_e in steady states (the working point of the process): The substrate concentration S_e of the Monod model is determined by the formula:

$$\mu_e = \mu_m \frac{S_e}{K_s + S_e} \quad (\text{optimal set point}) \Rightarrow \quad S_e = \frac{K_s \mu_e}{\mu_m - \mu_e}. \tag{9.4.21}$$

When *Wang-Yerusalimsky* kinetic model is used the situation is different. The substrate concentration S_e depends both from the growth rate μ_e and the biomass concentration E_x.

$$\mu_e = \mu_m \frac{S_e}{(Ks + S_e)} \frac{k_E}{(k_E + X_e)}. \tag{9.4.22}$$

The growth rate μ_e and the biomass concentration X_e are dependent now on the moment of interception with the manifold (9.4.9). Possible way out of this situation is replacement of the biomass concentration X_e with $X(t_1)$ and calculation of the manifold in any step of the equation solver.

$$S_e \approx \frac{Ks\mu_e(k_E + X(t_1))}{(k_E \mu_m - \mu_e(k_E + X(t_1))}. \tag{9.4.23}$$

In all cases, this particularity will provoke augmentation of the calculation. In this chapter, a control design for optimal control and stabilization of the specific growth rate is presented. The usage of the control via growth rate of the biomass is up to date branch in the scientific research in the area of the bioprocess control systems. The control design is based both on the Brunovsky normal form of the Wang-Yerusalimsky kinetic model and on a chattering optimal control design. The Wang-Monod kinetic model is a restricted form of this model (k_E=0). That is why the Wang-Yerusalimsky kinetic model could be accepted as a general model in different functional state regimes for some investigations.

The ethanol concentration and the acetate concentration are determined analytically as functions of the biomass concentration, substrate concentration and the specific growth rate, regarding Wang-Yerusalimsky model. The optimal profile and the control law for optimal control and stabilization of the specific growth rate of Wang-Yerusalimsky kinetic model remain the same as this of Wang-Monod kinetic model.

When the system is led to the initial position in a point from the optimal profile the stabilization of the process at this stage can be done using other suitable methods as sliding regimes.

By now we have reached a full mathematical description of the complex system "Biotechnologist-fed-batch process." We have overcome the restrictions connected with the observability of the Monod kinetics; we have overcome the obstacles with the singularities of the optimal control via finding and using of the Brunovsky normal form of the differential equation. The system was led to the working point from the "equivalent control in sliding mode" smoothly and stabilized in the optimal specific growth rate position. The solution and the determination of the optimal profile was done via synchronized usage of several mathematical approaches for modeling, reduction of nonlinear system, application of the Pontryagin maximum principle taking into account of the singularities of the optimal control as well as the dependence of the input control space on the initial conditions and the time.

5. ITERATIVE CONTROL DESIGN

The analytic construction of the utility function is an iterative machine-learning process. This interactivity allows a new strategy in the process of control design and in the control of the system with human participation in the final solution. The process of the iteration itself can unfold in several directions.

The first is increasing the number of training points. Using this approach continually, the representation of the Biotechnologist's opinion for the character of the optimal process could be approximated more accurately. By continually increasing the number of training points, details of his thinking can spring up and new details could be found. This process of iteration is natural for the stochastic recurrence procedure and it can be easily achieved with the pseudo random Lp_τ sequences (Sobol, 1979). It is also natural for the "machine-learning" process itself (Ayzerman, 1970).

Another possibility is hidden in the character of the pseudo random sequences. Using these sequences anytime, we can go back and return to previous solution if it is not satisfactory or a mistake in the process of evaluation of the utility function has been made. This process can be used in both directions-computer learning with machine learning procedures and learning of the tutor. The process of the correction itself can be accelerated by using the built until this moment utility approximations.

A third possibility is some combination of the previous two approaches, which allows flexible iterative mutual learning process of the control solution construction. At any moment here mathematical computer processing and use of the powerful analytical approaches is feasible. In fact, it can be said that mathematics is practically included in the process of forming of the control of the biotechnological process based on expert preferences.

We are going to demonstrate what is said above by starting with the iterative improvement of the utility function approximation via increasing the number of the learning points. The process of learning with 64 "learning answers" is shown on Figure 18. Here the best learning point is defined about the value of 0.31.

As we mentioned before starting from the process of recognition, approximation can be achieved using two approaches: regression or stochastic approximation. Each of the two approaches has its advantages. On Figure 19, regression approach has been demonstrated.

This approach can be very useful when investigating the uncertainty or mistakes in the answers of the Biotechnologist. In this approach specific for the expert or the decision maker solution statistical estimates can be determined. Such estimates can be even specific subjective characteristic of the personality of the expert or for his attitude or his confidentiality toward the considered process. In this regard, the regression model is preferable compare to the approximation via stochastic programming. Yet stochastic programming is always first in the training process, at least in the process of "pattern recognition" (Ayzerman, 1970).

If we approach the stochastic approximation with polynomials of lower degree, we can achieve stochastic utility function approximation. This approach is shown on Figure 20, compared with the process of "pattern recognition."

Now we increase the number of the "training points" from 64 to 128. This approach via inter-segment is obligatory for the pseudo random Lp_τ sequences. We always choose powers of 2. The reason is in the Sobol's theory of the pseudo random sequence.

The first effect we achieve is detailisation of the analytic representation of the expert's opinion about the process. This is shown on Figure 21.

We observe the process via using the system information for individual utility evaluations (Keeney, 1988). In the information system itself there are options which allow one to check the contradictions in the expert DM's answers. The

Figure 18. Pattern recognition with 64 "learning points" (maximum 0.31)

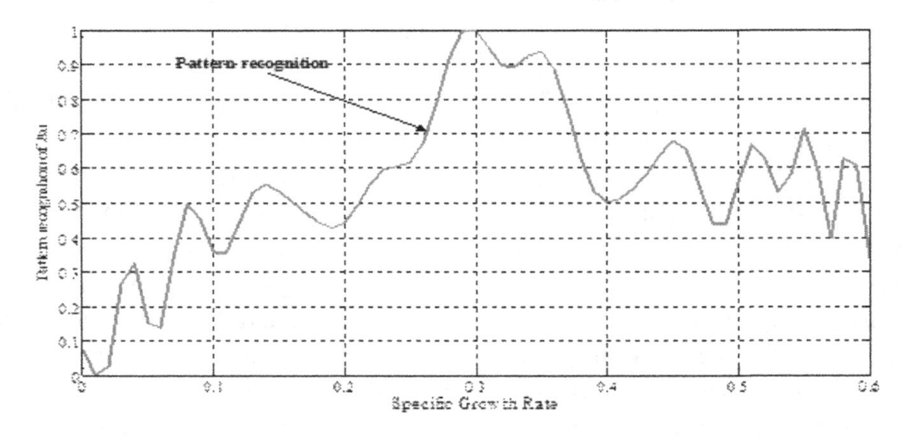

Figure 19. Utility regression approximation with 64 "learning points" (maximum 0.31)

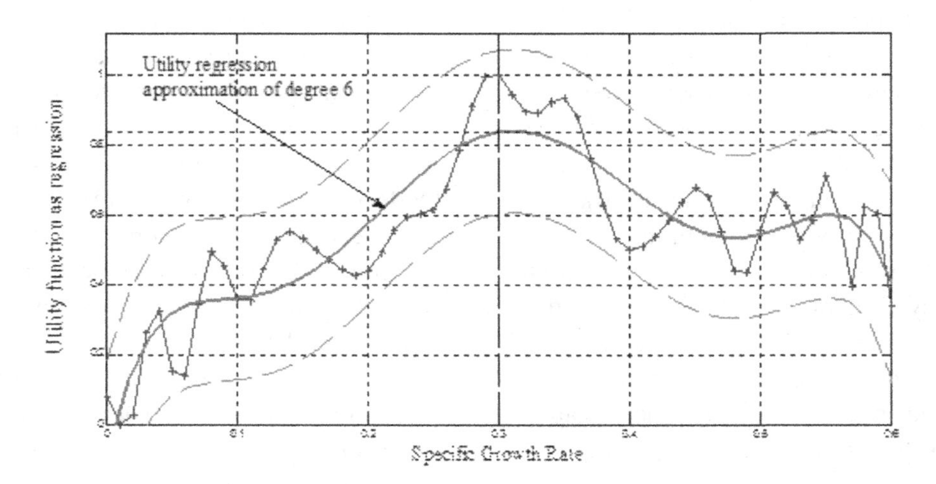

Figure 20. Stochastic approximation with 64 "learning points" (maximum 0.31)

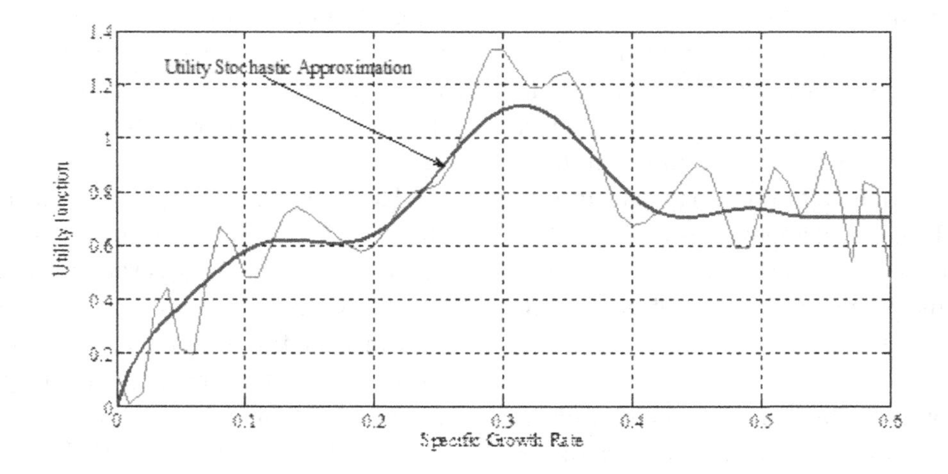

Figure 21. Pattern recognition with 128 "learning points" (maximum 0.31)

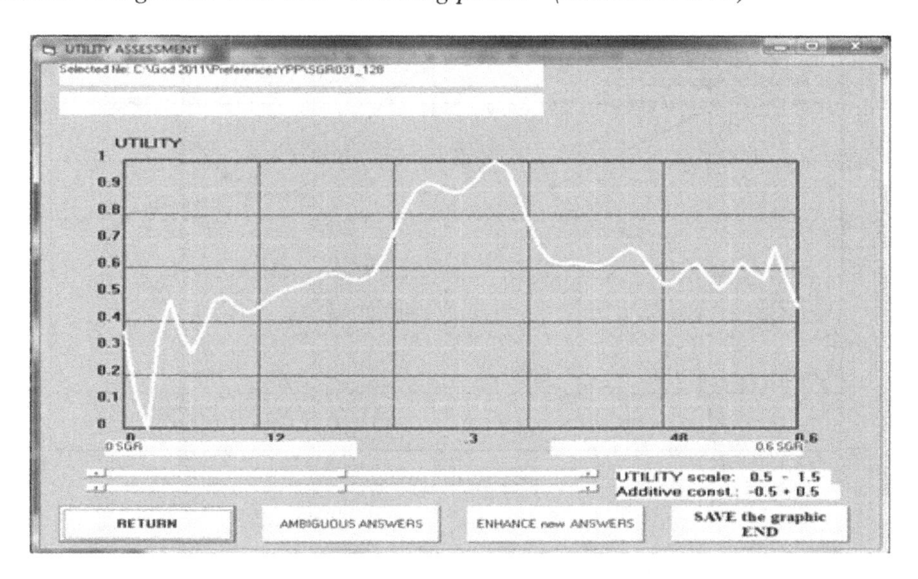

contradictions are the wrong preferences of the technologist caused by the threshold of the indetermination in his preferences and the uncertainty of his preferences ($\boldsymbol{A}_{u*} \cap \boldsymbol{B}_{u*} \neq \varnothing$).

Now we increase the number of the training points. On the next Figure 22 stochastic approximation of the utility expert function, using the new 128 training points is shown.

There are changes about the maximum of the recognition broken curve; also, there are changes at the ends of the interval. In fact, the utility function starts to develop without the necessity of additional processing.

We increase again the number of the training points to 512. The result is shown on Figure 23. In the figure it is seen that the pattern recognition process decreases the noise caused by the wrong answers, the mistakes are eliminated.

In this chapter, we demonstrated the flexibility and the diversity of the potential functions method and its conjunction with the utility theory when we completely analytically described the complex system "expert, biotechnologist-dynamical process." The first and the most important effect is the possibility for the analytical description of such complex systems. This has been achieved for the first time in scientific practice. The second effect is the introduction of the iterativity in the process of forming of the control as we use naturally and harmonically computer and analytical mathematical techniques. The third effect is the fact that the process of training can be reversed towards the trainer technologist expert with the aim of additional analysis and corrections.

At the end, we are going to summarize. By now, we have reached the mathematical description of the complex process "Biotechnologist-fed-batch process." Utility analytical mathematical descriptions have been built concerning the attitude of the technologist toward the dynamical process. Using this approach factors as ecology, financial perspective, social effect can be taken into account. They are included in the expert preferences via the expert attitude towards them.

We have overcome the restrictions connected with the observability of the Mono kinetics; we have overcome the obstacles connected with the singularities of the optimal control via finding and using the Brunovsky normal form of the differential equations. The system was derived smoothly to the working point in the equivalent control regime of sliding mode and stabilized there.

Figure 22. Pattern recognition and stochastic approximation with 128 "learning points" (maximum 0.31)

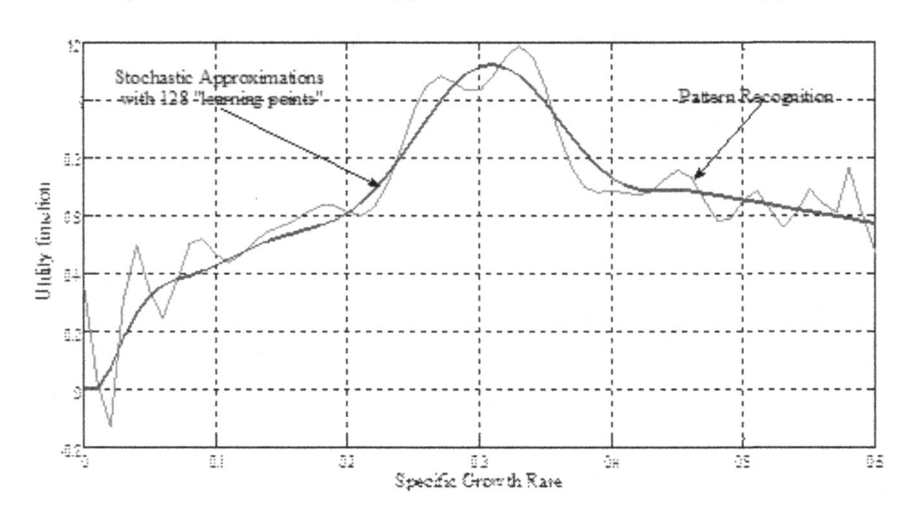

Figure 23. Pattern recognition and stochastic approximation with 512 "learning points" (maximum 0.31)

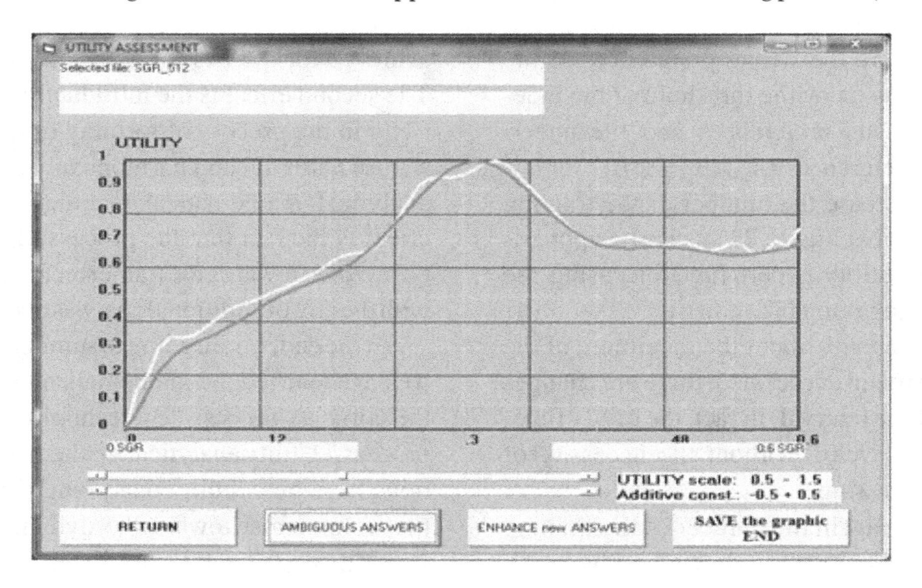

It can be seen that solution for complex processes is attained in practice by the synchronous use of several mathematical approaches, which are combined in synergy to achieve the main objective of the research. In general, this is characteristic of every practical task. By definition value driven design is a systems engineering strategy, which enables multidisciplinary design optimization by providing designers with an objective function. The preferences based description of the objective function input all the important attributes of the system being designed, and outputs a score. It this chapter the Value driven design was demonstrated in details, from the modeling to the optimal control. At the whole system level, the objective function which performs this assessment of value is called a value model and here this is the complex model "biotechnologist-fed-batch process."

REFERENCES

Aizerman, M., Braverman, E., & Rozonoer, L. (1970). *Potential function method in the theory of machine learning*. Moscow, Russia: Nauka.

Aleksiev, V., Tihomirov, V., & Fomin, S. (1979). *Optimal control*. Moscow, Russia: Nauka.

Cohen, M., & Jaffray, J.-Y. (1988). Certainty effect versus probability distortion: An experimental analysis of decision making under risk. *Journal of Experimental Psychology. Human Perception and Performance*, *14*(4), 554–560. doi:10.1037/0096-1523.14.4.554

Elkin, V. (1999). *Reduction of non-linear control systems: A differential geometric approach*. Dordrecht, The Netherlands: Kluwer.

Emelyanov, S., Korovin, S., & Levant, A. (1993). Higher-order sliding modes in control systems. *Differential Equations*, *29*(11), 1627–1647.

Farquhar, P. (1984). Utility assessment methods. *Management Science*, *30*, 1283–1300. doi:10.1287/mnsc.30.11.1283

Fishburn, P. (1970). *Utility theory for decision-making*. New York, NY: Wiley.

Hsu, J., & Meyer, A. (1972). *Modern control principles and applications*. New York, NY: McGraw-Hill.

Kahneman, D., & Tversky, A. (1979). Prospect theory: An analysis of decision under risk. *Econometrica*, *47*, 263–291. doi:10.2307/1914185

Keeney, R. (1988). Value-driven expert systems for decision support. *Decision Support Systems*, *4*(4), 405–412. doi:10.1016/0167-9236(88)90003-6

Keeney, R., & Raiffa, H. (1993). *Decision with multiple objectives: Preferences and value trade-offs* (2nd ed.). Cambridge, UK: Cambridge University Press. doi:10.1109/TSMC.1979.4310245

Krotov, V., & Gurman, V. (1973). *Methods and problems in the optimal control*. Moscow, Russia: Nauka.

Mengov, G. (2010). *Decision making under risk and uncertainty*. Sofia, Bulgaria: Publishing house JANET 45.

Neeleman, R. (2002). *Biomass performance: Monitoring and control in biopharmaceutical production*. (Thesis). Wageningen University. Wageningen, The Netherlands. Retrieved from http://edepot.wur.nl/121354

Pavlov, Y. (2001). Exact linearization of a non linear biotechnological model/Brunovsky model. *Proceedings of Bulgarian Academy of Sciences*, *10*, 25–30. Retrieved from http://adsabs.harvard.edu/abs/2001CRABS.54j.25P

Pavlov, Y. (2005). Subjective preferences, values and decisions: Stochastic approximation approach. *Proceedings of the Bulgarian Academy of Sciences*, *58*(4), 367–372.

Pavlov, Y. (2007). Brunovsky normal form of monod kinetics and growth rate control of a fed-batch cultivation process. *Online Journal Bioautomation*, *8*, 13–26.

Pavlov, Y. (2008). Equivalent form of Wang -Yerusalimsky kinetic model and optimal growth rate control of fed-batch cultivation processes. *Bioautomation*, *11*, 1–13.

Pavlov, Y. (2010). Preferences, utility function and control design of complex process. *Proceedings in Manufacturing Systems*, *5*(4), 225–231.

Pavlov, Y. (2011). Preferences based stochastic value and utility function evaluation. In *Proceedings of the International Conference InSITE 2011*. Novi Sad, Serbia: InSITE. Retrieved from http://proceedings.informingscience.org/InSITE2011/index.htm

Pavlov, Y., & Ljakova, K. (2004). Equivalent models and exact linearization by the optimal control of monod kinetics models. *Bioautomation, 1*, 42–56.

Raiffa, H. (1968). *Decision analysis*. Reading, MA: Addison-Wesley.

Sobol, I. (1979). On the systematic search in a hypercube. *SIAM Journal on Numerical Analysis, 16*, 790–793. doi:10.1137/0716058

Tzonkov, S., & Hitzmann, B. (2006). *Functional state approach to fermentation process modeling*. Sofia, Bulgaria: Prof. Marin Drinov Academic Publishing House.

Wang, T., Moore, C., & Birdwell, D. (1987). Application of a robust multivariable controller to non-linear bacterial growth systems. In *Proceedings of the 10th IFAC Congress on Automatic Control*, (vol. 39). Munich, Germany: IFAC. Retrieved from http://www.ifac-control.org/events/congresses

Chapter 10
Personalized E-Learning Systems

ABSTRACT

The chapter considers institutional learning in a framework that consists of four components—education system, teaching/learning system, learning environment, and learning theory. The teaching/learning system is the kernel of institutional learning. Nevertheless, it is a subsystem of the education system, which can be described as an autonomous (self-dependent) system that covers the major task of education—learning activity. A model of this system is presented. It describes integration of the learner and teacher and determines a teaching-learning process in its two forms. The realization is possible only in learning environment. The main characteristic of the learning environment is that it provides resources to the teaching-learning process. They are divided into two groups named technical means of teaching and learning, and learning and teaching resources. When the learning environment is developed on the basis of computer, information, and communication technologies, it is known as technology enhanced e-learning environment.

The usability of this environment is one of its important characteristics, since, in this way, the integration of the teaching-learning process is guaranteed. A framework for description of e-learning usability is constructed. The achievement of two of the usability attributes, adaptability and learnability, is necessary, since they ensure the development of personalized learning environment. Personalized learning resources can be developed by teachers on the base of adaptability and learners' representation. Learner modeling requires information about the learner's attributes. One of the approaches to information discovery is through measurement of the learner's preferences that reveal the learner's competences. The learner model can be determined indirectly through evaluation of teacher's preferences that present an opinion about existing learner types, which is based on teacher experience.

After presentation of a technique for measurement of human preferences and its implementation in an instrument for measurement, several attempts to reveal the great potential of evaluation of human preferences in learner modeling are described.

DOI: 10.4018/978-1-4666-2967-7.ch010

1. LEARNING SYSTEM AND LEARNING ENVIRONMENT

We consider institutional form of learning activity named *institutional learning*. It presents the performance of learning activity in the framework of an educational organization (institution) like school, university, college, etc. Its framework consists of four components—*education system*, *teaching/learning system*, *learning environment*, and *learning theory* (Figure 1). Institutional learning is a complex activity that is carried out as teaching-learning process and applies results of the work of different education-oriented experts in accordance with the learning abilities of the students. Usually the curriculum encompasses different types of learning (learning approaches), which bases are various learning theories. They demand different types of teaching (pedagogical approaches), on which the learning theories are mapped. No single teaching method can be applied for all occasions. An optimal program would be some mixture of diverse pedagogical approaches. To achieve an effective teaching-learning process a teacher has to collaborate with different specialists that do not participate directly in this process like other teachers, methodologists, speech therapists, psychologists, educators and specialists for design of pedagogy.

The education system envelops the other components of institutional learning. It ensures the collaboration of education-oriented specialists and management of the entire educational process that concerns the development, control, assessment, investigation, and implementation of the teaching-learning processes. Under the influence of the education system, the learning environment is regarded as an education environment where the main actors of an education system (methodologists, pedagogues, psychologists, educators, and others) participate in mini-societies devoted to specific educational objectives. They interact with each other to produce the overall, complex service to the teaching-learning process and perform specialized problem solving (decision making) activities.

Figure 1. Framework of institutional learning

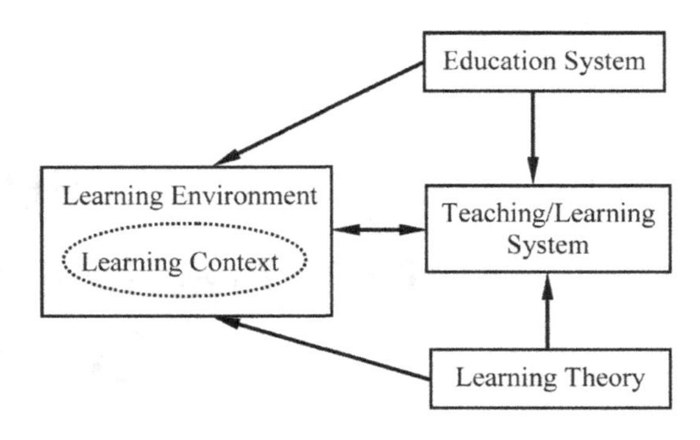

The teaching/learning system is the kernel of institutional learning. Nevertheless, it is a subsystem of the education system, which can be described as an autonomous (self-dependent) system that covers the major task of education—learning activity. The main components of this system are teacher(s) and learner(s). They are considered as agents, since they have intelligent functions. In broad terms the learner gains knowledge (knowledge acquisition) and the teacher generates knowledge and takes decision about the teaching-learning process. We present a model of the simplest teaching/learning system that consists of a teacher and a learner. They work together to achieve a shared objective. The following statement represents an agent-based view of this system (Andreev, 2006).

Teacher $\wedge$ Learner $\rightarrow$ shared learning objective (object).

This system consists of two kinds of agents: a *selfish agent* interested in his own need (he tries to maximize his own individual return) and an *agent-factor* helping to bring about a result. The teacher is an agent-factor in teaching-learning process. The learner is a selfish agent that is a necessary and sufficient condition for the existence of a learning activity. From this analysis, we come to the following conclusions:

- Learner is the cause of teaching-learning process;
- Learning activity is one of the innate human's functions, i.e. a characteristic of human being;
- Teacher is necessary condition only for realization of institutional learning.
- The presented description of teaching/learning systems is an opportunity to reveal the major characteristics of a teaching/learning system;
- The relation between the teacher and learner is symmetrical;
- The teaching/learning system is non-linear;
- The teaching/learning system that is based on the collaboration between teachers and learners is an integrated system;
- A collaborative system is framework for a real process that can be seen as a complex autonomous activity (Andreev, 2008): Teaching-learning process is such an activity that is shared between teacher and learner and can be view as an extension of an autonomous learning activity.

The complex autonomous activity is a real process, since it presents a continuous succession of activities. The continuity is a necessary consequence of the wholeness of the teaching/learning system due to its integrity and autonomous learning activity.

The teaching-learning process that exists in the collaborative teaching/learning system can be presented in two ways. The first of them is a result of the transformation of the non-linear agent-based model in the following linear form:

Teacher → (Learner → shared learning objective).

In this process-oriented form of representation of teaching/learning system, it is possible to do the following conclusions:

1. The teacher has to supply the learner activity with satisfactory products, i.e. he plays the role of a decision maker that has to decide "what" learning resources to ensure for the learner. The learner uses the prepared resources for achievement of a desired outcome.
2. The symmetry of the collaborative teaching/learning system is ensured through a harmonious relation between the teacher and learner. The former has to satisfy the learner requirements.
3. The teacher prepares the learning resources and controls the efficiency of the teaching-learning process—the functioning of the teaching/learning system is presented by a "teaching activity".

The effectiveness of the teaching-learning process depends on the *capacity of teaching/learning system for harmonization*. The latter is an important system parameter that presents the ability of a teacher to ensure learning resources that are suitable for learners. In this case, the teacher is decision maker, but the learner is problem owner that causes problems in system integration. The capacity for harmonization guarantees the symmetry of the non-linear teaching/learning system. An approach to increase the system capacity for harmonization is the building of learner-centered (problem-oriented) teaching/learning systems in the place of the instructor-led ones.

The second approach to process-oriented representation of the teaching/learning system results in the following linear model of the system:

Learner → (Teacher → shared learning objective).

According to this process-oriented description, the learner controls the system behavior. As a result of this authorization of the learner to manage the teaching-learning process, the teacher becomes a necessary condition that has to satisfy the learner's needs. This organization of teaching-learning process is service-oriented due to the following characteristics that it has:

- The teacher is a servant: he is treated as a means in the teaching-learning process;
- The symmetry is ensured by interaction between the learner and teacher: in this case the symmetry is presented by its quality of balance;
- The learner is not only problem owner, but decision maker, as well;
- The teacher is recommender in this form of teaching-learning process: he supports the process;
- The teacher concerns the working condition (situation) of a learner—he is responsible for support of learning environment;
- The teacher is a reactive agent and his actions are analyzed into the context of "stimulus and response": he does not take decisions.

Learning theories address how people learn. Different theories focus on different factors that influence the learning activity and determine learning models. They are basis for construction of pedagogical frameworks that describe the broad principles through which a theory is applied to a teaching-learning process. The theories are derived from different perspectives about the nature of learning that has psychological fundaments, since learning activity is a human characteristic. They are classified in several categories in correspondence with the main psychological perspectives that are used for research and explanation of the psychological nature of human being at this moment. We differentiate the following groups of learning models that are related to certain psychological theory: behaviorism-oriented theories, cognitive theories, constructive theories, and humanist theories. It is necessary to note that these categories are intersection groups of learning theories. Many of them belong to more than one group.

Behaviorism operates on a principle of "stimulus-response." The behavior caused by external stimuli can be explained without the need to consider internal mental states or consciousness. Although behaviorism is currently widely dismissed as a serious theoretical basis for education, this view is seriously wide of the mark. The learning models that are based on behaviorism are concerned with emphasizing active learning-by-doing with immediate feedback on success, the careful analysis of learning outcomes, and with the alignment of learning objectives, instructional strategies and methods used to assess learning outcomes. The cognitive perspective considers the teaching-learning process as an information process. It is a response to behaviorism, people are not "programmed animals" that merely respond to environmental stimuli. People are rational beings that require active participation in order to learn, and whose actions are a consequence of thinking. Changes in behavior are observed, but only as an indication of what is occurring in the learner's head. Cognitivism uses the metaphor of the mind as computer: information comes in, is being processed, and leads to certain outcomes (Marx, 1987).

Constructivism is a paradigm for learning where it is considered as an active, constructive process. The learner is an information constructor. People actively construct or create their own subjective representations of objective reality. New information is linked to prior knowledge, thus mental representations are subjective. The concepts are regarded as tools, to be understood through use rather than as self-contained entities to be delivered through instruction. The learner's search for meaning through activity is the essence of these learning theories. Knowledge is constructed based on personal experiences and hypotheses of the environment. Learners continuously test these hypotheses through social negotiation. Each person has a different interpretation and construction of knowledge process. The learner is not a blank slate (*tabula rasa*) but brings past experiences and cultural factors to a situation (Vygotsky, 1978; Bruner, 1967).

The learning theories that conform to humanistic perspective of psychology follow the belief that it is necessary to study the person as a whole, especially as an individual grows and develops over the lifespan. The learning is a self-motivated activity. A central assumption of humanism is that people act with intentionality and values. This is in contrast to the behaviorist notion of operant conditioning (which argues that all behavior is the result of the application of consequences) and the cognitive psychologist belief that the discovering knowledge or constructing meaning is central to learning (DeCarvalho, 1991). Very important variations of humanistic perspective are Activity Theory (Leontiev, 1978) and John Keller's ARCS Model of Motivational Design (Keller, 1983). According to the latter, there are four steps for promoting and sustaining motivation in the learning activity: Attention, Relevance, Confidence, and Satisfaction (ARCS).

Learning environment supports different learning styles that depend not only on the organization of the education system, but on the accepted learning models (theories), as well. If some behaviorism-oriented learning model is performed, the learning environment has to guarantee repetitive performance of certain tasks that build specific behavior and skills in learners. Depending on the selected variant of constructive learning model, the learner's environment has to provide circumstances for cooperative and collective learning, group learning that is based on communications among learners and so on. An important attribute of learning environment is the learning context that it contains. The learning context is determined by two factors—learning subject and the type of learners. It is possible to differentiate several types of learners—learners with specific learning difficulties (learners with dyslexia or autism), learners with difficulties to gain knowledge in certain subject domain, learners with talent and others. The specific type of learners and concrete subject domain determine specific learning situation (learning context) that requires selection of the most suitable learning style that has to be realized by the teaching-learning process.

The realization of teaching-learning process represented by its two forms is possible only in learning environment. The main components of this process realization and relationships between them are presented on Figure 2.

Figure 2. Components of realization of teaching-learning process

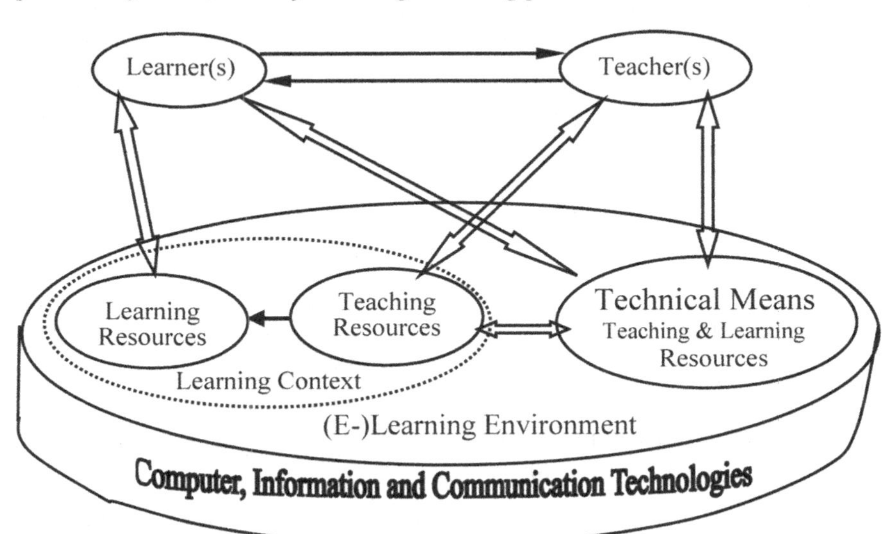

The main characteristic of the learning environment is that it provides resources to the teaching-learning process. Since they help and support this process, they are divided into two groups named *technical means of teaching and learning*, and *learning and teaching resources*. The learning and teaching resources support the teaching-learning process. They are the main factor for achievement of integration of teaching/learning system, since they can guarantee efficient teaching-learning interaction. If the learning resources satisfy the needs of the learners and the learning outcomes correspond to the shared objectives and satisfy the teachers' requirements, it brings the system to equilibrium. Since the learning resources are products of teaching activities, they are elements of the learning- teaching interface that tries to integrate teacher's and learner's activities in a united teaching-learning process. Several types of learning resources are distinguished: learning objects, units of learning, learning facilities and tests. Learning object is result of teacher's activity (f) presented in the following way:

Learning object = f (instr. resources, instr. strategy).

The construction of this type of learning resources depends on two parameters: instructional (teaching) resources that compose the contents of a learning object and instructional strategy that determines the way the teacher presents the content of the learning resource. The tests are learning resources that are derivates of learning objects. The production of tests has the following presentation:

Learning resource = f (learning objects).

The units of learning contain not only learning content, but pedagogical pattern and representation of learners, for which the learning unit is suitable for use (Koper, 2004).

Technical means help not only the teachers and learners, but the actors of education system, as well. They guarantee increasing capabilities of the learners, teachers, and educators in performance of the teaching-learning process. The necessary teaching instruments that support a teacher in performance of teaching activity can be divided in two groups—tools for performance of various pedagogical techniques and tools for authoring of learning resources. The pedagogical techniques concern the way of realization of learning models. Some of the most used pedagogical techniques are lectures, assignments, case studies, simulation, discussions, debates, forums, collaboration, role-play, personal consultations, etc. Some of the tools concerning the implementation of pedagogical techniques have relation to classroom management, mentoring, realization of different group communications patterns like discussion or conferencing, and others.

The educators' capabilities are ensured by tools for management of teaching-learning process, for assessment of learning outcomes and certification, for organization of education and information of other stakeholders that observe the results of teaching-learning process and for help of various administrative activities. Another group of instruments that learning environment has to provide concerns the augmentation of learner's capabilities in his/her access to learning content. In this group are included tools for both content sharing and accessing information—learners identify teaching-produced materials (learning objects) relevant to their educational objectives and subject domain, and attempt to access. Other tools are for:

- Scanning information, learners search for particular headings, information items or instructions related to their problem representation;
- Transferring information during a discussion or exercises;
- Knowledge understanding.

An aspect of institutional learning is social learning. It is based on the fact that understanding of content is socially constructed, through conversations about that content and through interactions concerning problems or actions. The focus is not only on *what is learned* but on *how it is learned*. In addition, social learning concerns not only *learning about* the subject matter but also *learning to be* full participants in the domain.

There is big difference between the cognitive learning environment that supports cognitive learning models and the constructive one, since the learner learns from the former while he learns with the constructive environment. The constructive learning environment supports the principles of constructive learning theories that determine the learning activities: Learning (activity) is a search for meaning; Learning occurs in a context; Learning activity complies with the learners' mental model; Constructing knowledge is the purpose of learning (not "right" vs. "wrong"). In the perspective of behaviorism, the learning activity is presented by learner's reaction or tasks that have to be carried out. These types of learning activities are usually supported by learning environment presenting some reality.

Since the learning environment is a common place for all participants in the teaching-learning process, it is regarded as a collaborative work environment. For that reason, it has to provide collaborative tools for establishment of "communities of practice" that guarantee sharing of learning objects that are products of teaching activity and achievement of interoperability among various teaching activities (Lave, 1998). Some of these tools serve for conferencing, shared workplaces, work grouping, content management, context management, and others.

If the basis, on which a learning environment is developed, consists of computer, information, and communication technologies, it is considered as *e-learning environment* or *technology-enhanced learning environment*. It gives opportunities for realization of new tendencies not only in searching, storing, producing, and implementation of learning and teaching resources, but in realization of various technical means for aiding learning and teaching, as well. We make mention of some of them by way of illustration:

1. Convergence of audio, video, and data into unified conferencing. Audio and data conferencing, screen sharing, and remote access are based on one and the same technology, so they are integrated at the infrastructure level in order to be transparent and easy to use.
2. Participation in virtual social networks, distance meeting, and conferencing. Consolidation of synchronous and asynchronous tools.
3. Consolidation of collaboration technologies. Because there are available vast amount of collaborative tools, IT specialists have to consolidate them by evaluation and choosing only one tool, having in mind the idea of the overall functions needed.
4. Aggregation of all people, content, conversations around content, keywords/tags to filter content, etc. Ultimately, aggregation affects social networks, portals, communities, streams of information, even relationships that could be addressed and dealt in one screen.
5. Mobile collaboration. These days' mobile phones become smarter and have more power processors, more memory, and therefore they could be used for more elaborated tasks.
6. Collaboration moving to a virtual world. One of the latest applications of 3D-technology is Virtual worlds (simulations). In some cases, 3D is better to use, but in other cases it is not because it is resource demanding—enough RAM, processing power, and graphics processing memory are required. Nevertheless, training and education using simulations could benefit from 3D-technology.

The collaborative character of e-learning environment is due not only to the fact that it is common for learners and teachers, but to the capabilities that are ensured by new technologies, as well. This characteristic of the environment is precondition for implementation of integrated teaching/learning system, but it cannot guarantee the achievement of this objective when a continuous teaching-learning process is put into effect. The implementation of a collaborative e-learning environment requires development of an e-learning environment that provides with usable resources the main agents of the teaching-learning process and in such a way guarantees usefulness and unity of this process. The resources are usable, if they meet the needs of teachers and learners and their shared learning objectives, i.e. they are adapted to their requirements.

The development of a usable e-learning environment concerns *e-learning usability* that is one of the main objectives of research on education technology and technology-enhanced learning. With respect to the learner, this objective is interpreted as *development of personalized e-learning*. Its achievement that is based on usage of adaptation techniques and description of individual learner depends on the decisions that the teacher has to take. The main difficulty (problem) in support of personalized e-learning is the description of individual learner, i.e. to construct his/her model. It is sub-problem of a decision making process that has to solve the problem of integration the teaching-learning process through optimization of teaching-learning interactions (relations). Since e-learning environment supports these interactions and it is possible to regard the teacher as a component of the environment that takes part in development of learning resources, the efficient and balanced teacher/learner interactions depends on the usability of learning resources that is achieved by their personalization.

2. E-LEARNING USABILITY

The term usability typically refers to technical issues, whether a product or system is comfortable, bug-free, and intuitively operable. We find this connotation is suitable for computer programs and information seeking environments where performance of specific tasks and access to information is the key objective. The results of usability studies are typically incorporated into several stages of the software life cycle, from early analysis and design through final testing and follow-up studies. For software engineers' perspective, usability concerns center on the user interface and the degree to which it meets various usability heuristics (Nielson, 1993).

We consider usability as an attribute of the way that a person interacts with a product rather being something, which can be assessed independently of usage. Usability is a concept that it is very difficult to define, since many aspects of a product contribute to how it is perceived. Instead of defining e-learning usability, we prefer to specify it in a framework that involves:

- **Context of Use of e-Learning:** Users, technology-enhanced learning environment (learning/teaching resources, technical means for teaching and learning, technological basis), context of learning (subject domain, learning objectives);
- **Usability Criteria:** They determine usability attributes (abstract properties) like effectiveness, efficiency, satisfaction, ease of use;
- **Implementation of Usability Attributes:** It bases on specification and determination of usability-carrying properties that have to be taken into consideration in design of e-learning environment and its resources.

Usually usability of some product is evaluated a posterior. It is designed and constructed first, and then subjected to token usability testing just prior to delivery. The specialists of usability test the already created product whether it satisfies some heuristics or tests. In case of the implementation of a teaching-learning process we assert that this is not appropriate because if one discovers at the end of some learning course that the learning materials are not usable, which means learners could not perform their tasks and goals successfully, it would be total waste of time and would discourage learners from attending courses. No one has approved this situation—neither teachers and students nor stakeholders. At this late stage in the development cycle, little can be done easily to improve the usability This is the reason why we claim that if teachers want to achieve e-learning usability of learning resources they must have a priory a determination about learners. Therefore, usability achievement has to be a requirement for the development of e-learning environment and its resources that is taken into consideration at the stage of their design. The latter requires not only to model technology-based teaching, learning and educational tools, but to ensure that they possess the necessary properties, as well. Hence, the design cycle does not finish with the modeling of a technology-enhanced learning environment, but it has to guarantee that the realization of the environment model will satisfies the needs of its users, i.e. harmonious relationship between e-learning environment and its users. The design evaluation of e-learning environment is important phase of the whole design cycle that requires the determination of design criterion. The latter can be determining in a framework that supports the construction of an e-learning usability model considered as design criterion.

The construction of an e-learning usability model for assessment and direction of the design of an e-learning environment in correspondence with usability requirements is based on the following design principles (Dormey, 1995, 1996):

- Association of abstract properties called usability attributes with e-learning environment;
- Usability attributes correspond to a set of teaching, learning and education-dependent tasks or uses of an e-learning environment;
- Usability attributes should be sufficient to represent the satisfaction of the needs of main groups of people using e-learning environment through determination of their requirements by means of usability goals;
- The properties of learning/teaching resources and technical means for teaching, learning, and education management must have in order to satisfy the usability goals.

The second principle (see above) reflects the observation that we associate the usability attributes of a technical means or a learning/teaching resource either with actions in or uses of an e-learning environment (Figure 3).

Environments contain the resources and references to resources needed to carry out an activity or set of activities. The e-learning environment contains teaching/learning resources that are necessary for teacher activity, learner activity, and activity of educational organization that controls the relationships among all participants in institutional learning (learners, teachers, parents). At the same time, it provides technical means that are necessary for carrying out of these activities. The computer, information, and communication technologies have great impact on learners and teachers in carrying out their activities in the integrated teaching-learning process. These components of context of e-learning usability determine two usability attributes—*ease of use* and *operability*.

The teaching and learning resources and technical means not only must be easy to use and suitable for operation, but they also should serve achievement of teaching/learning objectives, i.e. ensure useful activity. In the design of e-learning

Figure 3. Multiple criteria for assessing and design of e-learning usability

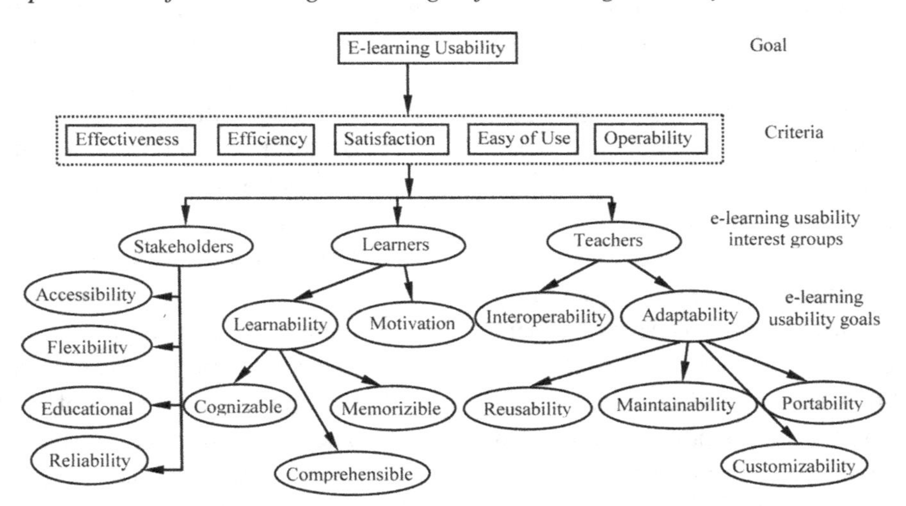

environment usability usefulness is necessary usable attribute that corresponds with learning situation. The usefulness has two aspects—effectiveness and efficiency. The effectiveness of e-learning environment is an indication how well it supports users activities in achievement of their goals and whether or not it is accurate in its working. Efficiency is an indicator how well learning environment contributes to the performance of some activity in achievement of satisfactory or desired results. Satisfaction is also very important aspect of e-learning usability. It addresses subjective responses to use of e-learning environment and concerns user's desire to continue his work with the help of e-learning environment (Mehlenbacher, 2005).

The users of e-learning environment are divided into three categories—learners, teachers, and stakeholders. The learners and teachers are of primary importance for the technology-enhanced learning environment. The set of stakeholders that are of secondary importance includes instructional designers, technical assistance, administrators, program coordinator, teaching assistance, and support staff and research. Highlighting the multiple users of e-learning environment is a critical first step in beginning to understand the complexity of e-learning usability challenge (Storey, 2002). Each one of these groups has specific viewpoint to e-learning usability that binds the members of the group to certain usability goals. Usability goals represent what the user is trying to achieve with the e-learning environment in correspondence with the usability attributes. It is also necessary to represent these goals through more tangible terms that concern the determination of usability-carrying properties of the environment and its resources.

The teachers are considered as producers of learning resources. That is why an e-learning environment is usable for them, if it supports effective and efficient teachers' activities. From their perspective, the usability goals concern interoperability and adaptability of available learning and teaching resources. An interoperability-supported learning environment has a relation with collaborative teaching activities that ensure sharing of teaching resources. The adaptability can be achieved through learning environment that has the following components (properties)—reusability, maintainability, portability, customizability. Maintainability is achieved through usage of various information systems and modifiability. A learning environment has the property reusability, if it guarantees data encapsulation, representation independence, and application independence.

The definition of learnability from learners' perspective should be considered to include the ability of learners to effectively and efficiently learn. It is possible at the level of using of learning resources that have to possess the following properties—easy to perceive (cognizable), easy to be comprehended (comprehensible), and easy to memorize (memorizable). The learnability is result of conforming learning resources to learner's competences. The latter have to be the argumentation why the learning resources are designed and developed by the teachers in this or other way. Satisfaction as usability attribute is interpreted by the learners from psychological viewpoint as motivation. The usability goals that have to satisfy the needs of stakeholders for usable e-learning environment are the following: accessibility, reliability, flexibility, and possibilities for performance of educational activities.

The personalization of an e-learning environment is possible, if the usability goals adaptability and learnability are achieved in its development. The adaptability of e-learning environment is very important usability goal, since it is precondition for effective production of learning resources that are qualified as learnable. The harmonization of the features of learning resources with the characteristics of learners is necessary for the integration of the teaching-learning process, i.e. useful interactions between the teacher and the learner.

3. ADAPTIVE AND ADAPTABLE E-LEARNING ENVIRONMENT

We say that there are two forms of existing of teaching-learning process. According to the first form the teacher develops (produces) the learning resources and controls the usefulness of the teaching-learning process—the functioning of the teaching/learning system is presented by a teaching activity. He plays the role of decision maker and producer. The second form authorizes the learner to manage the teaching-learning process, where the teacher becomes a necessary condition that has to satisfy the learner's needs. This organization of teaching-learning process is regarded as service-oriented. The learner has an active role in determination which learning resources meet his needs. Therefore, he is a more intelligent participant in the teaching-learning process that can take decisions.

If in realization of the second form the teacher is considered as a component of the learning environment, the latter could be qualified as a server-based learning environment. Its relationships with the learner can be described in the framework of a client-server architecture, where the learner is client (user). The adaptive behavior of a server-based e-learning environment is an aspect of its automatic adaptation. The client-server architecture is a natural setting for realization of automatic adaptation due to the symmetry of client-server interaction. The client (learner) goals and preferences depend on the circumstances, in which he is, i.e. the resources that a server (technology-enhanced learning environment) provides. On the other hand, the server has to satisfy the user requirements. There is correspondence between user (learner) characteristics, and attributes and the features of server's resources. The server has to take account of its relation with concrete users.

The client-server symmetry is the fundamental reason for the development of a model of automatic adaptation that is used in the design of adaptive e-learning environment. This model represents the concept of automatic adaptation that coincides with the concept of tailorability. It presents a way of binding up the two (complementary) sides that define an adaptation. The first of them describes *what* is to be adapted (what are the variables of learning environment's resources). The second part represents *to what* the server is ready to adapt (learner model or context model). The way of realization of adaptation model (*how* the adaptation is performed) is responsibility of the adaptation technology (Brusilovsky, 2001).

An adaptation technology can be dissected into adaptation method and adaptation technique. The adaptation method specifies the adaptation logic. The adaptation technique realizes an approach to implementation of an adaptation method in a particular adaptation dimension. There are several adaptation dimensions. Their definitions depend on certain variables of learning environment's resources and corresponding learner's characteristics or attributes, to which the resource variables adapt. There are various adaptation technologies that ensure servers adaptation in different adaptation dimensions. Most of them guarantee adaptive presentation of learning objects (content adaptation). For the other adaptation dimensions, there are the following technologies—adaptive delivery of learning objects, adaptive learning activity selection, and adaptive learning service provision and so on.

The adaptive servers that provide a learner with learning resources do not present a complete adaptive e-learning environment, since the latter has to consist of servers that are representatives of the two server types. Adaptive e-learning environments that support learners with learning instruments can be personalized, as well. An approach to personalization of adaptive learning instruments is described in Kay (2001). A model of personalization of a complete adaptive e-learning environment that provides learning resources and services (learning instruments) simultaneously is presented in Aroyo (2006).

According to the first form of the teaching-learning process, the e-learning environment has to include production components. A production component performs or supports the work of a teacher that develops learning resources. The development of learning resources is considered from two different viewpoints that determine the representation of a production component. According to the first of them, the development depends on its outcomes, i.e. the products of teaching activity (type of learning resources). In correspondences with the second aspect, the development is regarding as a process of producing that coincides with the teaching activity. These two aspects of a production component require the use of holistic and systematic approach to its design and development. An e-learning environment that includes production components could be considered as productive e-learning environment.

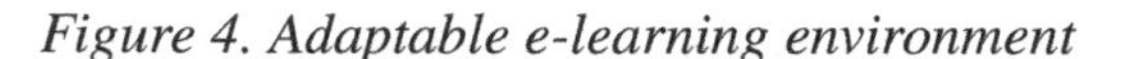
Figure 4. Adaptable e-learning environment

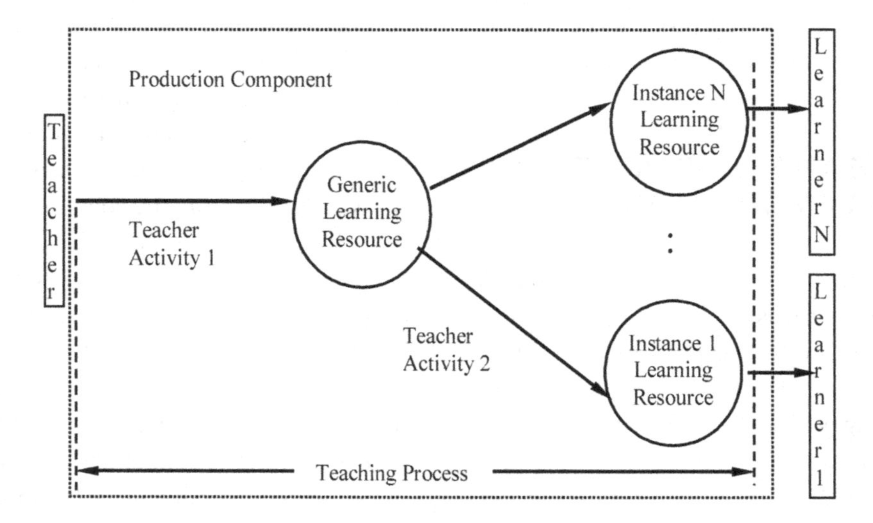

Learning resources are any explicit, physical result of a teaching activity realized by a process of producing that has two phases (Figure 4):

- Development of an initial generic learning resource concerning a subject domain – this generalized learning resource is an outcome of the teacher activity 1;
- Maintenance, reuse and customization of generic learning resources (teacher activity 2) – during this phase the teacher adapts the generic learning resource to a particular learner or context.

The adaptation of generic learning resources to the characteristics and attributes of a learner is one of the main tasks of a teacher during the maintenance phase.

The adaptation of learning resources to individual learners needs concerns the personalization of a productive e-learning environment and development of e-learning environment that meet one of the usability goals—adaptability. It distinguishes from the personalization of adaptive e-learning environment, since it transforms generic learning resources into specific learning resources and exists as a function of the teaching activity. A dynamic productive e-learning environment, which adaptation to individual learner bases on the transformation of generic learning resources into specific ones, is considered as "adaptable" e-learning environment (Stephanidis, 2001). The main components of adaptation model are: *learning resource description, learner model,* and *adaptation technology*. The learner and learning resource models are decoupled. It is due to the asymmetric nature of the production component. The adaptation logic of this technology is based on a *generalization/specification* method of reasoning. The general representation of learning resources has a great potential for reusing. Therefore, the adaptation technique that presents a way of adaptation performance could realize a reusing method (Barnes, 1991). This approach to personalization of a productive e-learning environment that is based on the process of development and maintenance of learning resources entail an engineering perspective on adaptation.

4. LEARNER MODEL AND LEARNER'S PREFERENCES

Learner modeling involves inferring unobservable information about a learner from observable information about him/her, e.g. his/her actions, utterances, and preferences. Its realization bases on modeling techniques that can be classified in accordance with three aspects—*type of modeling method, approach to model creation,* and *modeling objectives*. There are two types of learner modeling methods: empirical methods and reasoning-based (theoretic) methods. An empirical user modeling techniques use both explicit and implicit knowledge and data acquisition to determine the learner's characteristics or attributes (Zukerman, 2001). The reasoning-based methods realize substantiated subjective learner modeling on the basis of concepts.

There are two main approaches for model construction: *individual-oriented* and *socio-cultural* (Zukerman, 2001). The socio-cultural approach uses the observation that people within a particular group tend to conform under a given set of circumstances. The former approach is based on the tenet that each learner exhibits a particular behavior under circumstances.

The classification of modeling techniques in correspondence with the modeling objectives depends on the type of learner models that the corresponding technique creates. The classification of modeling objectives depends on the following aspects:

- Models of direct and indirect learners;
- Models of individual learner's competences—learning style and learner's skill, abilities, background, knowledge, interests, and individual traits;

- Stereotypical learner model or updating learner model.

The direct learner model has two aspects. According to the first viewpoint, the main container of actual information about a concrete learner is the learner. He knows his week and strong features, psychological problems and competences. If the learner is a problem of institutional learning, he is the problem owner. The other aspect of direct modeling of learners is the acquisition of data about them during their actual learning activity (interaction with learning environment and teachers) and from activity outcomes. The second aspect of direct modeling of learners concerns the creation of an adaptive e-learning environment and guarantees its adaptive control. The indirect modeling of learners is based on the opinion of experts or teachers about a concrete learner.

A stereotypical model of a learner is a result of a modeling approach that qualifies the behavior or preferences of a group of users in certain circumstances as similar, and they are represented by a fixed (constant) learner model. The updating learner model bases on a preliminary learner model that is updated in accordance with the actual learner behavior. In many cases, we are not interested in learner modeling in any general sense, but only in correspondence with the purposes for which the learner models are constructed. For this reason, there are different models for the representation of learner's characteristics, competences, interests, or background.

Independent of modeling techniques is the framework, in which a learner is described. Building of a learner model begins with the determination of this framework that represents for which learner's attributes and features it is necessary to gather information. We present a common pattern for learner model structure that is usually used. All learners of e-learning environments bring the following general attributes to the learning experience, which are included in the pattern of the learner model. They are divided into several major sections—general biological description, abilities, skills, subject-oriented knowledge:

Biological

- **Age:** Children, adults, seniors
- **Gender:** Male, female, masculine, feminine
- **Race:** Caucasian, African-American, Asian, etc.

Abilities

- **Cognitive:** Perceive abilities memorizing (short-term and long-term memory), absorption capacity, intelligence scores, observing ability, visual, auditory, and others.
- **Physical:** Ambulatory, haptic.
- **Personal Psychological:** Attention, concentration, motivation (self-definition, esteem levels), self-monitoring.

Skills

- **Communication:** Personal, group.
- **Computer:** Training or education with technology, platform (application) specific familiarity.
- **Task Experience:** Novice or expert.
- **Textual:** Knowledge of application area, education, testing capability, accredited expertise.
- **Visual:** Experience with scientific and technical data visualization, spatial systems.
- **Knowledge**
- **Domain:** Knowledge of application area, education, testing capability.

Socioeconomic Context

- **Geographic:** Rural, urban, low, or concentrated populations.

- **Organizational:** Large or small, private, public, educational, or production.

Personal Attributes

- **Learning Style:** Reflective, sequential, deductive, inductive.
- **Attitude:** Theoretical or practical, orientation towards task, affective expectations.

Usually the necessary information for determination of different learner attributes is gathered from different information sources. A great part of learner attributes can be interpreted under a learner's characteristic—learning competence. It is possible, since its meaning takes in learning abilities (cognitive, physical, psychological), skills (learning experience of the learner), learner's knowledge and personal attributes. Therefore, learning competence can be determined in four dimensions. On the other hand, the learning competence is on the basis of learner's preferences. The people prefer to do something in a manner that is in correspondence with their knowledge, skills, abilities, and personal attributes. For that reason, human preferences are multi-dimensional and they can be evaluated in different perspectives. In other words, human preferences are context-dependent. This fact gives great advantage to an instrument that can measure human preferences, since it can be used for accurate evaluation of different learner's attribute. In this case, it is not necessary to collect data from different sources. All needed information that can be extracted from the learner is gathered by an instrument for measurement of preferences.

One of the factors that influence the increase of human competence is experience. The latter has impact on the mental attitude of people, i.e. on their opinion. As the personal experience (attitude) is one of the aspects of people's competence, the human preferences can serve for subjective evaluation of the object of experience. This evaluation is an approximation of the real representation of this object. The measurement of the teacher's preferences that are determined by his experience can provide information about learners that are objects of his experience.

5. MEASUREMENT OF LEARNER'S AND TEACHER'S PREFERENCES

A man *(DM, problem owner, or expert)* can evaluate an object through explicit expression of his preferences with respect to its attributes and way of behavior, if the framework of evaluation is determined. When the object attributes correspond with man needs concerning some of his basic functions, the man's preferences evaluate his function's characteristics, i.e. the man self-evaluates his characteristics (abilities). In this case, the measurement of subjective preferences serves for measurement of human's characteristics that objectively are represented by object's attributes. For example, our preferences about the parameters of presentation of learning material reflect characteristics of our learning activity. If we prefer the educational material to be presented predominantly in text form and not in mathematical form, it reflects characteristics of our learning abilities, i.e. our learning competence. The main objective of this evaluation is to keep the correlation between the expressed preferences and the DM's utility function determined in the framework of the above-considered system. The theoretical basis, which leads to such a solution, is the so-called axiomatic approach and its main component the expected utility theory.

Standard description of the utility function is presented by formula (10.5.1.1). There are a variety of possible final results ($x \in \mathbf{X}$) that are consequence of a DM activity. A utility function U(.) assesses each of these results. The DM judgment is measured quantitatively by the following formula:

$$U(\mathbf{p}) = \sum_i P_i(U(x_i),$$
$$\mathbf{p} \text{ is probability distribution } \sum_i P_i = 1 \quad . \tag{10.5.1.1}$$

We denote with P_i ($i=1\div n$) subjective or objective probabilities, which reflect the uncertainty of the final results. Strong mathematical formulation of the utility function is the following: Let **X** be the set of alternatives (possible results) and ***Y*** be a convex subset of the set of probability distributions over **X**. The DM's preferences over ***Y*** are described by the binary *"preference"* relation ($\succ$) including those over **X** (**X** $\subseteq$ ***Y***). Its induced *"indifference"* relation ($\approx$) is defined thus $((x \approx y) \Leftrightarrow \neg((x \succ y) \vee (x \succ y)))$. A utility function is any function U(.) for which is fulfilled:

$$((p \succ q), (p,q) \in Y^2) \Leftrightarrow (\int U(.)dp > \int U(.)dq). \tag{10.5.1.2}$$

Thus, the mathematical expectation of the utility U(.) is a quantitative measure concerning the DM preferences about the probability distributions over **X**. In practice the set **P** is a set of finite probability distribution. There are different systems of axioms that give satisfactory conditions of utility existence. The most famous of them is the system of Von Neumann and Morgenstern's axioms exposed in chapter 3 of the book (Fishburn, 1978).

It is known that the expected utility function is determined with precision up to an affine transformation and that the utility function determines the "preference" relation ($\succ$) over **X** as "negatively transitive" and "asymmetric" one (Chapter 3). Consequently the presumption of existence of an utility function U(.) leads to the presence of: asymmetry $((x \prec y) \Rightarrow (\neg(x \succ y)))$, transitivity $((x \succ y),(y \succ z)) \Rightarrow (x \succ z))$ and transitivity of the "indifference" relation ($\approx$). We know that the transitivity of the relations ($\succ$) and ($\approx$) is the most breached in the practice. The violation of the transitivity of ($\succ$) could be interpreted as a lack of information or as an expert's subjective mistake. The violation of the transitivity of the relation ($\approx$) is due to the natural "indistinguishability" threshold of every expert and the qualitative nature of the subjective evaluation. In addition, this is a bottleneck in the expected utility theory applications.

There are many different evaluation methods of the utility functions (Keeney, 1993; Raiffa, 1968) that are based prevailingly on the "lottery" approach (gambling approach). A "lottery" is called every discrete probability distribution over X. We mark as <x,y,α> the lottery: α is the probability of the appearance of the alternative x and (1-α)—the probability of the alternative y. The most used evaluation approach is the assessment: $(z \approx \langle x,y,\alpha \rangle)$, where $(x \succ z \succ y)$, $\alpha \in [0,1]$, $(x,y,z) \in X^3$. Weak points of this approach are the violations of the transitivity and the so called "certainty effect" and "probability distortion" identified by Kahneman and Tversky (Kahneman, 1979). Additionally, the determination of alternatives x and y on condition that $(x \succ z \succ y)$ where z is the analyzed alternative is not easy.

The elicitation of utility for prescriptive analyses requires subjective judgments. We saw that this judgment fails as a descriptive choice under risk, especially concerning the independence axiom (Chapter 3). From practical point of view, there is a need for procedures to improve the descriptive validity of the gamble measurement. Several non-expected utility theories have been developed in response of the displayed violations (Kahneman, 1979; Bleichrodt, 1999). Among these theories, the rank dependent utility model and its derivative cumulative prospect theory are currently the most popular. In the Rank Dependence Utility (RDU) the decision weight of an outcome is not just probability associated with this outcome. It is a function of both probability and the rank x. For example the RDU of the lottery (p_1, x_1; p_2, x_2;

……..; p_n, x_n) is (Kahneman, 1979; Bleichrodt, 1999):

$$\mathbf{RDU} = \sum_{i=1}^{n} W(p_i)U(x_i) \qquad (10.5.1.3)$$

Based on empirical researches several authors have argued that the probability weighting function W(.) has the inverse S-shaped form, which starts on concave and then becomes convex (See Figure 3.4.1 in Chapter 3).

5.1. A Technique of Evaluation of Learner's Preferences

The uncertainty and the transitivity disturbances can be interpreted as probabilistic noise. In these conditions the problem of utility function evaluation can be considered in a context where the information regarding the preference ordering is of a probabilistic nature, a lottery being declared better (alternatively worse) than a certain alternative, with a probability.

Starting from the properties of the preference relation ($\succ$) and indifference relation ($\approx$) we propose a stochastic approximation procedure for evaluation of the utility function U(.). Let **X** be the set of alternatives (possible results) and ***Y*** be a convex subset of the set of probability distributions over **X**. In correspondence with Chapter 6 it is assumed that ($\mathbf{X} \subseteq \boldsymbol{Y}$), $((q,p)\in \boldsymbol{Y}^2 \Rightarrow (\alpha q+(1-\alpha)p)\in \boldsymbol{Y}$, for $\forall\alpha \in[0,1]$) and that the utility function U(.) exists. We define two sets:

$$\mathbf{A}_{u*}=\{(\alpha,x,y,z)/(\alpha U^*(x)+(1-\alpha)U^*(y))>U^*(z)\};$$

$$\boldsymbol{B}_{u*}=\{(\alpha,x,y,z)/(\alpha U^*(x)+(1\alpha)U^*(y))<U^*(z)\}.$$

The utility function is constructed by recognition of $\mathbf{A}_{u*}$ (Aizerman, 1970; Pavlov, 2005). The proposed assessment is machine learning based on the DM's preferences. The machine learning is probabilistic pattern recognition ($\mathbf{A}_{u*}\cap\mathbf{B}_{u*}\neq\varnothing$) and the utility evaluation is a stochastic approximation problem with noise elimination (Chapter 6).

The stochastic procedure is the following. The DM compares the "lottery" <x,y,α> with the alternative $z\in\mathbf{X}$:

"better - $\succ$, f(x,y,z,α)=1",

"worse - $\prec$, f(x,y,z,α)=-1" or

"can't answer or equivalent-$\approx$, f(x,y,z,α)=0."

He relates the "learning point" ((x,y,z,α), *f(x,y,z,α)*) to the set $\mathbf{A}_u$ with probability $D_1(x,y,z,\alpha)$ or to the set $\mathbf{B}_u$ with probability $D_2(x,y,z,\alpha)$. Then the probabilities $D_1(x,y,z,\alpha)$ and $D_2(x,y,z,\alpha)$ are a mathematical expectation of f(.) over $\mathbf{A}_u$ and $\mathbf{B}_u$, respectively. Let D'(x,y,z,α) is the random value: $D'(x,y,z,\alpha)=D_1(x,y,z,\alpha)$, if $M(f/x,y,z,\alpha)>0$; $D'(x,y,z,\alpha)=-D_2(x,y,z,\alpha)$, if $M(f/x,y,z,\alpha)<0$; $D'(x,y,z,\alpha)=0$, if $M(f/x,y,z,\alpha)=0$. We approximate D'(x,y,z,α) with $G(x,y,z,\alpha)=(\alpha g(x)+(1-\alpha)g(y)-g(z))$, where $g(x)=\sum_i c_i\Phi_i(x)$. Then *g(x) is an approximation of the utility U*(.)* as is demonstrated in Chapters 5 and 6. The following is fulfilled for *f*(.): $f=D'+\xi$, $M(\xi/x,y,z,\alpha)=0$, $M(\xi^2/x,y,z,\alpha)<d$, $d\in\mathbf{R}$ (Aizerman, 1970; Pavlov, 2005). It is assumed that U(.) is square summable function and $\mathbf{U(x)} \underset{L_2}{=} \sum_i r_i\Phi_i(\mathbf{x})$, where $r_i\in\mathbf{R}$, and $\boldsymbol{\Phi}_i(x)$ is a family of polynomials. We set $\mathbf{t}=(x,y,z,\alpha)$, $\psi_i(t)=\boldsymbol{\psi}_i(x,y,z,\alpha)=\alpha\Phi_i(x)+(1-\alpha)\Phi_i(y)-\Phi_i(z)$ on the presentation of $\mathbf{A}_u$. The stochastic recurrent algorithm is:

$$c_i^{n+1} = c_i^{n} + \gamma_n\left[\mathbf{D}'(\mathbf{t}^{n+1}) - \overline{(\mathbf{c}^n, \Psi(\mathbf{t}^{n+1}))} + \xi^{n+1}\right]\Psi_i(\mathbf{t}^{n+1})$$

$$\sum_{\mathbf{n}} \gamma_{\mathbf{n}} = +\infty, \sum_{\mathbf{n}} \gamma_{\mathbf{n}}^{2} < +\infty, \forall \mathbf{n}, \gamma_{\mathbf{n}} > 0\,. \qquad (10.5.1.4)$$

The coefficients c_i^n take part in the decomposition $g^n(x) = \sum_{i=1}^{N} c_i^n \Phi_i(x)$ and

$$(c^n, \Psi(t)) = \alpha g^n(x) + (1-\alpha) g^n(y) - g^n(z) = G^n(x, y, z, \alpha)$$

The line above $\overline{y = (c^n, \Psi(t))}$ means: $\bar{y} = 1$, if $y>1$, $\bar{y} = -1$ if $y<-1$ and $\bar{y} = y$ if $-1<y<1$. The learning sequence and the appropriate DM answers are posed by pseudo-random sequences (Lp_τ).

This procedure permits direct assessment of the utility function $U(x,\alpha)$ as a function of both probability and the rank (alternative) x, following the findings of Kahneman and Tversky. We noted $W(p)=p+\Delta W(p)$ in formula 10.5.1.3. We suppose that this cumulative function W(.) is symmetric in careless of the diagonal. If the cumulative function W(.) has a symmetric form (the inverse S-shaped form, which starts on concave and then becomes convex) then is true:

$$\int_0^1 \Delta \mathbf{W}(\mathbf{p})\mathbf{u}(\mathbf{x})\mathbf{dp} = 0\,. \qquad (10.5.1.5)$$

In this case the expected von Neuman and Morgenstern's utility function U(.) is exactly the integral because p is evenly distributed:

$$\int_0^1 \mathbf{W}(\mathbf{p})\mathbf{U}(\mathbf{x})\mathbf{dp} = \mathbf{U}(\mathbf{x})\,. \qquad (10.5.1.6)$$

Following Kahneman and Tversky, we can evaluate first $U(x,\alpha)$ and after that apply formula (10.5.1.6).

6. MEASUREMENT OF HUMAN PREFERENCES AND LEARNER'S MODEL CONSTRUCTION: USE CASES

In this section, we present our assertion that through measurement of human preferences it is possible to determine various learner attributes and in this way to build learner model. In that way, it is possible to ensure the realization of value based management in e-learning personalization. Using learner's preferences, we can construct objective learner model, since the most objective assessment of an attributes of a learner can be given by himself/herself. At the same time it is described how is constructed a learner model through the experience of a teacher. We present a usage of the above described technique for measurement of human's preferences for determination of individual learner's attributes. The mathematical representation of this technique is implemented by means of a decision support system developed in the environment of Visual Studio (Visual Basic 6.0) exposed in Chapter 12. The final calculations and graphics are performed in MATLAB (MathWorks Inc.) environment.

We want to determine an aspect of cognitive ability of a learner—comprehension of knowledge. This ability is determined in the following limits: theoretical presentation of knowledge and example-based presentation. We have made this attempt with two learners using the instrument for measurement of their specific preferences. The results of this measurement are presented on Figure 5. It concerns the choice of proportion of theoretical presentation to the example-based presentation of knowledge. The figure presents the preference of a learner for content exposition—% theoretical presentation of the whole content presentation.

The seesaw line of the left graphic in the figure recognizes correctly more than 95% of the learner's answers. The stochastic uncertainty makes the utility function resemble a seesaw line.

Figure 5. Learner's preferences to form of content presentation

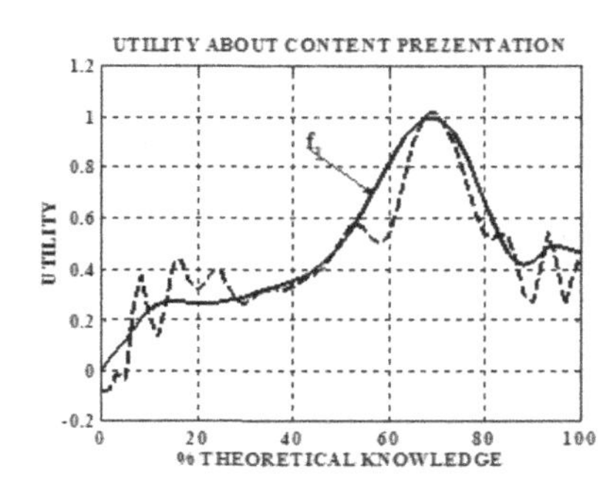

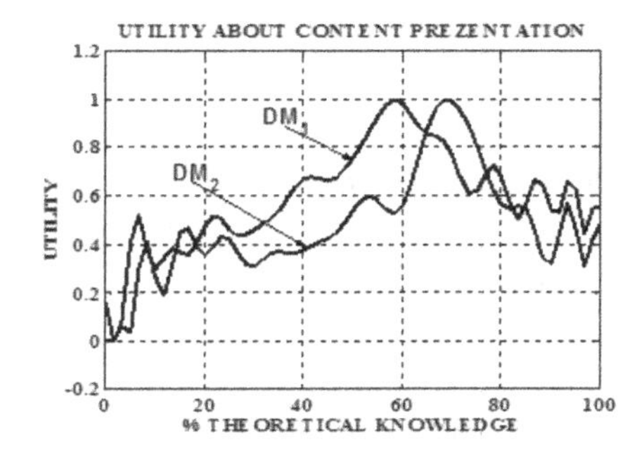

The first evaluation is based on 64 questions (learning points) – utility function $f_1(x)$. It is sufficient only for the first raw approximation. This assessment is fast, since the examination takes about 20 minutes. The utility function $f_1(x)$ clearly reveals the tendency of learner's preference. The right graphic shows estimation of the preferences of two learners (DM_1 and DM_2). It is noticeable that learners have different preferences for the percentage of theoretical knowledge presentation given in learning material (DM_1 chooses lower level of theoretical presentation than DM_2). Consequently, learners will need altered learning resources, adapted to their preference. Presented graphics are with stochastic uncertainty.

The second example depicts the measurement of learners' preference with respect to learning object delivery mode. A graphical representation of results is shown in Figure 6. The used criterion is the following—% synchronous learning objects delivery in relation to the entire course delivery. This concerns the proportion of synchronous delivery format to asynchronous one. Figure 6 shows the examination of preferred delivery mode of a learner in two ways with increasing precision. The utility function $f_2(x)$ is evaluated with 64 learning points, while the other utility function $f_1(x)$ (solid style)—with 128 points.

The used evaluation techniques allow learner's preferences to be measured more precisely, if it is necessary. The seesaw line on the figure is a pattern recognition based on 128 learning points and answers, while the solid line $f_2(.)$ is the utility function. The utility function $f_2(.)$ (solid style) reflects more precisely the learner's preference than the utility $f_2(.)$ (dot style) that is evaluated with 64 answers. As a result, we can assess specific learner's preference more precisely and use the corresponding "delivery mode."

A possible application of the technique for measurement of human preferences presented above is to indirectly determine the learner's competence

Figure 6. Preferences for learning objects delivery mode

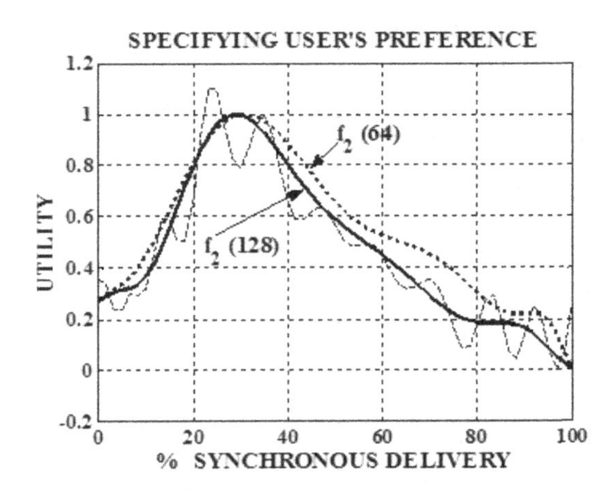

through measurement of teacher's (expert's) preferences. This technique is used for creation of a stereotypical representation of the learner's skill concerning the form and style of an examination. The learner modeling uses the teacher's opinion about learners' competence. It is a result of his/her experience. The form features of an examination (A) present the way of knowledge expression by a learner: test or free expression. The features of examination style (B) concern the oral and written style of examination. The possible criteria for the estimation of the preferences of learners which satisfaction is the objective of the teacher during an examination are the followings:

A. "% test in relation to the entire examination material" (0% to 100%) – illustrated in Figures 7 and 8:
B. "% time for written exam in relation to the whole time that is necessary for this exam" (0% to 100%) – illustrated in Figures 9 and 10.

The number of teacher answers is 64 for the Figure 7 and Figure 8. It is sufficient only for the first approximation. The seesaw lines in Figure 8 and Figure 10 recognize correctly more than 95% of the teacher answers. The utility dependence on probability can be accessed directly with the proposed stochastic procedure (formula 10.5.1.4). For this purpose we search for an approximation of the kind u(x, α), α∈[0,1], x∈**X** following Kahneman and Tversky. The utility function u(x,α) is shown on Figure 9. The explicit formula of the utility u(.) has the form:

$$\mathbf{u}(\mathbf{x}) = \int_0^1 \mathbf{u}(\mathbf{x}, \alpha) \mathbf{d}\alpha \,.$$

The utility functions on Figure 9 and Figure 10 are constructed by 1024 "learning points." Since the teacher accepts that the factors (A) and (B) are mutually independent in relation to "util-

Figure 7. Utility $f_1(x,\alpha)$

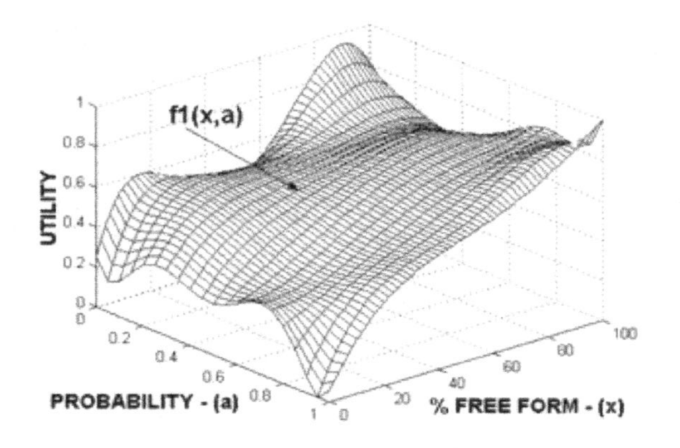

Figure 8. Utility $f_1(x)$

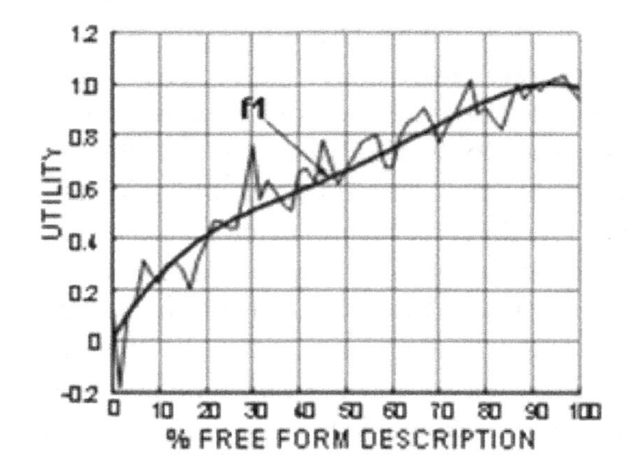

Figure 9. Utility $f_2(x,\alpha)$

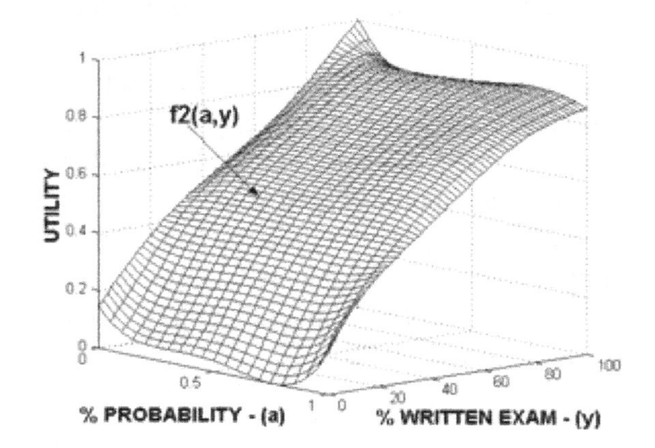

ity," the utility function has the following expression:

$$U(a,b)=K_1*f_1(a)+ K_2*f_2(b)+ (1- K_1- K_2)*(f_1(a)*f_2(b)), \qquad (10.6.6)$$

where a,b∈[0,100]%. The determination of the coefficients K_1 and K_2 depends on the determination of $f_1(.)$ and $f_2(.)$ and possible algorithm for their evaluations is described in chapter 7. This utility function is presented by Figure 11 and Figure 12.

The Figure 12 illustrates the lines of identical preferences that show a way for partitioning the group of learners in subgroups in accordance with their identical preferences. We can determine to which subgroup a learner belongs through the construction of his $f_1(.)$ and $f_2(.)$. They are a source for the determination of a_{max} and b_{max} and $U(a_{max},b_{max})$ that shows the position of the learner in the space presented by Figure 12, i.e. the subgroup to which the learner belongs.

Figure 10. Utility f_2 (x)

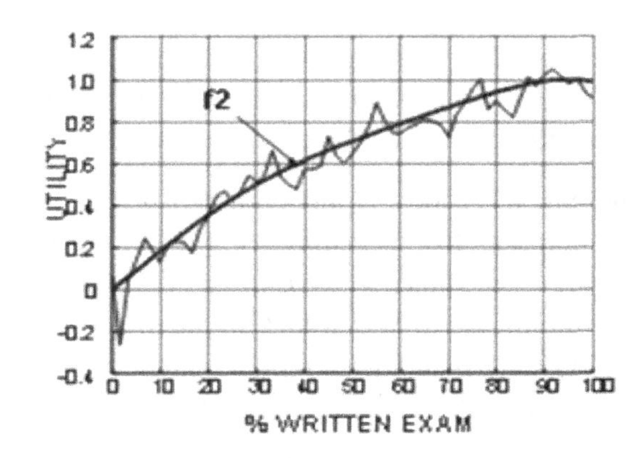

Figure 11. Teacher utility

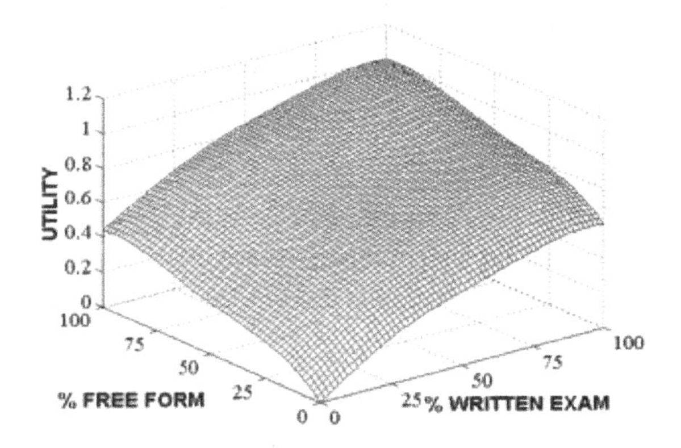

Figure 12. Partition of e-learner's group

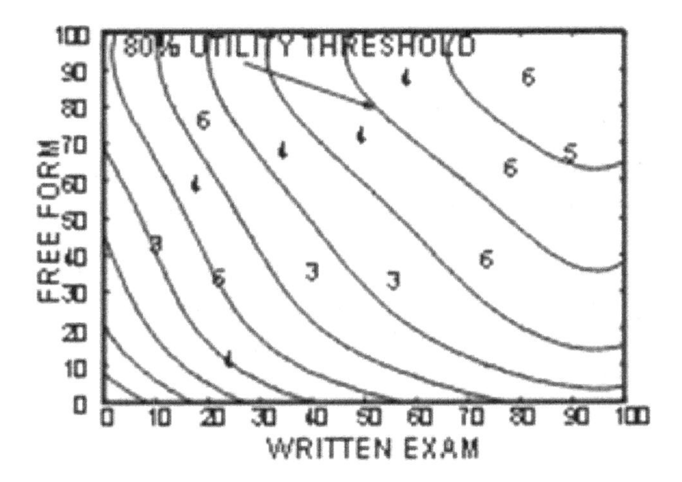

REFERENCES

Aizerman, M., Braverman, E., & Rozonoer, L. (1970). *Potential function method in the theory of machine learning*. Moscow, Russia: Nauka.

Andreev, R. (2008). A model of self-organizing collaboration. In *Proceedings of IFIP WCC 2008 Conference on Biologically Inspired Collaborative Computing, Book Series in Computer Science: IFIP Advances in Information and Communication Technology*, (pp. 223-231). Springer.

Andreev, R. D., & Troyanova, N. V. (2006). E-learning design: An integrated agent-grid service architecture. In *Proceedings of IEEE John Vincent Atanasoff' 2006 International Symposium on Modern Computing*, (pp. 208-213). Los Alamitos, CA: IEEE Computer Society.

Aroyo, L., Dolog, P., Houben, G.-J., Kravcik, M., Naeve, A., Nilsson, M., & Wild, F. (2006). Interoperability in personalized adaptive learning. *Journal of Educational Technology & Society*, *9*(2), 4–18.

Barnes, B. H., & Bollinger, T. (1991). Making reuse cost-effective. *IEEE Software*, *1*, 13–24. doi:10.1109/52.62928

Bleichrodt, H., & van Rijn, J. (1999). Probability weighting and utility curvature in QALY-based decision making. *Journal of Mathematical Psychology*, *43*, 238–260. doi:10.1006/jmps.1999.1257

Bruner, J. S. (1967). *On knowing: Essays for the left hand*. Cambridge, MA: Harvard University Press.

Brusilovsky, P. (2001). Adaptive hypermedia. *User Modeling and User-Adapted Interaction*, *11*(1-2), 87–110. doi:10.1023/A:1011143116306

DeCarvalho, R. (1991). The humanistic paradigm in education. *The Humanistic Psychologist*, *19*(1), 88–104. doi:10.1080/08873267.1991.9986754

Fishburn, P. (1978). Utility theory. In *Handbook of Operational Research: Foundations and Fundamentals*. New York, NY: Van Nostrand Reinhold Company.

Kahneman, D., & Tversky, A. (1979). Prospect theory: An analysis of decision under risk. *Econometrica*, *47*, 263–291. doi:10.2307/1914185

Kay, J. (2001). Learner control. *User Modeling and User-Adapted Interaction*, *11*(1-2), 111–127. doi:10.1023/A:1011194803800

Keeney, R., & Raiffa, H. (1993). *Decision with multiple objectives: Preferences and value trade-offs* (2nd ed.). Cambridge, UK: Cambridge University Press. doi:10.1109/TSMC.1979.4310245

Keller, J. M. (1983). Motivational design of instruction. In Reigeluth, C. M. (Ed.), *Instructional-Design Theories and Models: An Overview of their Current Status*. Hillsdale, NJ: Lawrence Erlbaum Associates.

Koper, R., & Oliver, B. (2004). Representing the learning design of units of learning. *Journal of Educational Technology & Society*, *7*(3), 97–111.

Lave, J., & Wenger, E. (1998). *Communities of practice: Learning, meaning, and identity*. Cambridge, UK: Cambridge University Press.

Leontiev, A. N. (1978). *Activity, consciousness, personality*. Englewood Cliffs, NJ: Prentice Hall.

Marx, M. H., & Cronan-Hillix, W. A. (1987). *Systems and theories in psychology*. New York, NY: McGraw-Hill.

Nielsen, J. (1993). *Usability engineering*. Boston, MA: Academic Press.

Pavlov, Y. (2005). Subjective preferences, values and decisions: Stochastic approximation approach. *Comptes rendus de L'Academie bulgare des Sciences, 58*(4), 367-372.

Raiffa, H. (1968). *Decision analysis*. Reading, MA: Addison-Wesley.

Stephanidis, C. (2001). Adaptive techniques for universal access. *User Modeling and User-Adapted Interaction*, *11*(1-2), 159–179. doi:10.1023/A:1011144232235

Vygotsky, L. S. (1978). *Mind and society: The development of higher mental processes*. Cambridge, MA: Harvard University Press.

Zukerman, I., & Albrecht, D. W. (2001). Predicative statistical models for user modeling. *User Modeling and User-Adapted Interaction*, *11*(1-2), 5–18. doi:10.1023/A:1011175525451

Zukerman, I., & Litman, D. (2001). Natural language processing and user modeling: Synergies and limitations. *User Modeling and User-Adapted Interaction*, *11*(1-2), 129–158. doi:10.1023/A:1011174108613

Chapter 11
Decision Support in Bird's Production Farms

ABSTRACT

In this chapter, the authors demonstrate the flexibility and the possibilities of the Multiattribute Utility Theory and of the Multiple Linear Extrapolations Theory together with Stochastic Programming for Modeling of Complex Systems in difficult to be modeled areas like poultry farming. The flexibility of the methodology allows designing information decision support systems or even autonomous expert systems for such complex processes and objects.

The aim of this chapter is to develop models and methodology for forecasting the development of chickens and broilers (the growth curve) as well as production of eggs which will help the rational nourishing and developing the chickens of different breeds and obtaining more and better production. This means that this methodology allows taking into account the special features of the concrete population of birds in the concrete bird-producing place. It would be very perceptive for modern bird breeding to create information systems, which cannot only prognosticate for the forthcoming period based on obtained data and also giving advice to the bird breeder for control of the observed processes.

In order to achieve this aim, methods and theoretical studies in the area of expert or decision support systems are used in the area of bird production. The investigation includes development of the methodology (approach and algorithms) for such systems and its testing on the level of modeling and data simulation.

DOI: 10.4018/978-1-4666-2967-7.ch011

1. INTRODUCTION TO THE PROBLEM

Modern bird production requires application of modern approach and using effective methods and means for producing cheaper products and protein. One of the basic scientist tasks is to find as many effective conditions for increasing the quality of the production of the level of modern standards as possible. The aspiration is for observing the standards for ecologically clean production and preservation of the environment.

The effectiveness of the modern bird production is shown by the fact that the countries from the European Union have increased their production by 80% in the recent ten years. At the moment the main problems for the Bulgarian bird production are to preserve the current level and to gradually reach the European level of bird production (Ljakova, 2003). This is naturally related to the increasing funds for scientific needs, which are currently insufficient. For raising the bird production level to the modern European standards, it is necessary to apply modern technological and scientific approaches (Technical Documentation ISABROWN, 1995).

The aim of this research is to develop models and methodology for prognosis of the development of chickens and broilers (the growth curve) as well as production of eggs, which will help for rational nourishing and developing of the chickens of different breeds and obtaining more and better production. In order to achieve this aim methods and theoretical research in the area of expert or Decision Support Systems (DSS) can be used in the area of bird production (Pavlov, 2001; Ljakova, 2002).

The research is done for Shaver breed broilers and ISA-brown breed chickens (Technical Documentation ISABROWN, 1995). The period for development of the broilers is 0 to 6 weeks and for the chicken 0 to 32 weeks. For egg production the period of research is 16 to 76 weeks (divided in 16 intervals).

The task includes making prognosis for the change of the weight for the next 3 weeks after the last measurement, under the initial feeding conditions. The prognosis is based on the particular curve for measurement of the mean weight of the population until the moment. The same is valid for the percent of the laying hens per day in the considered interval of time, with regard to the production of eggs. In the technical documentation concerning laying hens of the breed ISA-brown and broilers of the breed Shaver, data of the optimal intervals in which considered parameters can take values in the different periods of development are given (Technical Documentation ISABROWN, 1995).

It would be very perspective for the modern bird breeding to create information systems, which can not only prognosticate for the forthcoming period based on obtained data but also give advice to the bird breeder for control of the observed processes (Ljakova, 2002, 2003, 2005; Pavlov, 2001). The task also includes development of the methodology (approach and algorithms) for such systems and its testing on the level of modeling and data simulation. For this purpose, methods for processing of expert information are included and their estimating capabilities for the light program influence over the egg production have been demonstrated.

2. FACTOR ANALYSIS

The modern market requires effective methods and means for production of cheap animal products. One of the main scientific tasks is to investigate the essential factors and their interdependencies that can allow such a production. For production on the level of modern standards, it is necessary to use modern technologies including DSS in this specific area (Keeney, 1993). In all cases, the aspiration for obeying the standards for ecologically clean production and preserving the environment remains (Ljakova, 2003). Such scientific investigations can allow quick start for large-scale and mass production even on the level of the individual producer. Of course, the use of DSS requires corresponding knowledge of the mass producer, as well as some capabilities in the area of the information technologies.

Before we approach the mathematical description and development of the models, it is necessary to seek for structural description of the task (Gig, 1981; Diday, 1979). The aims are two. The first is the reduction of the dimension of the model via investigating the mutual interdependencies in the given task from the point of view of the factor analysis. The second aim is structural description of the reduced system.

Reducing the dimension of the used models via investigation of the mutual interdependencies is a basic moment for the development of the mathematical model (Enstein, 1977; Keeney, 1993). The aim is the correct determination of the basic factors from which the investigated processes are dependent and reduction of the information overlapping which occurs when large numbers of information criteria are used. For the development of models of information support systems, the choice of these factors determines the structure of the system. This stage is the most difficult for formalization and in its base may be placed the factor analysis (Lloyd, 1984). On the base of closeness matrices, mutual dependencies between the separate features are estimated. These matrices are computed based on closeness measure.

The choice of such measure must take into account the scales in which the measurements are made (Krantz, 1971, 1989, 1990). Not taking into account the variety of scaling corresponding to the different criteria is a mistake often occurring when complex systems are investigated. It can predetermine the success or the difficulties in developing such systems. Not taking into account the variety of scales and their correspondence with the empirical data will lead to the fact that the final numerical result will depend on the concrete measurement values and on the concrete scales (Braverman, 1983). This is not admissible in the investigations. In the concrete research, we are using information for the following important for the bird breading factors (Pavlov, 2001):

1. Temperature;
2. Humidity;
3. CO_2;
4. Ammonia;
5. Speed of the air;
6. Days;
7. Average layingness;
8. Average mortality.

The research is done via closeness matrix based on the coefficient of Goodman (Pavlov, 2001; Ljakova, 2002, 2003).

$$\mathbf{GOOD} = \left|\frac{\mathbf{P}-\mathbf{Q}}{\mathbf{P}+\mathbf{Q}}\right|.$$

Each feature determines one vector. For each pair of vectors the coefficient of Goodman is computed. The data for the features is computed in pairs. Here P means that the measurements are done in one direction and Q means that the measurements are done in the opposite direction. The Goodman coefficient is used as information input for the next step, the empirical algorithm SPECTRE (Braverman, 1983). It rearranges the features seeking maximization using the following formula:

$$K(G,p_l) = \frac{1}{m}\sum_{i\in G} a_l^i .$$

In this case, the group of the ordered features is denoted with G, which number until the moment we denote by m. By p_l we denote the following feature which does not belong to the group of ordered by the moment features, by a^i_l we denote the element of the matrix of Goodman, with regard to the "i"-th feature of G and the "l"-th feature which has not been classified yet. We optimize using the given formula and to the last place for the moment, we add the feature for which the formula has minimal value. For each subsequent computation with the augmentation of the number of elements of G the coefficient $1/m$ decreases. After the initial grouping the set of features is divided in L groups $G_1, G_2, \ldots G_L$. The grouping is obtained by following the moments of sharp decrease of the instant value of $K(G,p_l)$. This is a sign that we pass to the new subgroup. The order obtained in this way is shown in Figure 1.

The order obtained via SPECTRE with the Goodman coefficient is the following:

(1 3 4 7 2 5 6 8)

One can see that the following features are close: (1, 3), (4, 7), (5, 6, 8). According the coefficient of Goodman the temperature and CO2 have equal informativeness.

Of course, if we use other coefficients, as correlation coefficient, for example, we can receive other interdependencies (Enstein, 1977; Lloyd, 1984; Montgomery, 1980). The shape of the curve based on SPECTRE for correlation coefficients is shown in Figure 2. The order is:

(1, 7), (3, 4), (5, 2), (8, 6).

Independently of the bounded excerpt, the features (5, 8, and 6) are close again. The same is true for (1, 7, and 3). Yet, (5, 2), (8, 6) have to be placed in one group.

It makes impression that independently of the bounded excerpt the features (or the observed quantities in control theory terms) (5, 8, and 6) are close again. The same is true for (1, 7, and 3). Yet we have to place in one group (5 and 2) and (8 and 6).

Then, for example, the reduction for basic features in the research can be the following:

1. Temperature (CO_2, average laying eggs);
2. Humidity (speed of the air);
3. Ammonium;
4. Days (average mortality).

The previous stage can be viewed as a transitional for another more complex factor analysis. On the next level, the groups themselves are optimized. Algorithms such as the algorithm for "extreme grouping of the parameters" can be applied (Braverman, 1983). For this purpose, the results of the SPECTRE with matrix based on

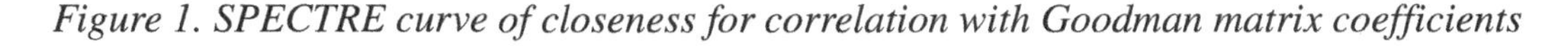

Figure 1. SPECTRE curve of closeness for correlation with Goodman matrix coefficients

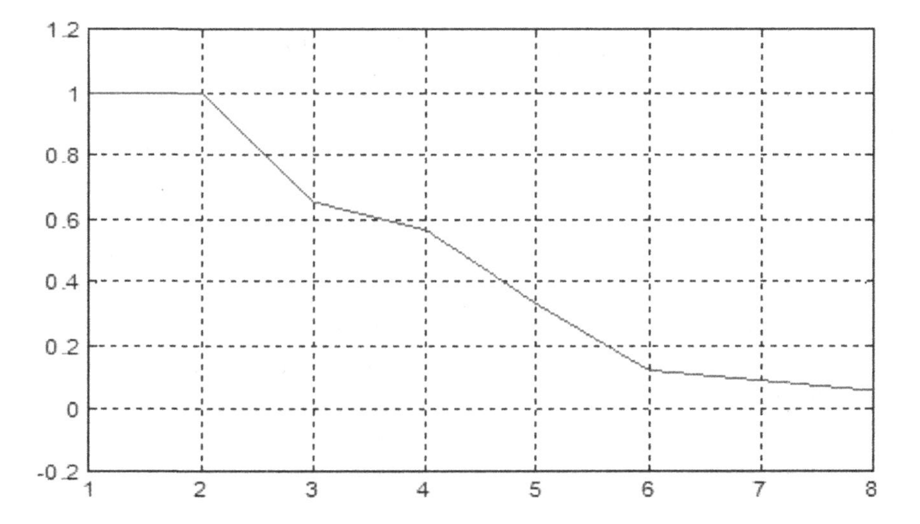

Figure 2. Curve of closeness for correlation closeness matrix

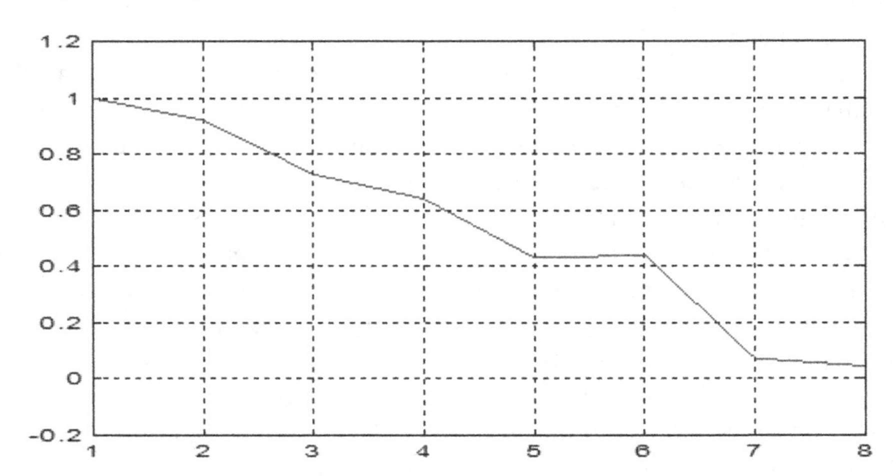

Goodman coefficients are included as initial data for the "extreme grouping of the parameters." For each subgroup $G_1, G_2, \ldots G_L$, a factor $(f_1, f_2, .. f_L)$, is computed via solving the problem:

$$\sum_{i \in G_k} (x_i, f_k)^2 = \max, (f_k, f_k) = N, k = 1 \div L.$$

After that, the optimization is done using the criteria:

$$I = \frac{1}{N^2} \times \sum_{i \in G_1} (x_i, f_1)^2 + \frac{1}{N^2} \times \sum_{i \in G_2} (x_i, f_2)^2 + \frac{1}{N^2} \times \sum_{i \in G_L} (x_i, f_L)^2.$$

Based on the achieved by the moment grouping $G_1, G_2, .. G_L$ for each feature x_i the following value is computed.

$$\frac{1}{N^2} \times \sum_{i \in G_k} (x_i, f_k)^2 \quad .$$

After this, x_i is placed in the group with maximal value. This proceeds until the groups stop changing. It is guaranteed that a local extremum can be found. In this way, we can receive new optimal reordering of the groups. When we use the correlation matrix the result is:

(1, 3, 7), 4, (2, 5), (6, 8),

- Temperature (CO_2, average laying eggs);
- Humidity (the speed of the air);
- Ammonia
- Days (average mortality).

Groups are finally determined as an intersection of the all received until here (summable) results, from all calculations with chosen different type coefficients. On this stage of the research the factor analysis ends, but if necessary it can continue iteratively.

The choice of the type of the coefficients of nearness is very important and this is a stage with very low formalization concerning mainly "the art of taking decisions." Here leading methodology can be the theory of measurement (scaling) (Gig, 1981; Krantz, 1971, 1989, 1990). The closeness coefficient has to be considered with the singularities of the process and the aim of the research. We have to point out again that the excerpt is relatively bounded. We can certainly say that there exist enough data from the literature for serious research. Such data can be gathered from bird producing farms without any effort.

These researches are very important for complex systems, because the result of them can predetermine the end result of the investigations.

If we are looking for adequate model, which is not additionally complicated, one feature has to be chosen from any of the formed groups. In this case we are not losing information, because features from one group carry one and the same informativeness. For example, if we are developing a DSS we must determine the factors that indeed influence the process (Keeney, 1993; Braverman, 1983). The use of this approach in forecasting and consultation model diminishes the noise from the overlapping criteria.

3. PROGNOSIS WITH MULTIATTRIBUTE LINEAR EXTRAPOLATION

We start with the prognosis of the average weight of the population of the bird-producing farm. The aim of this work is to investigate and develop method for prognosis of the development of the population of the Isa-brown egg laying hens (curve of the growth) which may help for rational nutrition and development of the birds (Ljakova, 2002, 2003). It is perspective to develop information system, which not only makes prognosis, but also gives an advice and controls and manages the production (Pavlov, 2001). It is possible to use several methodologies for prognosis depending on the quantity of available information. Such methodologies are regression analysis, different optimization techniques from operation researches, the method of Multiaatribute Linear Extrapolation (MLE) and others.

In this research, we are going to use the method of MLE. The period of the development is 0 to 32 weeks. In this period, the birds reach sexual maturity for given required weight. The task includes the prognosis of the development of the weight during the next several weeks. The prognoses are made under the assumption that the conditions of feeding have preserved the conditions used for the specific measurement curve for the specific average weight of the population until the moment. In the literature for Isa-brown egg laying hens the optimal borders in which the weight can deviate are given. In Figure 3 are given the borders of the weight in the period 0-32 weeks (in the relative units from 0 to 16).

It is possible to use several methodologies depending on the quantity of information which

Figure 3. Optimal borders for development of the weight according to the literature

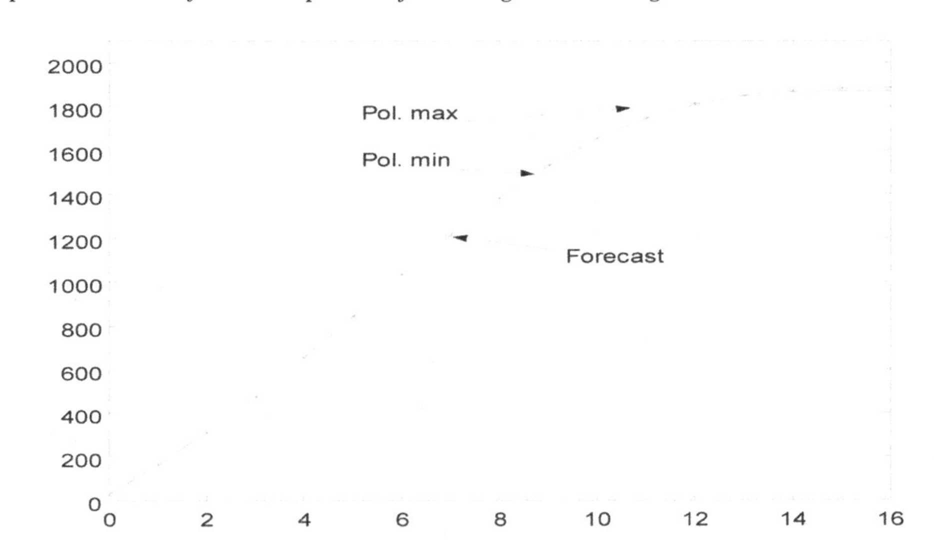

we have (measured average weight (Ljakova, 2002)). In this case, we have used the methodology and algorithms of MLE, because of the fact that it requires relatively small number of measurements and it has shown good result in practice (Rastrigin, 1981; Pavlov, 2001). By T we denote the "situations," in this case, the chosen moment of time in which the measurements are done, and with A we denote the values of the measurements for the "object"—population of birds. We suppose that the functional dependency between T and A is denoted as A=F(T). The mathematical essence of the problem is starting from a not very large number of situations T_i and the known decisions for them A_i (in general vectors) to estimate F(.):

$(T_i \Rightarrow A_i, i=1 \div k) \Rightarrow F,$

$Q(T_i, A_i)=\text{extremum}_{Ai\in\Omega(Ti)},$

Where Q is an optimization criterion, using which we choose A_i or in fact the description of the experimental data for the concrete estimated object for the moment and $\Omega(T_i)$ is the set of potential solution decisions for the moment T_i. The aim is to prognoses A_{k+1}, A_{k+2}, A_{k+3} for chosen T_{k+1}, T_{k+2}, T_{k+3}. The methodology is the following:

1. Under the hypothesis that the transformation A=F(T) is close to the linear we can use the method of MLE.
2. We make the projection A* in the m-dimensional space with basis (A_j, $j=1\div k$), caused by the known teaching situations F_i, $i=1\div k$, where A* are the empirical data for the process and (A_j, $j=1\div k$) are created by F_i, $i=1\div k$.
3. We represent the looked for solution F in the space generated of the teaching situations F_i, $i=1\div k$ with the same coefficients with which the projection Pr(A*) is represented.

We are going to interpret the methodology in the means of the concrete practical formulation. We are going to denote by t_1 the initial, and with t_m the current moments. The curve of growth is approximated in the interval $[t_1, t_m]$ with one of the appropriate for the purpose of the investigation approximation methods (Ljakova, 2002, 2003). In the case, the methodology of stochastic approximation is used (Aizerman, 1970; Ermolev, 1970). The reason is that for these classes of methods we clean the noise of the measurements and the degree of the approximating polynomial is easily changed until we reach good approximation. Let the approximating polynomials have the type:

$$P(t) = \sum_{i=0}^{n} x_i t^i .$$

We choose (κ) training polynomials considering the character of the investigation task or by general theoretical formulation. In the case, these polynomials are obtained from the polynomial description the above optimal upper border of hens growth as the coefficients (x_i, $i=0\div n$) are noised with some noise. Let their values at the moment t_i, $i=1\div m$ are a_{ij}, for $j=1\div k$:

$$A_j = (a_{1j}, a_{2j}, ..., a_{ij}, ..., a_{mj}),\ j = 1 \div k.$$

The coefficients of the polynomial form the following (n+1) dimensional vectors:

$$X_j = (x_{0j}, x_{1j}, ..., x_{nj}),\ j = 1 \div k.$$

The aim is to estimate the vector X* of the unknown coefficient of the polynomial of the real curve of growth and to receive good prognosis. If we multiply the corresponding coefficients of the concrete j^{th} polynomials with the corresponding degree of the chosen moment of time t_i and add them ($j=1\div k$, $i=1\div m$) we are going to receive the following system of linear equations (Rastrigin, 1981):

$$a_1^* = a_{11} + \sum_{j=1}^{k-1} \lambda_j \times (a_{1(j+1)} - a_{11})$$

$$\text{..} ,$$

$$a_m^* = a_{m1} + \sum_{j=1}^{k-1} \lambda_j \times (a_{m(j+1)} - a_{m1})$$

$$\vec{\lambda} = PINV(A_{ij}) \times (\vec{a}^* - \vec{a_1}),\ PINV(A_{ij}) = A^+,$$

$$\left\| (\vec{a}^* - \vec{a_1}) - [A_{ij}] \vec{\lambda} \right\| = \min .$$

To solve the system we are using the pseudo-inverse Moore-Penrose matrices (Arthur, 1972). In the notation A^+ is the Moor-Penrouse pseudo-inverse matrix. Their usage makes the approach independent of the number of measurements during the period of observation of the given process. Moreover, in the general case the measurements can be done in the different intervals of time, which makes the procedure useful in practice. In this case, $A^*=(a_1^*,a_2^*,...,a_m^*)$ are the measured values of the average weight of the investigated population of birds till the moment t_m. Solving the system of linear equations with respect to λ_j and applying the decisions rules using the approach described in point (3) of the methodology we find the solution X* of the given problem. In the common case $(\kappa \neq (m+1))$ and in practice often $(m < (n+1))$. In some cases we look for solutions by applying additional conditions to the initial problem. The properties of the pseudo–inverse matrices are used as it is in the case. In the practical problem polynomial of degree 7 is used. The shape of the given training system of polynomial is shown in Figure 4.

The training polynomials are 5. One example of the teaching excerpt is shown. The number of the measurements in the observed period is 5. In order to check the capabilities of the MLE we noise the data for the upper bound with 10% noise. The results are shown in Figure 5.

It can be easily seen that the prognosis data almost coincide with the literature data for the upper border of development of the weight in the population of birds, which demonstrates the capabilities of the MLE method. Prognoses are made for weeks 6, 7, 8, 9, and for 25, 26, 27, and 28 weeks. The data is shown on Tables 1 and 2.

Figure 4. Training excerpt of polynomials (-the generating polynomial)*

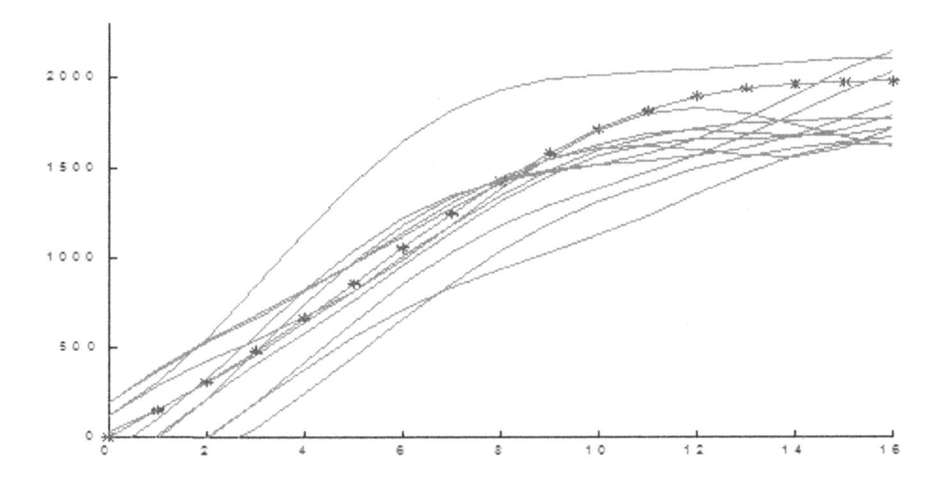

Figure 5. Optimal borders for development of the weight and prognosis for development using the lower border

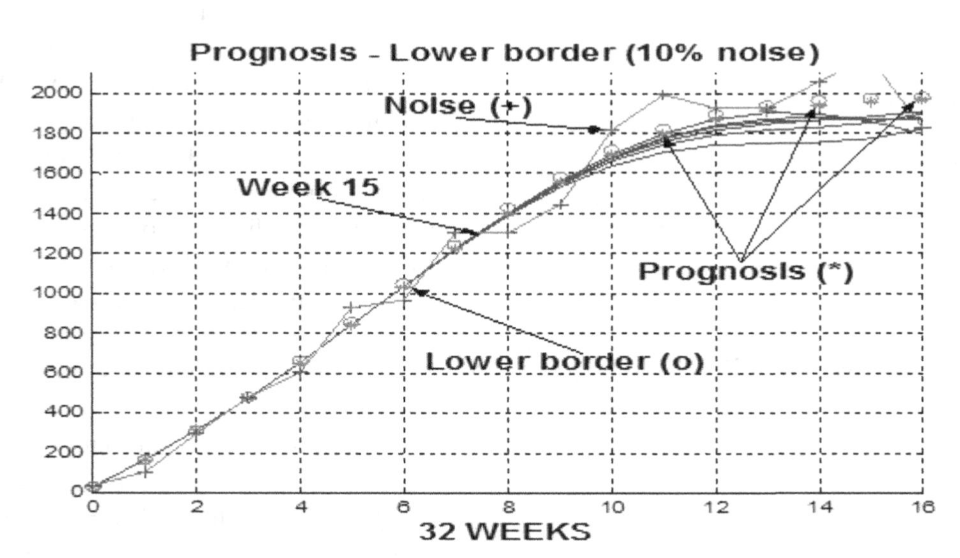

Table 1. Data

week	6, 7, 8, and 9 week			
exact data	312.71	393.24	477.55	565.31
prognosis data	312.82	393.65	478.54	567.27
error %	0.035	0.104	0.21	0.325

Table 2. Data

week	25, 26, 27, and 28 week			
Exact data	1782.43	1810.57	1832.118	1847.72
prognosis data	1782.77	1811.48	1833.794	1850.35
error. %	0.018	0.0604	0.0873	0.184

The research is conducted and prognosis is made for different moments of time, different number of measurements and different level of noise. For example, the prognosis after the 5-week using the upper border has the shape shown in Figure 6.

It can be seen that the prognosis is good even for week 32. The prognosis for the development of the weight of the population can allow a coordination of the way of feeding with the corresponding population of birds with aim to receive better results and better usage of the available resources. This method allows paying attention on the special features of the concrete population of birds on the concrete bird producing farm. Of course, it requires gathering the necessary information. In this direction the MLE method requires minimum restrictions (Rastrigin, 1981).

When the real information contains noise in order to receive training excerpt (set) of polynomials the method of regression analysis or set regression or stochastic recurrence approximation can be used (Pavlov, 2001; Ljakova, 2002, 2003; Ermolev, 1970). The aim is to clear the noise in the real data in order to receive better training polynomials. For this purpose could be used and could be convenient the following stochastic recurrent procedure for high level of noise (Braverman, 1983; Aizerman, 1970):

$$f^{n+1}(x) = f^{n}(x) + \gamma_n \left[f^*(x) + \xi_n - f^{n}(x^{n+1}) \right] [\Psi(x), \Psi(x^{n+1})]$$

$$\sum_{i=1}^{\infty} \gamma_i = \infty, \sum_{i=1}^{\infty} \gamma_i^2 < \infty, \gamma_i \geq 0, \forall i.$$

Figure 6. Prognosis after the 5th week

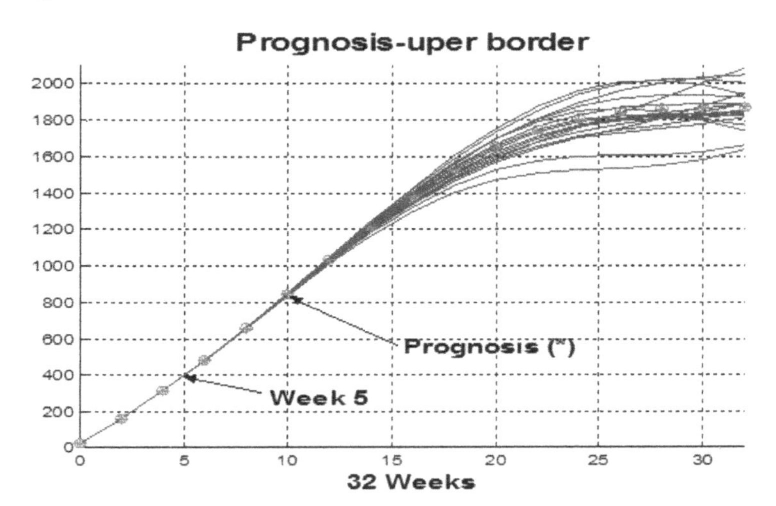

$$M_x\{(f * (x) - f^n(x))^2 \xrightarrow{e.w.} \min_\Psi.$$

In the case, we denote by ξ_n the noise in the measurements, and by the brackets at the end of the formula, we denote the inner product. M_x has the meaning of mathematical expectation, according to the distribution with which the training point's x^n appear. The stochastic procedure $M_x\{(f * (x) - f^n(x))^2 \xrightarrow{e.w.} min_\Psi$ looks for the best quadratic approximation, which bases on the family of functions $\{\Psi\}$. This stochastic procedure is used in the process shown in Figure 5.

In the next stage, we are going to design on the base of MLE method for prognosis egg production process. The results are shown in Figure 7. The polynomials from the training excerpt are built on the first stage with the approach of stochastic approximation and are of degree 25, because of the special feature of the process. The reason is that the polynomial describes process, which at the beginning is with large dynamics, and after the 8 week is described with curve, which is close to line. This makes the polynomial description sensitive toward the digital data for the coefficient of the polynomial. Such a description does not give as good results, and the approximation and the prognosis were not acceptable. The space spanned from the training excerpt does not reflect well the essence of the process in this case. One of the main reasons is the very high degree of the teaching excerpt of polynomials. In the Figure 7, one can see that the approximation is a bit set in motion in the straight line after 8th week. Such deviations in the coefficients of the polynomials combined with the high degree of the polynomials, which generates the approximation space with high dimension, are the reason that the results are not good enough. Under these conditions, the computations are done with matrix of dimension higher than 20 and this effect the final results.

Because of this reason, we approach this description again with the methods of regression analysis. The built new training excerpt of polynomials is generated by an approximating model done with regression of lower degree. The results are shown in Figure 8.

The good correspondence between the model and the prognosis giving polynomial is obvious. The error is under 0.5%. For the best

Figure 7. Egg weight ((g.)hens/day): (exact data-the midle line; (*)- prognosis) and two training polynomials

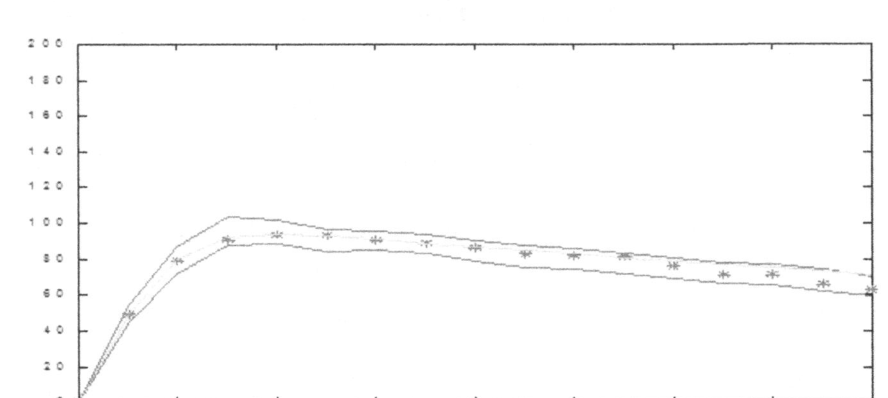

Figure 8. Prognosis (regression model)

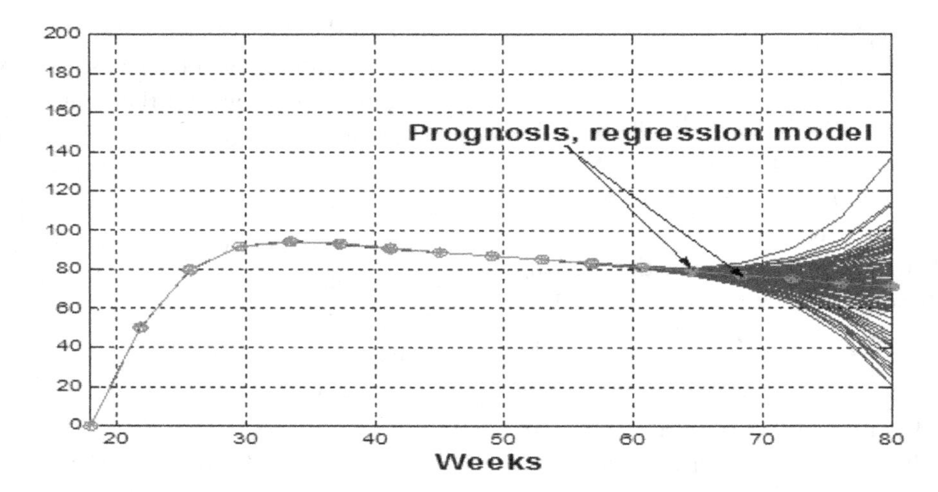

approximations, the error is under 0.05% for the whole interval of prognosis. The prognosis is done using polynomial built from the data from the literature (Technical Documentation ISA-BROWN, 1995). In practice in order to achieve such low error good initial data with enough quantity are needed.

Now the results are good. In the Figure 9, the training excerpt of polynomials is shown.

The conducted model building confirmed again the good capabilities of the MLE method. In this approach the feature that the estimate of the coefficients achieved via MLE methods are statistically not shifted. The obtained results base the viewpoint to admit that the method gives good prognosis and is good in practice. At the end, we have to say that MLE is usable independently of the values of the dimensions (k, m, n), but in certain cases, it is good to combine it with other methods and approaches (Rastrigin, 1981; Pavlov 2001).

Figure 9. Teaching excerpt, egg production (regression model)

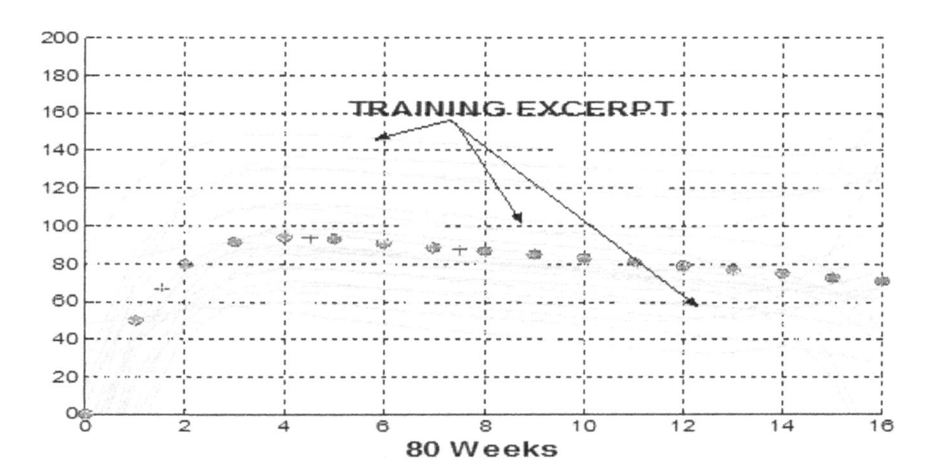

4. PROGNOSIS WITH UTILITY FUNCTION DESCRIPTION AND POLYNOMIAL PRESENTATION

It is possible that other approaches and methods are used for the prognosis in animal husbandry. For example, for complex problems or difficulties in gathering empirical data expert knowledge can be used and the theory for decision-making and Utility theory can be used (Fishburn, 1970; Keeney, 1993). Because of the difficult quantity estimate of the influence of the illumination on the process of egg production and because of the lack of data in the literature, we used expert analysis and Utility theory for measurement of the influence of the illumination (Pavlov, 2001, 2005, 2011).

The first stage is the building of utility function for the decision maker. In the case this is the farmer or the bird breeder who has experience for the concrete bird producing place. We have used the approach described in the chapter 5. The construction of the utility function $f_2(.)$ for the illumination influence proceeded 2-3 days (Pavlov, 2001). We have analyzed the errors in the expert answers and the indetermination in the expert opinion. We have built iteratively ten variants of the utility functions. The final result is shown on Figure 10. It is built based on 128 learning points and expert answers.

In these estimates, the difficult to formalize factors as the price of the eggs in the market at the moment can be taken into account. The expectations of the farmer or another decision maker for the change in the market can be mathematically taken into account (Pavlov, 2001, 2005, 2011).

The following expert estimate is connected with the influence of the illumination in the egg production, but as a function of the average age in the population. We suppose that the influence of the illumination during the first one third of the interval of eggs laying is minimal and after that with their development it gets stronger till it reaches its total capabilities. This effect depends on the concrete population, on the concrete features of the farm as the way of feeding, type of mixture and others. This influence is also expert estimated and we have built with 64 learning points an utility function f1(.), Figure 11.

The accuracy can be increased as we increase the number of the learning points and expert answers, but this is done in dependence if the aim of the processing. We have to point out that this is analytical utility description of the expectations of the decision maker (the farmer) for the way of

Figure 10. Expert utility for the influence of the illumination depending on the deviation from average time (in hours)

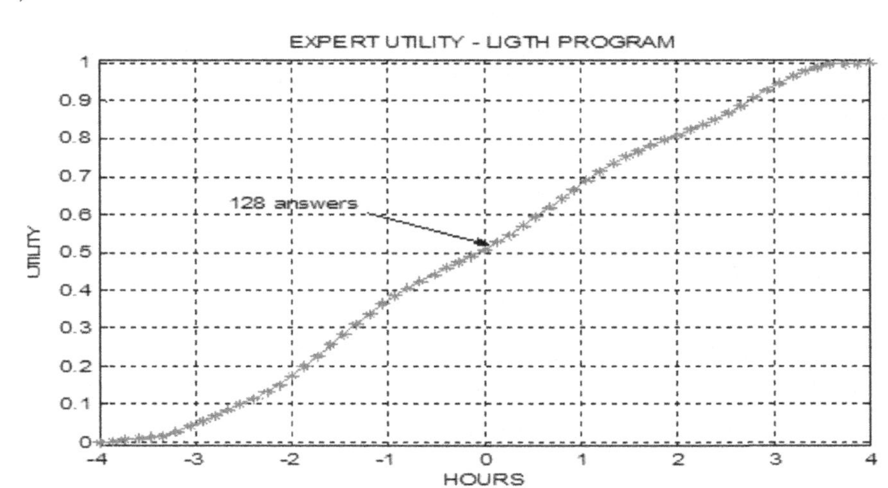

Figure 11. Change of the effect of the eggs production as a function of the development of the population in weeks (64 training points)

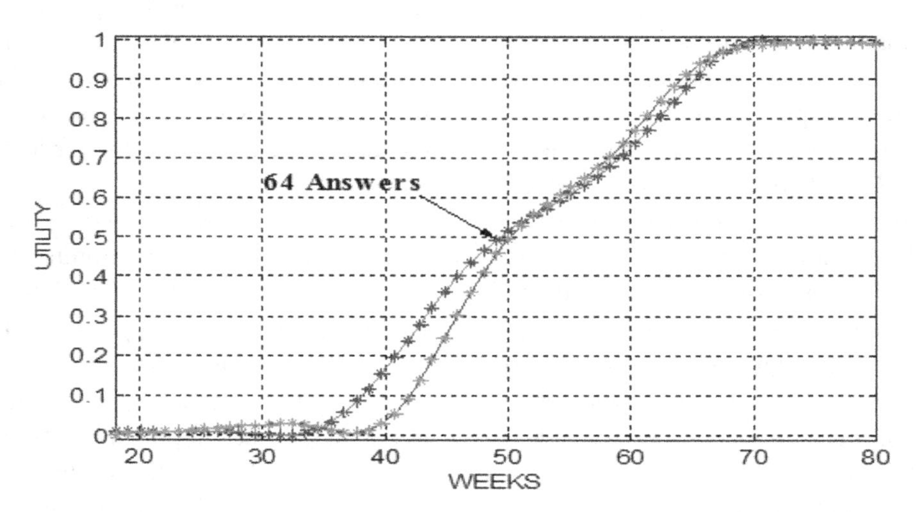

the process flowing in the farm. In fact, this is a description of his empirical knowledge and is well used when there are no other descriptions of the important processes and factors as the illumination in the egg production.

We are going to describe the process of the estimate of the expert's knowledge again shortly. We already know from the description in chapter 5 that we are looking for approximation estimate of the set A_{u^*} of the positive DM's preferences and of the set B_{u^*} of the negative DM's preferences:

$$A_{u^*} = \{(x,y,z,\alpha)/\ (\alpha u^*(x)+(1-\alpha)u^*(y))>u^*(z)\},$$

$$B_{u^*} = \{(x,y,z,\alpha)/\ (\alpha u^*(x)+(1-\alpha)u^*(y))<u^*(z)\}.$$

The star in the notations means empirical estimate of the utility of the decision maker. The utility itself is built as a recurrence procedure for the recognition of the set A_{u*} as is shown in chapter 5 and chapter 6.

Moreover, this procedure in its essence can be considered, as a computer training of the same preferences as these of the expert (Pavlov, 2011). The violations of the transitivity of the relations $(\succ)$ и $(\approx)$ are considered as a noise generated by the uncertainty in the answers of the expert in the correspondence with the prescriptive approach for decision-making.

In the foundation of the chosen approach DM compares "lottery" $(\alpha x+(1-\alpha)y,\ x,y\in\mathbf{Z}, \alpha\in[0,1])$ with simple alternatives $z\in\mathbf{Z}$ and the answer is determined by him ("better," "worse," and "indifferent-equivalency or impossibility for clear delimitation"). The DM compares the "lottery" $<x,y,\alpha>$ with z " training point" in the recurrent stochastic procedure denoted by (x,y,z,α) and he assigns it with probability $D_1(x,y,z,\alpha)$ to the set A_u and with probability $D_2(x,y,z,\alpha)$ to the set B_u.

To each point (x,y,z,α) corresponds the following random function $f(x,y,z,\alpha)=1$ for $(\succ)$, $f(x,y,z,\alpha)=-1$ for $(\succ)$ and $f(x,y,z,\alpha)=0$ for $(\approx)$ (this correspondence is subjective characteristic describing the intuition in the decision maker of the BP in which subjective uncertainty and probability uncertainty are included.) In this way we form the training points $(x,y,z,\alpha, f(x,y,z,\alpha))$. We suppose that (x,y,z,α) are given by the probability distribution $F(x,y,z,\alpha)$. In fact in the information system this is one pseudo random Lp_τ sequence (Pavlov, 2005, 2011). Then the probabilities $D_1(x,y,z,\alpha)$ and $D_2(x,y,z,\alpha)$ are the conditional mathematical expectations of $f(.)$ over the sets A_u and B_u correspondingly (chapter 5). By M we denote the mathematical expectation, in the case the conditional mathematical expectation of $f(.)$ for fixed (x,y,z,α). By $D'(x,y,z,\alpha)$ we denote the random quantity:

$$D'(x,y,z,\alpha) = \begin{cases} = D_1(x,y,z,\alpha), & \text{when } M(f/x,y,z,\alpha)>0, \\ = -D_2(x,y,z,\alpha), & \text{when } M(f/x,y,z,\alpha)<0, \\ = 0, & \text{when } M(f/x,y,z,\alpha)=0. \end{cases}$$

The measurable function $D'(x,y,z,\alpha)$ is approximated by a function from the type $G(x,y,z,\alpha)=(\alpha g(x)+(1-\alpha)g(y)-g(z))$. The function $G(x,y,z,\alpha)$ is positive on $\boldsymbol{A}_u$ and negative on $\boldsymbol{B}_u$ according to the degree of approximation till D'(x,y,z,α). In the case g(x) can be accepted as approximation of the utility function $u(.)$. In fact full recognition of the both sets is impossible since $(\boldsymbol{A}_{u*}\cap\boldsymbol{B}_{u*}\neq\varnothing)$ because of the mistakes and the uncertainty in the preferences of the technologist and the violation of the transitivity generated by the threshold of the uncertainty in his imagination, etc. (Keeney, 1993).

5. EGG-PRODUCTION MODEL DECISION SUPPORT AND PROGNOSES

In the current research we suppose that the joint effect is described by multiplication f(x,y)=f1(x) xf2(y). This effect, if necessary, can be analyzed in details in other investigations and to look for more complicated description of the process (Keeney, 1993; Pavlov, 2001). The built function f(x,y) is shown on Figure 12.

The model of the complete process W(t,x) is the sum of the average eggs production as a function of time in weeks g(t) and the total effect f(t,x) as a function of the time of developing of the population in weeks and the deviation from the illumination (in hours):

W(t,x)= g(t)+c.f(t,x).

In this case, we denote the normalizing constant with (c). In Figure 13, the current prognosis model is graphically shown.

In Figure 14, the fragment concerning 50-80 weeks is shown. In the model, expert estimates are included as utility functions in the complex multiplicative- additive system.

This detail from the methodic of the determining the structure of the function cam require an additional deeper analysis with the aim of exact

Figure 12. Model of the influence: the illumination in dependence of the development in weeks and the deviation from the average illumination in hours

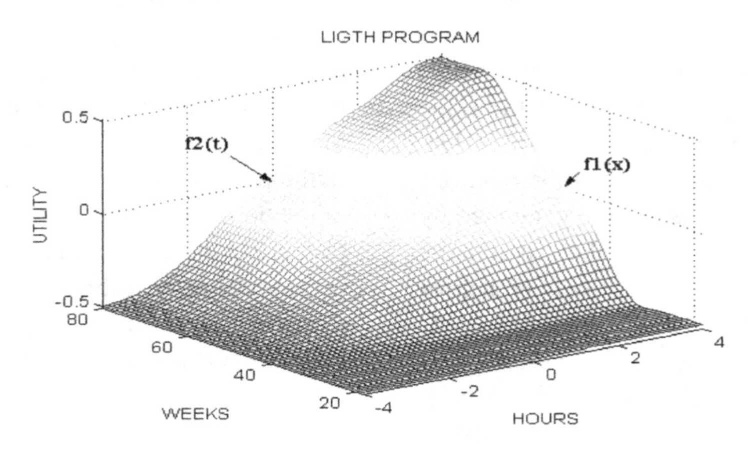

Figure 13. General model of the influence of the illumination

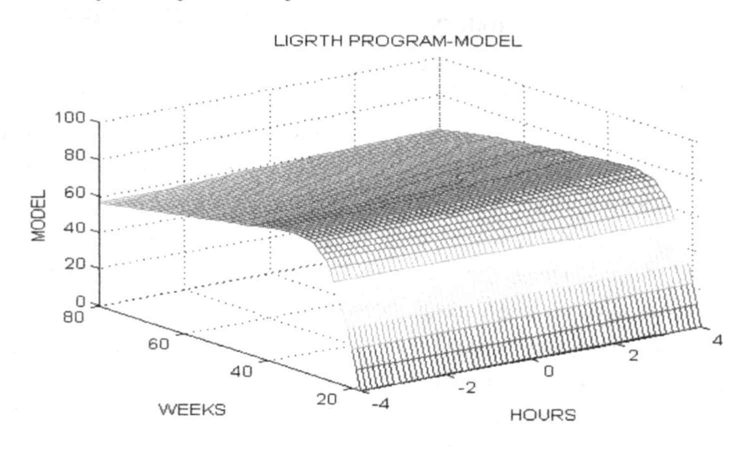

Figure 14. Changes in the model after week 50

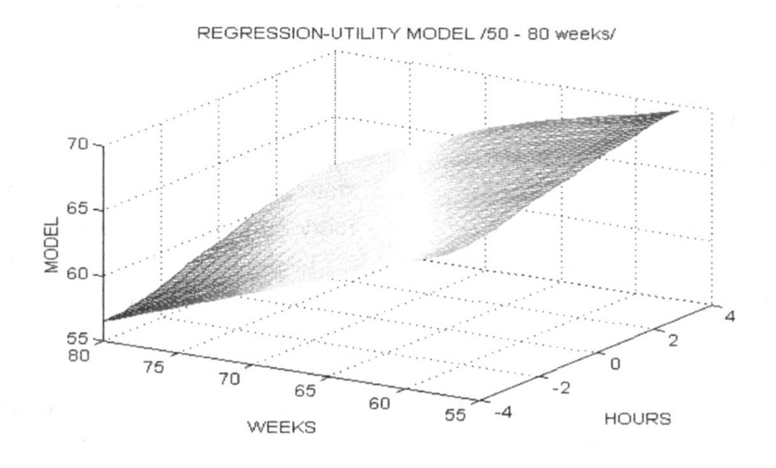

depicturing of the mutual dependencies between the investigated factors. Here we can use the theory developed by Ralph Keeney and Howard Raiffa (Keeney, 1993).

6. EGG-PRODUCTION DECISION SUPPORT INFORMATION SYSTEMS

There are investigations concerning the decision support in agriculture. The investigations debate type, category, and fields of applications of Decision Support Systems (DSS) and their contribution to decision-making in agriculture, mainly in the fields of planning and management of farms, farm regions, and agricultural resources. For example, Manos look through taxonomy of the decision support systems in agriculture (Manos, 2004). The taxonomy is based on an analysis of all published works on applications of DSS from 1987 to 2001, as well as a categorized presentation of these applications. The relevant classification of DSS is made: by treating subject (theoretical DSS or applicative), the source of publication (proceedings, scientific journal, or dissertation), the type of DSS (Decision Support System, Expert System, or Knowledge-Based DSS), the operational research model that each one DSS uses, the operational field (Diagnostic—Forecasting, Advisory, Control; Educational—Informational, Operational), the category of decisions (Strategic or Tactic planning decision). The information system for supporting the decisions in its conception can be self-training, depending on the way of organizing and the determination of the training excerpts (set) of polynomials. We are going to model the most important computational actions of such decision support system based on the directions in Keeney (1993) and Ljakova (2003, 2005).

It is known that the illumination programs can be used as in the growing up as well as in the maturity of the hens laying eggs (Technical Documentation ISABROWN, 1995). The influence of the illumination period is mainly on the consummation of fodder which from its side affects the quantity and the quality of the eggs. The aim is to make prognosis of the development based on the accepted model (Figures 13 and 14) as well as to offer actions and improve the process, even managing the process if the result of the measurements and the process flow does not satisfy the producer.

There is lack of digital data in used from as stocks, but from common considerations, we suppose the following mathematical model (Figure 15). Following the shown above literature stock the illumination programs can be used in growing up period as well as in the period of the maturity of birds. The influence of the illumination period is, as we mentioned above, mainly on the consumption of fodder, which from its side affects the quantity and the quality of the eggs. The increase in time of the illumination makes birds more active and this affects the process of feed up. The maximal deviations are feasible in the borders of $\left(\pm 5\%\right)$. The aim is not only to make prognosis of the development based on the corresponding model but also to offer actions, which improve the process in case the farmer is not satisfied. In this setting, we suppose the following; until the moment we have done N measurements; the illumination program for the development until the moment has been close to the average value, despite the fact that the last is not a restriction under the condition that we track the time for the illumination. This means that each measurement contains two factors—age and the time for the illumination. In Figure 16, we have shown the change in the production of eggs for the minimal and for the maximal illumination.

The aim is to make prognosis for the next three measurement units (~12 weeks) and to see how the process will change with the change of the illumination with the aim to improve the process. The training polynomials have been computed based on set regression (Ljakova, 2002, 2003).

Figure 15. Production of eggs as a function of illumination period and the time in weeks (1 unit = 4.7 weeks)

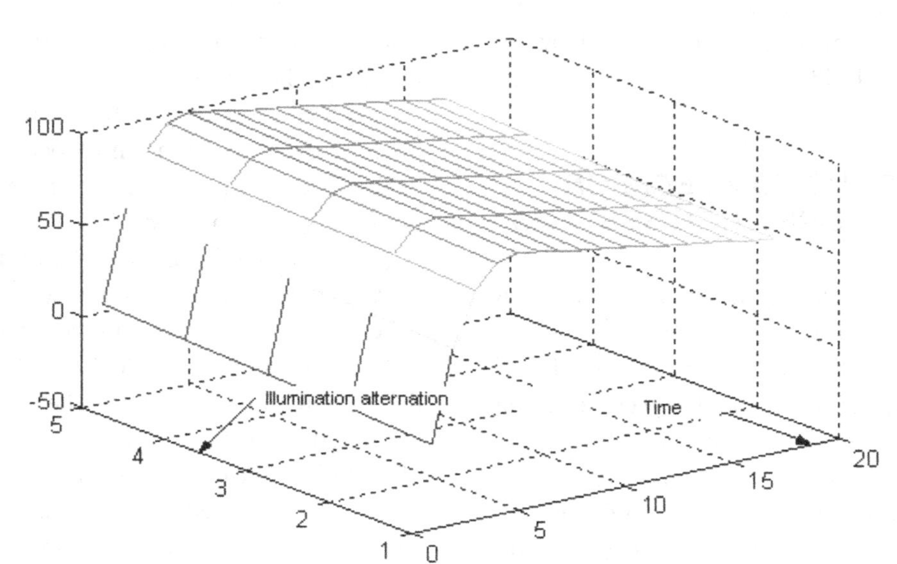

Figure 16. Light program prognoses and the curves for the minimal and the maximal illumination

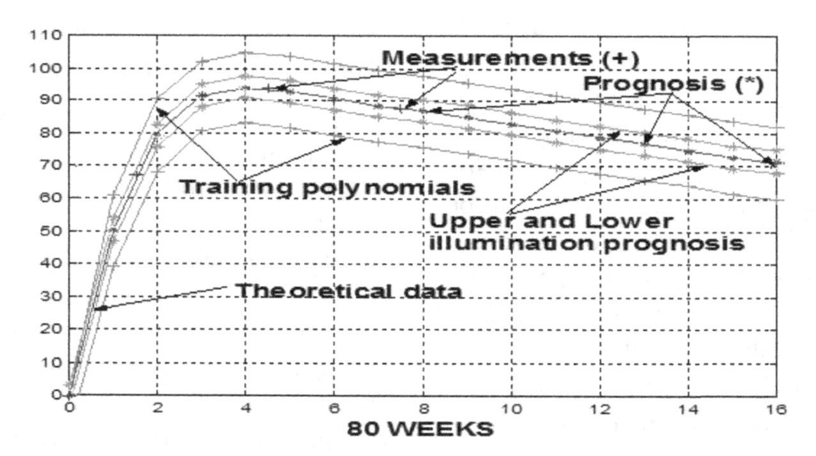

The type of the prognosis as well as the expected changes with the increase or decrease of the illumination can be followed in Figure 17. The modeling proved the applicability of the MLE methods and the accompanying algorithms for developing information decision support systems. At the same time, we discovered some of the special features of its applications. Mainly this is the fact that the main difficulties are connected basically with the determination of the training family of polynomials. The obligatory first stage in the development of DSS is deep factor analysis. This is basic principle for every modeling of complex systems.

For the investigated process, we have received good result for the modeling and the prognosis with the available information. The prognosis and the development of the weight of the population allows to coordinate the way of feeding with the concrete population with the aim to achieve bet-

Figure 17. Prognosis for normal illumination (o) and expected process for maximal illumination (*)

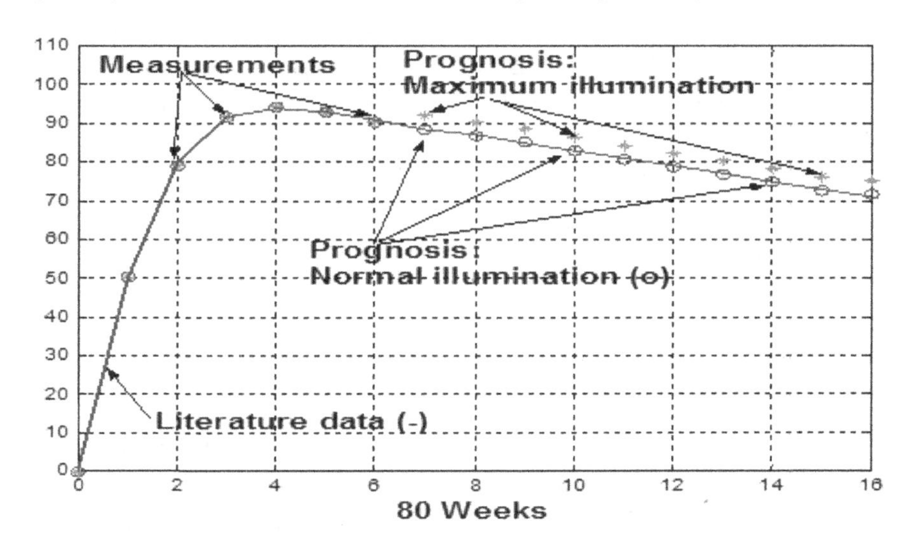

ter results and more efficient usage of the available raw materials. This means that this methodology allows taking into account the special features of the concrete population of birds in the concrete bird-producing place. Of course, this requires gathering the necessary information. However, in this direction MLE method imposes minimal restrictions.

The obtained results give us a reason to accept that MLE gives to us good prognosis and it is applicable in practice. As a final result, the approach is applicable independently of the values of the dimensions (k, m, n), but in certain cases it is better to use other approaches or they can be combined with the proposed approach. The quality of the prognosis depends on the degree of the correspondence between the experimental data and their approximating polynomial (in fact the training family of polynomials). For some of the processes an additional processing of the empirical information may be necessary in order to obtain better excerpt (training family) (Rastrigin, 1981; Ljakova, 2002). A great convenience of the proposed approach is that it is applicable for small number of practical measurements.

It is possible that other approaches can be used in animal stockbreeding. For example, where there are difficulties with gathering empirical information expert knowledge can be used and the theory of decision-making and the utility theory (Aizerman, 1970; Keeney, 1993; Pavlov, 2001, 2011). This theory can be used in order to create a training family, excerpt of polynomials, based on which prognosis is made, or to determine future changes in the parameters.

The flexibility of the methodology allows designing information decision support systems or even autonomous expert systems. In this chapter were demonstrated the flexibility and the possibilities of the Multiattribute Utility theory and of the Multiple linear extrapolations theory together with Stochastic programming for modeling of complex systems. The flexibility of the methodology allows designing information decision support systems or even autonomous expert systems.

This methodology allows taking into account special features of the complex area of the investigation, to model empirical knowledges and to include them in mathematical models for control or management. It would be very perspective for the modern bird breeding to create information systems, which can not only prognosticate for the forthcoming period based on obtained data but also give advice to the bird breeder for control of the observed processes.

REFERENCES

Aizerman, M., Braverman, E., & Rozonoer, L. (1964). Theoretical foundations of the potential functions method in pattern recognition learning. *Automation and Remote Control*, *25*, 821–837.

Arthur, A. (1972). *Regression and the moor-penrose pseudo-inverse*. New York, NY: New Academic Press.

Braverman, E., & Muchnik, I. (1983). *Structure methods for the processing of empirical data*. Moscow, Russia: Nauka.

Diday, E. (1979). *Optimization en classification automatique*. Paris, France: Institut National de Recherche en Informatique et en Automatique.

Enstein, K., Ralston, A., & Herbert, S. (Eds.). (1977). *Statistical methods for digital computers*. New York, NY: John Wiley & Sons.

Ermolev, Y. (1976). *Stochastic programming methods*. Moscow, Russia: Nauka.

Fishburn, P. (1970). *Utility theory for decision-making*. New York, NY: Wiley.

Gig, J. (1981). *Applied general systems theory*. Moscow, Russia: Mir.

Keeney, R., & Raiffa, H. (1993). *Decision with multiple objectives: Preferences and value trade-offs* (2nd ed.). Cambridge, UK: Cambridge University Press. doi:10.1109/TSMC.1979.4310245

Krantz, D. H., Luce, R. D., Suppes, P., & Tversky, A. (1971). *Foundations of measurement* (*Vol. 1*). New York, NY: Academic Press.

Krantz, D. H., Luce, R. D., Suppes, P., & Tversky, A. (1989). *Foundations of measurement* (*Vol. 2*). New York, NY: Academic Press.

Krantz, D. H., Luce, R. D., Suppes, P., & Tversky, A. (1990). *Foundations of measurement* (*Vol. 3*). New York, NY: Academic Press.

Ljakova, K. (2005). Knowledge base implementation in biotechnological processes. In *Proceedings of the International Symposium Bioprocess System*. Sofia, Bulgaria: Bioprocess System.

Ljakova, K., & Pavlov, Y. (2002). Modeling and prognoses in birdproduction decision support. *Automatica & Informatics*, *1*, 42–48.

Ljakova, K., & Pavlov, Y. (2003). Methodological and information provision for some aspects in poultry farming. *Ecological Engineering and Environment Protection*, *2*, 52–59.

Manos, B., Ciani, A., Bournaris, T., Vassiliadou, I., & Papathanasiou, J. (2004). A taxonomy survey of decision support systems in agriculture. *Agricultural Economics Research*, *5*(2), 80–94.

Montgomery, D. (1980). *Design and analysis of experiments*. New York, NY: John Wiley & Sons.

Pavlov, Y. (2005). Subjective preferences, values and decisions: Stochastic approximation approach. *Proceedings of the Bulgarian Academy of Sciences*, *58*(4).

Pavlov, Y. (2011). Preferences based stochastic value and utility function evaluation. In *Proceedings of the Conferences InSITE 2011*. Novi Sad, Serbia: InSITE. Retrieved from http://proceedings.informingscience.org/InSITE2011/index.htm

Pavlov, Y., & Ljakova, K. (2001). Expert information and decision support in complex bird production systems. In *Proceedings of the International Symposium and Young Scientists' School Bioprocess Systems*. Sofia, Bulgaria: Bioprocess Systems.

Pavlov, Y., & Ljakova, K. (2001). Utility assessment in complex systems. In *Proceedings of the International Conference Automatics and Informatics*. Sofia, Bulgaria: ICAI.

Rastrigin, L. (1981). *Adaptation of complex systems*. Riga, Russia: Zinatne.

Technical Documentation. (1995). *ISABROWN breed chickens*. ISABROWN.

Chapter 12
A Preference Utility-Based Approach for Qualitative Knowledge Discovery

ABSTRACT

The chapter presents a conceptual framework of decision-making domain. Decision-making is the major component of the problem-solving system that is based on the principles of decision theory. Important factors that influence decision-making activity are decision environment and decision situation (decision context). Decision-making is observed from several viewpoints that determine different cognitive frameworks.

This chapter gives a short description of a prototype Decision Support System (DSS), which assesses the value and/or utility functions of the individual user. This DSS allows de-facto training of the computer in the same preferences as that of the individual user without the need of additional participant or mediator in the process of utility evaluation. It is mathematically backed up by the methods described in the preceding chapters. The presented methodology and mathematical procedures allow for the creation of such individually oriented DSS for analytic representation of the preferences as objective function based on direct comparisons or on the gambling approach. Such systems may be autonomous or parts of a larger information decision support system.

1. DATA MINING AND KNOWLEDGE DISCOVERY

Data mining generates information about situation structure. Usually, the situation is interpreted to mean a problem, system, or condition. The systematic observation of a situation requires its consideration in certain framework that gives a generic view of its elements and system-based description. The framework guarantees representation and record of situation appearances. A system consists of interrelated objects. An object can be either a real entity or a mere concept (Andreev, 2001). It comprises a set of attributes leading to a precise definition of the system as a relation among attributes of objects. The value of an attribute may be either variable or constant.

DOI: 10.4018/978-1-4666-2967-7.ch012

An appearance of a situation is described by the knowledge about the system represented through objects' attributes. A situation's appearance is given by the values that its constituting attributes hold, when the situation is observed at a particular point in time *t*. Change of values of single attributes results in a modification of the situation's appearance. A sequence of appearances of a situation presents situation (system) behavior. The latter is restricted by the domains of values of the system attributes. However, the situation behavior does not generally imply subsequent appearances to depend on each other. The situation structure is determined by means of static relations and dynamic relations. The former represent constraints on the values of single attributes and interdependences among the attributes of an appearance. Dynamic relations specify the interdependence of the values of attributes regarding subsequent situation's appearances.

Observation of situation behavior is an elementary way of gaining knowledge about a situation. Recording of situation's appearances provides data representing this knowledge. However, situation determination (formulation) requires knowing of values of situation's attributes beyond the data available from observation. It might either not be possible to observe the values of certain attributes of a situation's appearance at hand or the value of a single attribute of a future appearance might be unknown. In both cases information about situation structure is needed for inference unknown attributes' values. Information about static relations allows for drawing conclusions on the value of an attribute on the base of the values of other attributes of the same situation's appearance. The dynamic relations give information necessary for making forecasts about values of attributes of future appearances.

Data mining is an information extraction activity that could be presented by actions with different goals. Such a goal is to search information that is necessary for problem formulation. This information could be derived not only from some information system, but from experts familiar with the situation, as well. Another goal that determines another type of action realizing data mining is to search large volumes of data for patterns and discover hidden facts contained in databases. A unique definition of data mining has not been established yet, since its meaning is not determined by the objective of an action. This activity can be realized by various actions.

From a scientific viewpoint, the data mining process is regarding as data evaluation that consists of two steps (actions): a preprocessing step and core data mining action that seek for a representation of either static or dynamic relations of a situation. It is concerned with secondary analysis of a huge amount of data typically being at hand in the form of recorded situation's appearances. The data is not collected based on experiments designed to answer a certain set of a priory known questions. It is result of either a situation simulation or actual acting of a situation. However, the data cannot be processed by data mining activity without hypothesis about situation (system) structure. Such an a priori hypothesis determines the type of information to be derived from data. The preprocessing step is denoted as a map from a set of recorded appearances into a target data set (Meisel, 2007). This map is implemented by various instances of the following types of preprocessing operations—selection of a subset of appearances, projection of situation's attributes into a subset of these attributes, modification of the domain of an attribute, construction of new attributes by aggregation of original attributes. Usually the preprocessing operations are implemented as search procedure that ensures an optimal target data set with respect to the performance criteria of the subsequent core data mining action.

The main characteristic of data mining process is "provision of information." Due to this characteristic, data mining is very useful for supporting problem determination in the boundaries of a problem-solving activity. Operating within the problem-solving paradigm, data mining is suitable to support decision-making. As decision-making support process, it has to provide with information decision-making process during its three stages—intelligence, design, and choice. With that end, in view our endeavor is to present data mining in the framework of decision problem solving. We accept Humphreys' approach to representation of a decision problem at five qualitative distinct levels through progressively employment of problem expressing, framing and fixing processes to strengthen the constraints on the way of problem description until a course of action is prescribed (Humphreys, 2007a). These are level 5, exploring "what needs to be thought about"; level 4 necessary for expressing the problem and identifying frames, in which it has to be represented; level 3, developing structure within a frame; level 2, determining the decision alternatives within the frame; and level 1, making assessments. The way that participants in DM process set the constraints at these levels progressively establishes their view about decision situation.

These levels have been presented in a point-down triangle, or "decision spine" (Humphreys, 2007b), indicating the progressive decrease in discretion in considering what knowledge can be included in the problem representation being developed as one moves downward from level 5 towards fixed structure, and zero discretion at level 1. Any decision situation is represented at all levels. In the actual decision-making process, the sequence movement through the levels is not linear, but correspond to a spiral through the circular logic of choice where a particular course of action is prescribed as the true solution to the decision situation. The decision spine is not intended to indicate a prescriptive process model for decision-making that starts at level 5, establishes constraints, then goes down one by one through the levels until an action is prescribed at level 1. The representation of decision situation at an upper level is mapped on the next lower level as its more detail representation. This presentation of decision-making as process of extraction of more detailed representation of a decision situation is suitable for presentation of the use of data mining in decision-making context. The Figure 1 presents the classical data mining process that support situation determination during decision-making process at the top three levels (level 5, level 4, level 3).

The preprocessing step of data mining process is considered at the level 5 and level 4 of decision spine. Data mining starts at the top level 5 with situation's appearances within a small world whose bounds are determined by what each of the participants in problem-solving activity is prepared to observed. The appearances of a situation are results of observations that constitute the contextual knowledge available in forming the content elements of situation (problem, system) representations that are manipulated in situation structur-

Figure 1. Model of data mining process

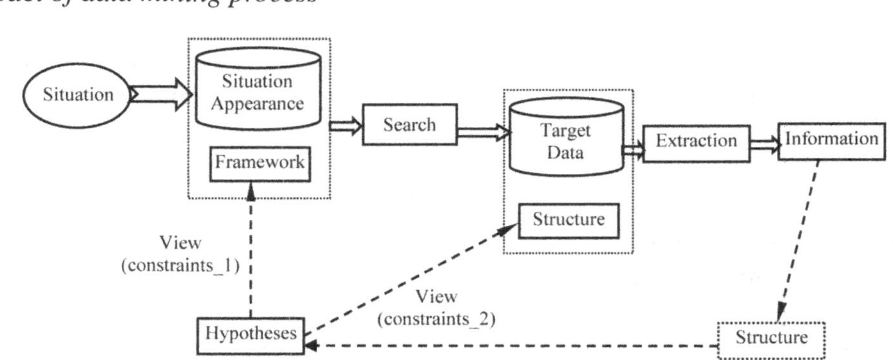

ing at lower levels. At the next level down (level 4), situation's appearances are regarding in suggested hypothetical framework to make claims that particular elements of what was observed should be included in the representation of the situation. The successful establishment of the claims of "what to include in the representation of the situation" is promoted and accepted through the use of the suggested framework. It proposes which elements (aspects) of situation's appearances have to be taken under consideration and to search for the data that present their state. These data is collected as *target data*. The framework is used to define the constraints on what elements can be observed as sources of data so that their implication for potential basis for extraction of information about situation structure can be explored.

The core data-mining step that performs at level 3 relies on both the target data set and a structure, to which the information that has to be derived must conform. The selection of such an information structure is based on a priory hypothesis about situation structure. The implementation of core data mining uses an extraction procedure that through gradually fitting the target data set to suggested situation structure discovers knowledge about the situation. There are two alternative types of knowledge that may be identified by data mining. They depend on the descriptive power of the structure that is suggested as the true structure of the situation (Meisel, 2011). It is distinguished between model representing structure and pattern representing structure. The latter represents exceptional situation behavior while models embrace a whole of a group of situations. The different types of knowledge that are discovered are categorized in correspondence with following core data mining tasks:

1. **Descriptive Modeling:** It results in models that describe an aggregative view of the set of situation's appearances. The descriptive model of a situation has not a central attribute and guarantees inference of a missing value of an attribute of any single situation's appearance (sample of a situation). The most popular technique that concerns this knowledge discovery is cluster analysis. Its objective is to aggregate appearances of data generating process within the target data set by grouping them according to a similarity measures. There are two main clustering-classification types: *supervised* (categorization), in which a fixed number of clusters are predetermined and the samples are categorized into these groups; *unsupervised* (clustering)—the preferred number of clusters is formed by an algorithm while processing the dataset (Bittmann, 2007). Unsupervised methods as hierarchical clustering methods provide visualization of the entire "clustering space" and in the same time enable predetermination of a fixed number of clusters.
2. **Predictive Modeling:** The discovered information presents models describing a relation between one specific situation attribute and a set of others. Such models either allow for forecasts of attribute values of future appearances or they allow for inference of an attribute value that could not be observed in the past. Common techniques for this type of modeling are decision trees and support vector machines for instance.
3. **Discovering Patterns and Rules:** This tasks result in patterns allowing for inference of attribute values of situation appearance subject to the presence of specific values of other attributes within the same appearance. Important techniques for carrying out this task are association rule algorithms, rule induction methods and rough set approaches.
4. **Exploratory Data Analysis:** It relies on visual or interactive technique. A visual or interactive model may provide useful insights into situation structure and support the generation of new hypotheses.

It is necessary to be determined to what extent the derived model or pattern actually represents situation structure, i.e. the information derived by data mining must be verified. Verification is one of the most difficult steps within the data mining process. In many cases verification is done by comparison of attribute values derived from a model or pattern respectively to attribute values derived from observed system appearances.

2. DATA MINING: DECISION-MAKING SUPPORT PROCESS

In the previous section, data mining is presented as basis of a process of situation (problem) determination, since it serves for extraction of knowledge about certain situation. Therefore, data mining process can be considered as decision-making support process that provides decision-making with information during its first stage—intelligence. If we can enlarge the limits of data mining process over the other stages of decision-making process, it gives an opportunity to integrate all stages of decision-making on the base of data mining. On Figure 2 is presented a way for covering of all levels of decision spine in the framework of data mining process. In the preprocessing step of data mining is included decision model structure that reveals decision maker's viewpoint on the situation, since the observed situation fixes the context for decision-making, i.e. it is a decision situation.

At the level 3 the decision situation is structured not only on the base of our hypotheses about the basic system structure that is in situation's foundation, but on the base of hypotheses about decision problem structuring (decision model structure), as well. Decision model structuring frames employed by decision makers discovering decision situation structure within a decision spine fit usually into the following basic categories (Chatjoulis, 2007):

- Frames for structuring future scenarios through modeling act-event-sequences with decision trees: they are basis for linking imagined acts and events in the future through potential consequences of immediate acts with the aim of establishing course of action and investigating side effects under conditions of uncertainty;
- Frames that structure preferences between alternatives through decomposition and re-composition within multi-attribute util-

Figure 2. Data mining-based decision-making process

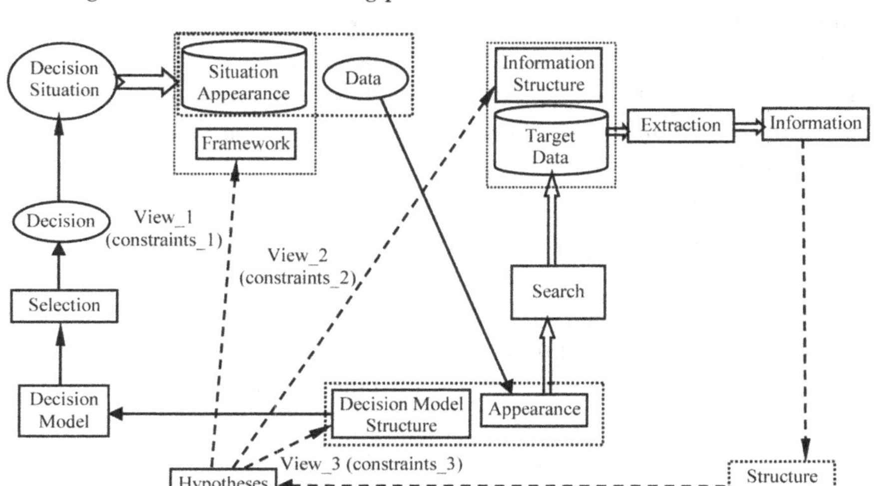

ity hierarchy: they serve to identify value-laden attributes on alternative courses of action under consideration;

- Rule-based structures used for reducing decision-making space: the aim is to constraint possibilities for determination different courses of an action by specifying a set of rules that prescribes only one course of action.

At level 2, the content manipulated with the structured decision models need to be represented as hypotheses, of which a decision model consists. At this level, it is explicitly recognized that probabilities can vary according to individual decision makers' views on future states of the situation and that utilities vary in reflecting the range of decision makers' interests and preferences. At level 1, the only degree of discretion left for the decision makers in developing decision problem (situation) structure within the decision spine is to decide on how to make a "best assessment" of the of "the most likely value" at those points in the problem representation that have been determined as "uncertain" within the constructed decision frames, such that a particular course of action is prescribed. The actions at this level are presented by *"selection"* on Figure 2.

From the perspective of the extended data mining process, the determination of decision model structure and values of the decision attributes of a decision situation is result of an optimization procedure. The latter imposes decision situation structure (knowledge) by definition of static relations between decision attributes and other situation attributes determined by the situation framework. The situation framework and structure are a critical precondition for carrying out this optimization. The model of extended data mining process that describes decision-making process starting with hypotheses about situation's structure and ending with determination of the values of decision attributes consists of three major steps:

1. Identification of basic framework of a situation. It is directly available from observation of situation's appearances. If a domain of an attribute cannot be fully observed, hypotheses about system structure must be consulted.
2. Refinement of basic decision situation model by decision attributes and evaluation attributes. Introduction of these additional attributes implies additional situation structure. A decision model is derived as an instance of a generic decision model structure. Identification of the decision model structure is a crucial step. If the generic structure defining the domains of decision attributes or evaluation attributes cannot be identified easily, hypotheses must be consulted. These hypotheses may have a significant impact on the set of possible search procedures to be applied in step three.
3. Application of a search procedure. The applied search procedure implements a map from the domains of regular situation's attributes into the domains of decision attributes. It seeks for those values of the decision attributes receiving the best evaluation possible. Limitations of the set of applicable search procedures are subject to the decision model structure.

Note that both step one and two may heavily depend on hypotheses about situation's framework and structure. If they do not originate from precise situation's appearances, the determination of structure and values of decision attributes can be pushed into absurdity.

The unified view of decision-making process and data mining brings to consideration of optimization problems in determination of a decision. Optimization derives the values of decision attributes, while data mining identifies information about situation structure. The Figure 2 highlights another important link between decision-making and data mining. Both procedures may heavily rely on hypotheses about situation's structure. The dependency on hypotheses shows the fundamental limitations of decision-making and data mining. Nevertheless, considering decision-making and data mining as an integral methodology affords significant synergies On the one hand, decision-making increases the efficiency of a large number of data mining operations using decision models for optimization of knowledge discovery. On the other hand, data mining leads to an increased effectiveness of many decision-making approaches.

Preprocessing and core data mining step present a proper application domain of carrying out the first stage (intelligence) of decision-making process. Both data mining steps may be conducted according to the model of decision-making model. It is necessary to distinguish clearly the data concerning framework-oriented representation of a decision situation and the process generating target data. This distinction of target data and framework-oriented data is important, since the data generating process of the preprocessing step of data mining searches for target data taking into consideration decision-oriented view on decision situation. This view is represented by the decision model structure that serves for optimization of preprocessing and knowledge discovery in application of data mining process on decision situation determination.

In one-shot decision-making single decision model (model structure) is used as basis for searching suitable target data set and extracting of an instance of the information structure for describing decision situation that is the final result of data mining process. In this case, the framework-oriented data present an appearance of decision situation and the collected data represents constant attribute value. Multi-stages decision-making bases on an incremental data mining process, in which the attributes of framework-oriented representation of decision situation are variables. In contrast with the non-incremental data mining, a suitable target data set and an instance of the information structure is derived for each decision situation appearance. To this end, the decision model of an appearance is determined as an update of the decision model of the previous appearance.

The effectiveness of optimization in determination of a decision depends on the decision model under consideration. The characteristic of a decision model determine the type of optimization technique that may be applied. A decision model allowing for use of an exact algorithm leads to a more effective decision than a model restricting the set of optimization techniques to heuristic search. The effectiveness of optimization vanishes if a decision model cannot be identified at all. The lack of a decision model may be caused in two ways. Either the available hypotheses about situation structure do not permit setting up a decision model structure or the decision model structure at hand cannot be mapped into an effective decision model. In both cases optimization remains ineffective but a decision may be effected with the aid of data mining. The use of data mining in case a decision model structure cannot be established at all is a useful practice. In such cases, a collection of information about situation structure is substituted for a decision model. Typically, this information is represented as a set of rules possibly derived by data mining. The rules are derived as an abstraction of situation appearances but at the same time they may be considered as an abstraction of the decision models that would have been consulted if a decision model structure could have been established.

3. PREFERENCES, ORDERING, GAMBLING

The orientation of many of the information systems, whose purpose is in the area of decision-making, is prescriptive. In other words, aiding the decision maker and not description or predetermining the final outcome. We will emphasize again that the decision as a result is manifestation of three main points, which are the available alternatives, the discrete distributions by which the alternatives are manifested as consequences of certain actions and desires of the decision maker, explicitly expressed as DM's preferences, with respect to the occurrence one or another alternative as consequence. Utmost attention in this book is devoted to the last two points: the desire as expression of utility (value) of the decision and partially the probability distributions of the alternatives as a consequence of the chosen decision (relative likelihoods for each alternative).

The main assumption in each solution is that the values of the subject making the decision are the locomotive force, and as such, they are the main moment in supporting the decisions (Keeney, 1988, 1993). This means that the values are the guiding force in supporting the decisions and due to this are determining for the time and forces, which are selected for the formation of the decision. In information systems in general and especially in the expert systems, the values are implicitly and heuristically included. It is meant that there is no explicit objective function to allow for flexible behavior of the decision maker when forming the decisions. Such objective value function allows for quantitative analysis and removal of logical inconsistencies and errors. It is meant that such objective function nuances the viewpoints with accounting the values with mathematical precision (Keeney, 1993; Chaudhury, 1995).

We will summarize that the decision support systems have additional properties and capabilities in concept, which separates them from the other information or in more narrow sense expert systems. One more informal description is that they reflect the philosophy of the theory of decision-making expressed in logical axioms, methodology and procedures based on these axioms, through which it is possible to analyze the internal complexity of the solutions in scientific way. The relative strength of the decision support systems lies in the fact that they are based on scientific knowledge expressed in forming a logically reasonable structure of the decision as an element of decision theory and scientific analysis expressed axiomatically and quantitatively by proven mathematical procedures based on the values as motivation (Keeney, 1988).

The complete application of this approach requires time and special knowledge in different scientific fields, necessitated mainly by problems in the evaluation of the values of the decision maker based on the expressed preferences (Keeney, 1988; Kersten, 1999; Zeleznikow, 2002). The reason is in the cardinal nature, uncertainty and errors in the particular expression of the preferences if the decision maker, as well as in that they are not explicitly and consciously related to the final decision in complex problems. This requires the design of decision support systems at the level of evaluating the individual preferences and their value representation as utility functions. This opens in application aspect the way for the concrete user to the achievements of the decision-making theory without the need of special knowledge and with the computer power to process the information.

The presented in this book approach, methodology and mathematical procedures allow for the design of individually oriented systems for analytic representation of the preferences and their value expression as objective function based on direct comparisons or on the gambling approach. Such systems may be autonomous or parts of larger information decision support system (Eom, 1990, 2006; Chaudhury, 1995). We will describe a prototype of such system developed based on the mathematical conclusions given in the preceding chapters of the book. We use the following formalization of the preferences and notions of the decision maker. At the lowest level of formalization, when the problem is described verbally, the DM is tasked with choosing from several variants the one that is more preferred by him. Usually the alternatives are compared two by two and is sought the one that is more preferred in all comparison with the others. There are two fundamental comparative value concepts, namely "better" (strict preference) and "equal in value to" (indifference). The relations of preference and indifference between alternatives are usually denoted by the symbols ($\succ$) and ($\approx$). In accordance with a long-standing philosophical tradition, $A \succ B$ is taken to represent "B is worse than A," as well as "A is better than B." There are commonly accepted postulates for choice at this level. The first postulate is the capability to realize comparison. This means that for every couple A and B *it is true* $A \succ B$ *or* $B \succ A$, or $A \approx B$ *(equivalent or cannot take solution).* The second postulate is transitivity of the preference, i.e. if $A \succ B$ *and* $B \succ C$ *then it is true* $A \succ C$. *The same is true for the indifference relation; if* $A \approx B$ *and* $B \approx C$ *then it is true* $A \approx C$. *In complex multifactor problems, the transitivity is the most often violated property. Often time the preferences are closed in cycles. In other words in practices there are preferences for which transitively,* $A \succ C$ *and* $C \succ A$. *In the indifference relation, additional natural obstacle is the threshold of non-distinguishability between two events, inherent for every person.*

In theoretical plan there are strict and logically sound mathematical theories, yielding theorems of existence or theorems for expression (representations) of preferences in stronger scales as the interval or the ratio scales or the absolute scale (Chapter 4*). The problem arises in practice with the application of the theoretical results. The Allais paradox for the comparison of lotteries in the gambling approach is widely known.* The associative transaction of theoretical results from the normative approach is not well founded. We include in the DSS numerical methods that would combine the beautiful theoretical results of the theory of measurement and utility theory with the qualitative nature of expression of empirical knowledge through preferences and taking into account the inherent uncertainty in the expression of the subjective knowledge as preferences.

We will give short formal description of the preferences. We suppose that the formal preference relation $(x \succ y)$ and its related indifference relation $((x \approx y) \Leftrightarrow \neg((x \succ y) \vee (x \prec y)))$ fulfill the following postulates:

- The preference and indifference relation are transitive: $((x \approx y) \wedge (y \approx t)) \Rightarrow (x \approx t))$, $((x \succ y) \wedge (y \succ t)) \Rightarrow (x \succ t))$, $((x \approx y) \wedge (y \succ t)) \Rightarrow (x \succ t))$, $((x \succ y) \wedge (y \approx t)) \Rightarrow (x \succ t))$;
- The indifference relation is equivalence;
- For any two alternatives (x, y) one and only one of the following three possibilities is fulfilled $(x \succ y)$ or $(x \prec y)$ or $(x \approx y)$.

We assume that the DM in the expressing of the preferences makes mistakes or has uncertainties in the choice. We suppose that this uncertainty or errors have random nature and may be represented as random variable with mathematical expectation zero and bounded variance, different in the different practical problems. We assume that this uncertainty is expressed in each comparison independently of the other comparisons. We define sets $\mathbf{A}_{u*}$ and $\mathbf{B}_{u*}$ over the set of alternative $\mathbf{X}$, of positive preferences and negative preferences. Let are the sets:

$$\mathbf{A}_{u*}=\{(x,y)\in\mathbf{X}^2 / (u^*(x))>u^*(y)\},$$

$$\mathbf{B}_{u*}=\{(x, y)\in\mathbf{X}^2 / (u^*(x))<u^*(y)\}.$$

In the conditions of exact preferences it is fulfilled $(\mathbf{A}_{u*}\cap\mathbf{B}_{u*})=\varnothing)$, which is in accordance with the normative theory (Pavlov, 2011). However, in practice, due to errors, the uncertainty in the expert preferences and the threshold of indistinguishability the two sets intersect $(\mathbf{A}_{u*}\cap\mathbf{B}_{u*})\neq\varnothing)$. Then arises the problem, to construct function of two variables (x,y), which divide in the best stochastic way the sets $\mathbf{A}_{u*}$ and $\mathbf{B}_{u*}$, $(\mathbf{A}_{u*}\cap\mathbf{B}_{u*})\neq\varnothing)$. Such function of two variables is one of the ways for evaluating the value functions in ordinal aspect. Evaluation of the value functions is analyzed in Chapter 5.

We know from Chapter 3 that the existence of a value function u(.) leads to the existence of: asymmetry $((x \prec y) \Rightarrow (\neg(x \succ y))$, axiom 4 in paragraph 3.1, Chapter 3), transitivity $((x \succ y) \wedge (y \succ z) \Rightarrow (x \succ z)$, axiom 5) and transitivity of the "indifference" relation (≈) (axiom 3). Therefore, the existence of a value function needs transitivity of the preferences and transitivity of the indifference relation. Therefore, the construction of a value function reorders the preferences as weak order. We remember the definition of "*weak order.*" This is an asymmetric and "negatively transitive" relation. The transitivity of the "*weak order*" ($\succ$) follows from the "asymmetry" and the "negative transitivity." The irreflexivity of the preferences and the negative transitivity of the preference relation split the set of alternatives $\boldsymbol{X}$ into non-crossing equivalence classes. The factorized set of these classes is marked by $\boldsymbol{X}/\approx$. It is proven in Chapter 3 that the existence of a "*weak order*" ($\succ$) over $\boldsymbol{X}$ leads to the existence of a "*strong order*" preference relation ($\succ$) over $\boldsymbol{X}/\approx$. So far, we are in the ordering scale, presentation with precision to monotonous functions (Chapter 3).

The construction of a utility function needs measurement and descriptions of the preferences in more powerful scales. We remember briefly the definition of a utility functions. Let $\boldsymbol{X}$ be the set of alternatives and $\boldsymbol{P}$ is a set of probability distributions over $\boldsymbol{X}$ and $\boldsymbol{X} \subseteq \boldsymbol{P}$. A utility function $u(.)$ will be any function for which the following is fulfilled:

$$(p \succ q,\ (p,q) \in \boldsymbol{P}^2) \Leftrightarrow (\textstyle\int u(.)dp > \int u(.)dq).$$

In keeping with von Neumann (Keeney, 1993) the interpretation is that the integral of the utility function $u(.)$ is a measure concerning the comparison of the probability distributions of alternatives and that the probability distributions are consequence of DM choice. Let the probability distributions p and q *are* defined over the set of alternatives $\boldsymbol{X}$. The notation ($\succ$) expresses the preferences of DM over $\boldsymbol{P}$ including those over $\boldsymbol{X}$ ($\boldsymbol{X} \subseteq \boldsymbol{P}$). The assumption of existence of a utility (value) function $u(.)$ leads to the "*negatively transitive*" and "*asymmetric*" relation ($\succ$) and to transitivity of the relation (≈). The proposition 3.4.1 from Chapter 3 and theorem 4.2.1 from Chapter 4 reveal that the utility measurement scale is equivalent to the temperature scale (interval scale). In Chapter 3 and 4 we saw that the gambling approach (comparisons of lotteries defined in Chapter 6) is necessary condition for the evaluation of the utility function in the sense of von Neumann (Figure 3).

Starting from the gambling approach for the definitions and the presentation of the expert's preferences we use the following sets, motivated by Proposition 6.2.1 of Chapter 6:

$$\boldsymbol{A}_{u^*}=\{(\alpha,x,y,z)/(\alpha u^*(x)+(1-\alpha)u^*(y))>u^*(z)\},$$

$$\boldsymbol{B}_{u^*}=\{(\alpha,x,y,z)/(\alpha u^*(x)+(1-\alpha)u^*(y))<u^*(z)\}.$$

The approach we are using in our investigation for the evaluation of the utility functions in its essence is the recognition of these sets. Through stochastic recurrent algorithms, we approximate

Figure 3. Gambling approach: comparisons of lotteries

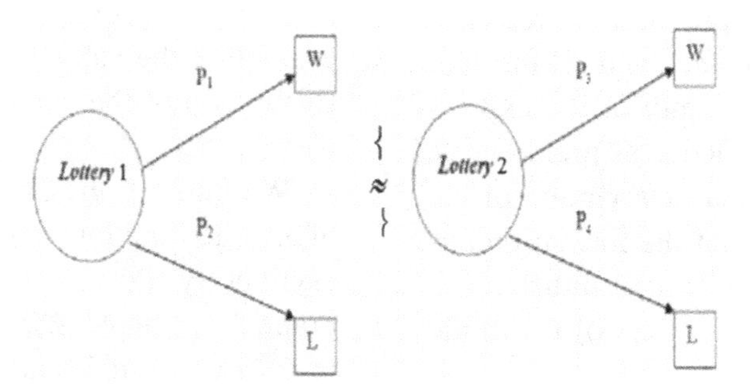

functions recognizing the above sets. And again in the conditions of exact preferences it is fulfilled $(\mathbf{A}_{u*}\cap\mathbf{B}_{u*})=\varnothing$), but in practice, due to errors and uncertainty in the expert preferences, the two sets intersect $(\mathbf{A}_{u*}\cap\mathbf{B}_{u*})\neq\varnothing$).

4. STOCHASTIC PATTERN RECOGNITION OF POSITIVE AND NEGATIVE ANSWERS

Following the ideas of Professor Ralph Keeney, the main assumption in each management or control decision is that the values of the subject making the decision are the determine force and due to this are determining for the formation of the decisions (Keeney, 1988, 1993). In the model driven design, the required human value considerations are engineered into practices, activities, and management via value functions. Such objective value functions allow for quantitative analysis and removal of logical inconsistencies and errors. It is meant that Value-driven design creates an environment that enables design optimization by providing designers with an objective function. The objective value function inputs all the important attributes of the system being designed, and outputs a score. The purpose is to enable the assessment of a value for every design option so that options can be rationally compared and a choice taken.

Value-based management concepts are prevalent in theory and practice since value creation is commonly considered the paramount business goal. Validate mathematical preferences value evaluation could be the first step in realization of a human-adapted design process and decision-making in value-based management. In this manner, we have posed the decision-making problem as a problem of constructing functions based on stochastic recurrent procedures, which can later be used in optimization problems (Pavlov, 2011). Here is described a prototype of an information decision support system for individual's utility evaluations. In DSS we include the gambling approach on the base of the stochastic recurrent procedures developed in Chapters 5, 6, and 7 (Figure 4).

Following the approach in (Keeney, 1993), DM compares the "lottery" $<x,y,\alpha>=(\alpha x+(1-\alpha)y)$ with the separate elements (alternatives) $z\in\mathbf{X}$. DM expresses his preferences in a qualitative aspect - "better," "worse," or a refusal to choose—"equivalent":

$$(\alpha x+(1-\alpha)y) \overset{\succ}{\underset{\prec}{\approx}} z$$

Proposition 4.1 of Chapter 3, proposition 2.1, and theorem 3.1 of Chapter 6 show that such comparisons of "lotteries" are of the simplest and sufficient possible type for individual's utility evaluation. The expressed preferences, the answers of DM and comparisons are of cardinal (qualitative) nature and contain the inherent DM's uncertainty and errors (Figure 5).

We already know that we are looking for the approximate evaluation of the set A_{u*} of the positive preferences and to the set B_{u*} of the negative preferences:

Figure 4. Evaluation of value, utility functions, or subjective probabilities

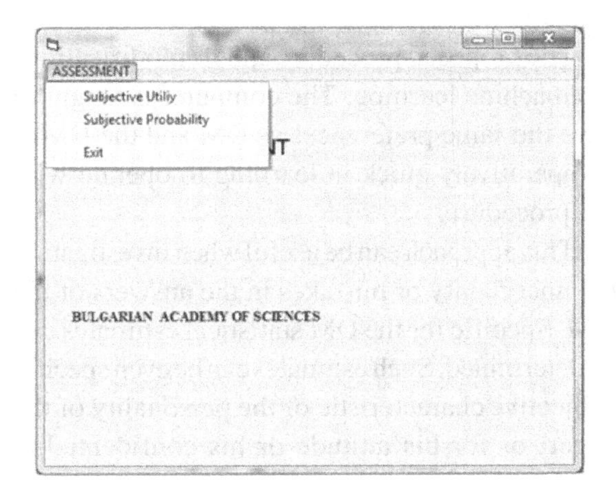

Figure 5. Evaluation of value, utility functions, or subjective probabilities

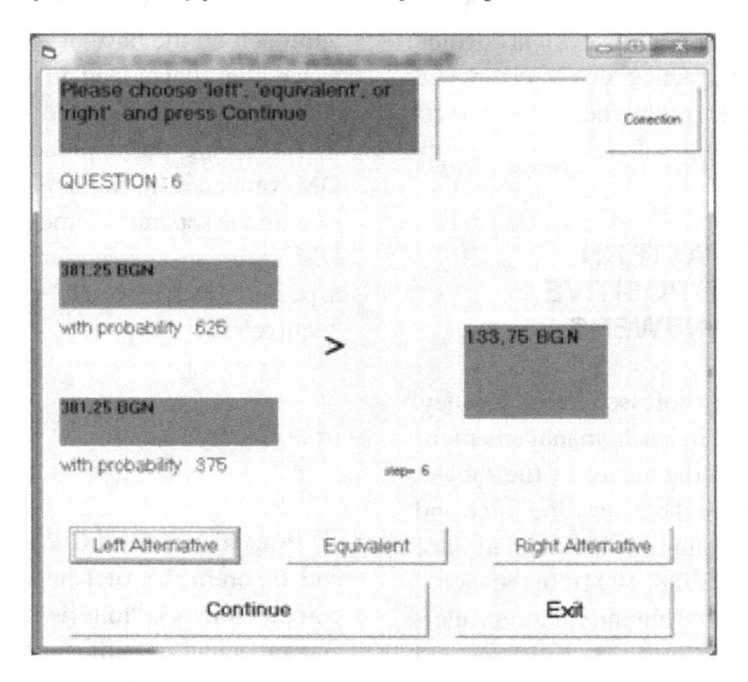

$A_{u*} = \{(x,y,z,\alpha)/ (\alpha u*(x)+(1-\alpha)u*(y))>u*(z)\}$,

$B_{u*} = \{(x,y,z,\alpha)/ (\alpha u*(x)+(1-\alpha)u*(y))<u*(z)\}$.

The star in the notations means an empirical estimate of the utility of the technologist.

This procedure in its essence can be considered as a machine-learning computer training on the same preferences as these of the expert (Ayzerman, 1970). The calculations in the DSS are based on the stochastic procedures in Chapter 5, 6, and 7 (Figure 6).

The proposed procedure and its modifications are machine learning. The computer is taught to have the same preferences as DM and the DM is comparatively quick in learning to operate with the procedure.

This approach can be useful when investigating the uncertainty or mistakes in the answers of the DM. Specific for the DM statistical estimates can be determined. Such estimates can be even specific subjective characteristic of the personality of the expert or for his attitude or his confidentiality toward the considered process. Other examples of evaluation of utility functions and subjective probabilities could be seen in Chapter 7, 9, 10, and 11.

5. ITERATIVE UTILITY EVALUATIONS AND CORRECTIONS OF THE DECISION MAKER'S WRONG ANSWERS

The analytic construction of the utility function is an iterative "machine-learning" process. The process of the iteration itself can unfold in several directions. The first is increasing the number of training points in any moment of the investigation. It is easy achieved iterative improvement of the utility function approximation via increasing the number of the learning points. Using this approach continually, the representation as utility function of the DM preferences could be approximated more accurately. By continually increasing the number of training points, details of his thinking can spring

Figure 6. Evaluation of utility function

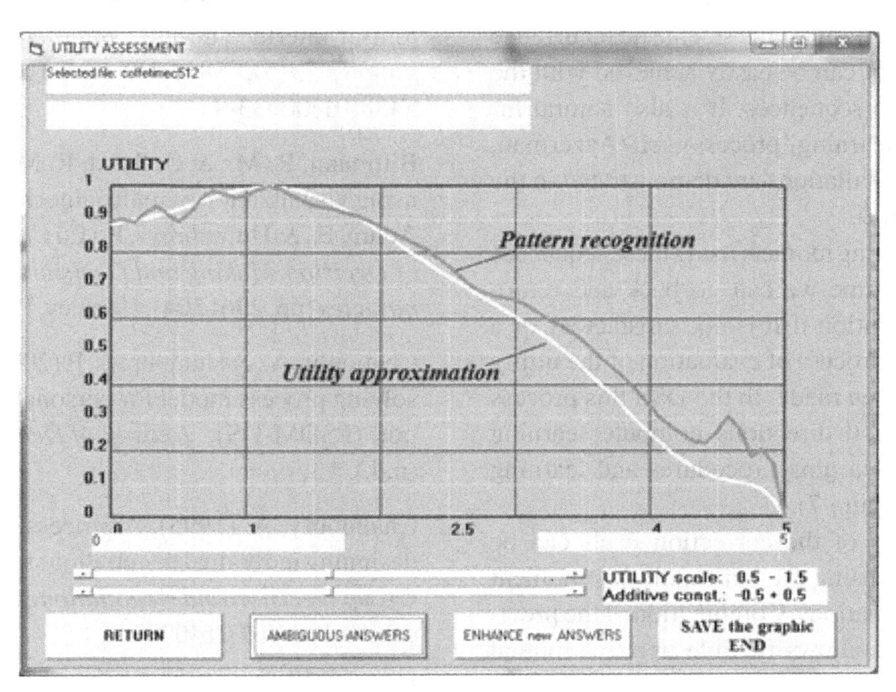

Figure 7. Qualitative comparison of the utility approximation with the DM preferences

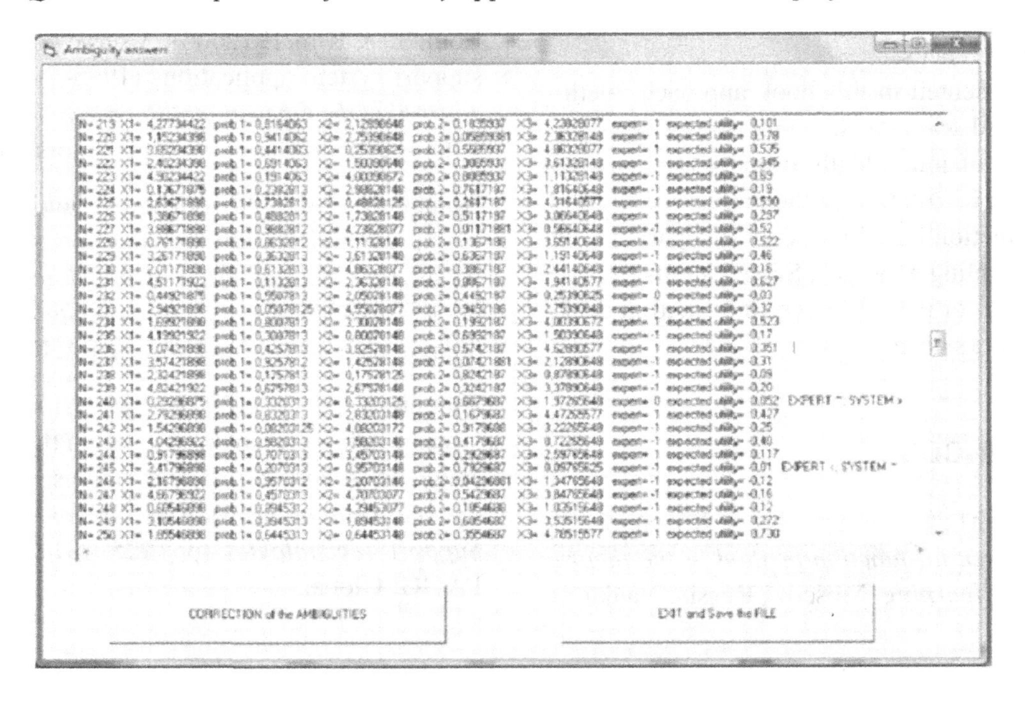

up and new details could be found. This process of iteration is natural for the stochastic recurrence procedure and it can be easily achieved with the pseudo random sequences. It is also natural for the "machine-learning" process itself (Ayzerman, 1970). Such calculations are demonstrated in the end of Chapter 9.

New interesting moment using these sequences is that at any time we can go back and return to previous solution if it is not satisfactory or a mistake in the process of evaluation of the utility function has been made. In the DSS this process can be used in both directions -computer learning with machine learning procedures and learning of the tutor (Figure 7).

The process of the correction itself can be accelerated by using the built until this moment utility approximations. Combination of the previous approaches allows flexible iterative mutual learning process of the control solution construction. At any moment here mathematical computer processing and use of the powerful analytical approaches is feasible. In fact, it can be said that mathematics is practically included in the process of forming of the decisions based on expert preferences.

The presented in this book approach, methodology and mathematical procedures allow for the design of individually oriented systems for analytic representation of the preferences as objective function based on direct comparisons or on the gambling approach. Such systems may be autonomous or parts of larger information decision support system.

REFERENCES

Aizerman, M., Braverman, E., & Rozonoer, L. (1970). *Potential function method in the theory of machine learning*. Moscow, Russia: Nauka.

Andreev, R. D. (2001). A linguistic approach to user interface design. *Interacting with Computers*, *13*(5), 581–599. doi:10.1016/S0953-5438(01)00033-9

Bittmann, R. M., & Gelbard, R. M. (2007). DSS using visualization of multi-algorithms voting. In Adam, F., & Humphreys, P. (Eds.), *Encyclopedia of Decision-Making and Decision Support Technologies* (pp. 296–304). Hershey, PA: IGI Global.

Chatjoulis, A., & Humphreys, P. (2007). A problem solving process model for personal decision support (PSDM-DS). *Journal of Decision Systems*, (n.d.), 13.

Chaudhury, A. (1995). A process perspective to designing individual negotiation support systems. *Group Decision and Negotiation*, *4*(6), 525–548. doi:10.1007/BF01409714

Eom, H., & Lee, S. (1990). Decision support systems and applications research: A bibliography (1971-1988). *European Journal of Operational Research*, *46*, 333–342. doi:10.1016/0377-2217(90)90008-Y

Eom, S., & Kim, E. (2006). A survey of decision support system applications (1995–2001). *The Journal of the Operational Research Society*, *57*, 1264–1278. doi:10.1057/palgrave.jors.2602140

Humphreys, P. C. (2007a). The decision hedgehog: Group communication and decision support. In Adam, F., & Humphreys, P. (Eds.), *Encyclopedia of Decision-Making and Decision Support Technologies* (pp. 148–154). Hershey, PA: IGI Global.

Humphreys, P. C. (2007b). Decision support systems and representation levels in the decision spine. In Adam, F., & Humphreys, P. (Eds.), *Encyclopedia of Decision-Making and Decision Support Technologies* (pp. 225–231). Hershey, PA: IGI Global.

Keeney, R. (1988). Value-driven expert systems for decision support. *Decision Support Systems*, *4*(4), 405–412. doi:10.1016/0167-9236(88)90003-6

Keeney, R., & Raiffa, H. (1993). *Decision with multiple objectives: Preferences and value trade-offs* (2nd ed.). Cambridge, UK: Cambridge University Press. doi:10.1109/TSMC.1979.4310245

Kersten, G. E., & Noronha, S. J. (1999). WWW-based negotiation support: Design, implementation, and use. *Decision Support Systems*, *25*(2), 135–154. doi:10.1016/S0167-9236(99)00012-3

Meisel, S. (2011). *Anticipatory optimization for dynamic decision-making*. New York, NY: Springer Science + Business Media.

Meisel, S., & Mattfeld, D. (2007). Synergies of data mining and operations research. In *Proceedings of the 40th Hawaii International Conference on System Sciences,* (pp. 54-58). Los Almos, CA: IEEE Press.

Pavlov, Y. (2011). Preferences based stochastic value and utility function evaluation. In *Proceedings of the Conferences InSITE 2011*. Novi Sad, Serbia: InSITE.

Zeleznikow, J. (2002). Risk, negotiation and argumentation—A decision support system based approach. *Law Probability and Risk*, *1*(1), 37–48. doi:10.1093/lpr/1.1.37

Compilation of References

Ackoff, R. L. (1981). The art and science of mess management. *Interfaces*, *11*(1), 20–26. doi:10.1287/inte.11.1.20

Ackoff, R. L. (1998). *Ackoff's best: His classic writings on management*. New York, NY: John Wiley & Sons.

Adam, F., & Pomerol, J.-C. (2008). Understanding the influence context on organizational decision-making. In Adam, F., & Humphreys, P. (Eds.), *Encyclopedia of Decision-Making and Decision Support Technologies* (pp. 922–929). Hershey, PA: IGI Global. doi:10.4018/978-1-59904-843-7.ch104

Aizerman, M. A., Braverman, E., & Rozonoer, L. (1970). *Potential function method in the theory of machine learning*. Moscow, Russia: Nauka.

Aizerman, M., Braverman, E., & Rozonoer, L. (1964). Theoretical foundations of the potential function method in pattern recognition learning. *Automation and Remote Control*, *25*, 821–837.

Aleksiev, V., Tihomirov, V., & Fomin, S. (1979). *Optimal control*. Moscow, Russia: Nauka.

Allais, M. (1953). Le comportement de l'homme rationnel devant le risque: critique des postulats et axiomes de l'école amèricaine. *Econometrica*, *21*, 503–546. doi:10.2307/1907921

Andreev, R. (2008). A model of self-organizing collaboration. In *Proceedings of IFIP WCC 2008 Conference on Biologically Inspired Collaborative Computing, Book Series in Computer Science: IFIP Advances in Information and Communication Technology*, (pp. 223-231). Springer.

Andreev, R. D., & Troyanova, N. V. (2006). E-learning design: An integrated agent-grid service architecture. In *Proceedings of IEEE John Vincent Atanasoff' 2006 International Symposium on Modern Computing*, (pp. 208-213). Los Alamitos, CA: IEEE Computer Society.

Andreev, R. D. (2001). A linguistic approach to user interface design. *Interacting with Computers*, *13*(5), 581–599. doi:10.1016/S0953-5438(01)00033-9

Aranda, G., Vizcaíno, A., Cechich, A., & Piattini, M. (2005). A cognitive-based approach to improve distributed requirement elicitation processes. In *Proceedings of the 4th IEEE Conference on Cognitive Informatics (ICCI 2005)*, (pp. 322-330). Irvine, CA: IEEE Press.

Aroyo, L., Dolog, P., Houben, G.-J., Kravcik, M., Naeve, A., Nilsson, M., & Wild, F. (2006). Interoperability in personalized adaptive learning. *Journal of Educational Technology & Society*, *9*(2), 4–18.

Arrow, K. (1950). A difficulty in the concept of social welfare. *The Journal of Political Economy*, *58*(4), 328–346. doi:10.1086/256963

Arthur, A. (1972). *Regression and the moor-penrose pseudo-inverse*. New York, NY: New Academic Press.

Bandura, A. (1986). *Social foundations of thought and action: A social cognitive theory*. Englewood Cliffs, NJ: Prentice-Hall.

Barnes, B. H., & Bollinger, T. (1991). Making reuse cost-effective. *IEEE Software*, *1*, 13–24. doi:10.1109/52.62928

Baron, J. (2008). *Thinking and deciding* (4th ed.). Cambridge, UK: Cambridge University Press.

Bazara, M., & Shetty, C. (1979). *Nonlinear programming: Theory and algorithms*. New York, NY: Wiley.

Bell, D. E., Raiffa, H., & Tversky, A. (1988). *Decision-making: Descriptive, normative and prescriptive interactions*. Cambridge, UK: Cambridge University Press. doi:10.1017/CBO9780511598951

Bennet, A., & Bennet, D. (2008). The decision-making process for complex situations in a complex environment. In Burstein, F., & Holsapple, C. W. (Eds.), *Handbook on Decision Support Systems* (pp. 1–14). New York, NY: Springer-Verlag. doi:10.1007/978-3-540-48713-5_1

Bertoni, M., Eres, H., & Isaksson, O. (2011). Criteria for assessing the value of product service system design alternatives: An aerospace investigation. In *Proceedings of Functional Thinking for Value Creation* (pp. 141–146). Springer. doi:10.1007/978-3-642-19689-8_26

Beynon, M., Rasmequan, R., & Russ, S. (2002). A new paradigm for computer-based decision support. *Decision Support Systems*, *33*, 127–142. doi:10.1016/S0167-9236(01)00140-3

Birnbaum, M. (2008). New paradoxes of risky decision-making. *Psychological Review*, *115*(2), 463–501. doi:10.1037/0033-295X.115.2.463

Bittmann, R. M., & Gelbard, R. M. (2007). DSS using visualization of multi-algorithms voting. In Adam, F., & Humphreys, P. (Eds.), *Encyclopedia of Decision-Making and Decision Support Technologies* (pp. 296–304). Hershey, PA: IGI Global.

Bleichrodt, H., & van Rijn, J. (1999). Probability weighting and utility curvature in QALY-based decision making. *Journal of Mathematical Psychology*, *43*, 238–260. doi:10.1006/jmps.1999.1257

Braverman, E., & Muchnik, I. (1983). *Structure methods for the processing of empirical data*. Moscow, Russia: Nauka.

Brown, R. (2008). Decision aiding research needs. In Adam, F., & Humphreys, P. (Eds.), *Encyclopedia of Decision-Making and Decision Support Technologies* (pp. 141–147). Hershey, PA: IGI Global.

Bruner, J. S. (1967). *On knowing: Essays for the left hand*. Cambridge, MA: Harvard University Press.

Brusilovsky, P. (2001). Adaptive hypermedia. *User Modeling and User-Adapted Interaction*, *11*(1-2), 87–110. doi:10.1023/A:1011143116306

Carlson, S. A. (2008). An attention based view on DSS. In Adam, F., & Humphreys, P. (Eds.), *Encyclopedia of Decision-Making and Decision Support Technologies* (pp. 38–45). Hershey, PA: IGI Global. doi:10.4018/978-1-59904-843-7.ch004

Castagne, S., Curran, R., & Collopy, P. (2009). Implementation of value-driven optimisation for the design of aircraft fuselage panels. *International Journal of Production Economics*, *117*(2), 381–388. doi:10.1016/j.ijpe.2008.12.005

Chatjoulis, A., & Humphreys, P. (2007). A problem solving process model for personal decision support (PSDM-DS). *Journal of Decision Systems*, (n.d.), 13.

Chaudhury, A. (1995). A process perspective to designing individual negotiation support systems. *Group Decision and Negotiation*, *4*(6), 525–548. doi:10.1007/BF01409714

Checkland, I. J. (1999). *Svstetns thinking, systems practice: Includes a 30-year retrospective*. Chichester, UK: John Wiley & Sons.

Checkland, P. (1981). *Systems thinking, systems practice*. New York, NY: Wiley.

Cohen, M., & Jaffray, J.-Y. (1988). Certainty effect versus probability distortion: An experimental analysis of decision making under risk. *Journal of Experimental Psychology. Human Perception and Performance*, *14*(4), 554–560. doi:10.1037/0096-1523.14.4.554

Collopy, P., & Hollingsworth, P. (2009). *Value-driven design. AIAA Paper 2009-7099*. Reston, VA: American Institute of Aeronautics and Astronautics.

Courtney, J. (2001). Decision-making and knowledge management in inquiring organizations: Toward a new decision-making paradigm for DSS. *Decision Support Systems*, *31*(1), 17–38. doi:10.1016/S0167-9236(00)00117-2

Courtney, J., & Paradice, D. B. (1993). Studies in managerial problem formulation systems. *Decision Support Systems*, *9*(4), 413–423. doi:10.1016/0167-9236(93)90050-D

Csaki, C. (2008). The mythical decision maker: Model of roles in decision-making. In Adam, F., & Humphreys, P. (Eds.), *Encyclopedia of Decision-Making and Decision Support Technologies* (pp. 653–660). Hershey, PA: IGI Global. doi:10.4018/978-1-59904-843-7.ch073

DeCarvalho, R. (1991). The humanistic paradigm in education. *The Humanistic Psychologist, 19*(1), 88–104. doi:10.1080/08873267.1991.9986754

Diday, E. (1979). *Optimization en classification automatique*. Paris, France: Institut National de Recherche en Informatique et en Automatique.

Dixmier, J. (1967). *Cours de mathematiques*. Paris, France: Gautier-Villars.

Eden, C., & Ackermann, F. J. (2001). SODA - The principles. In I. Rosenhead & Mingers (Eds.), *Rational Analysis for a Problematic World Revisited*. Chichester, UK: John Wiley & Sons, Ltd.

Ekeland, I. (1979). *Elements d'economie mathematique*. Paris, France: Hermann.

Elkin, V. (1999). *Reduction of non-linear control systems: A differential geometric approach*. Dordrecht, The Netherlands: Kluwer.

Elliott, R. J. (1982). *Stochastic calculus and applications*. Berlin, Germany: Springer-Verlag.

Emelyanov, S., Korovin, S., & Levant, A. (1993). Higher-order sliding modes in control systems. *Differential Equations, 29*(11), 1627–1647.

Engelking, R. (1985). *General topology*. Warsaw, Poland: Panstwove Widawnictwo Naukove.

Enstein, K., Ralston, A., & Herbert, S. (Eds.). (1977). *Statistical methods for digital computers*. New York, NY: John Wiley & Sons.

Eom, H., & Lee, S. (1990). Decision support systems and applications research: A bibliography (1971-1988). *European Journal of Operational Research, 46*, 333–342. doi:10.1016/0377-2217(90)90008-Y

Eom, S., & Kim, E. (2006). A survey of decision support system applications (1995–2001). *The Journal of the Operational Research Society, 57*, 1264–1278. doi:10.1057/palgrave.jors.2602140

Ermolev, Y. (1976). *Stochastic programming methods*. Moscow, Russia: Nauka.

Ermolev, Y. M. (1976). *Methods of stochastic programming*. Moscow, Russia: Nauka.

Farquhar, P. (1984). Utility assessment methods. *Management Science, 30*, 1283–1300. doi:10.1287/mnsc.30.11.1283

Fishburn, P. (1970). *Utility theory for decision-making*. New York, NY: Wiley.

Fishburn, P. (1988). Context-dependent choice with nonlinear and nontransitive preferences. *Econometrica, 56*(2), 1221–1239. doi:10.2307/1911365

Forgionne, G. A., & Kohli, R. (2000). Management support system effectiveness: Further empirical evidence. *Journal of the Association for Information Systems, 1*(3), 1–37.

Forgionne, G., & Russel, S. (2008). The evaluation of decision-making support systems' functionality. In Adam, F., & Humphreys, P. (Eds.), *Encyclopedia of Decision-Making and Decision Support Technologies* (pp. 329–338). Hershey, PA: IGI Global. doi:10.4018/978-1-59904-843-7.ch038

Fransella, F., Bell, R., & Bannister, D. (2003). *A manual for the repertory grid technique* (2nd ed.). Chichester, UK: John Wiley & Sons, Ltd.

Friedman, M., & Savage, L. (1948). Utility analysis of choices involving risk. *The Journal of Political Economy, 56*(4), 279–304. doi:10.1086/256692

Gatev, G. (1978). *Analysis and synthesis of automatic systems*. Sofia, Bulgaria: Tehnika.

Gig, J. (1981). *Applied general systems theory*. Moscow, Russia: Mir.

Gikhman, A., Skorokhod, V., & Yadrenko, M. (1988). *Probability theory and mathematical statistics*. Kiev, Russia: Vyshcha Shkola.

Greeno, J. G. (1978). Nature of problem-solving abilities. In Estes, W. K. (Ed.), *Handbook of Learning and Cognitive Processes*. Hillsdate, NJ: Erlbaum.

Grifits, T. L., & Tenenbaum, J. B. (2006). Optimal prediction in everyday cognition. *Psychological Science, 17*(9), 767–773. doi:10.1111/j.1467-9280.2006.01780.x

Groner, R., Groner, M., & Bischof, W. F. (1983). *Methods of heuristics*. Hillsdale, NJ: Erlbaum.

Hahn, G. J., & Kuhn, H. (2012). Designing decision support systems for value-based management: A survey and an architecture. *Decision Support Systems, 53*(3), 591–598. doi:10.1016/j.dss.2012.02.016

Hall, D., & Davis, R. (2007). Engaging multiple perspectives: A value-based decision-making model. *Decision Support Systems, 43*(4), 1588–1604. doi:10.1016/j.dss.2006.03.004

Hastie, R., & Dawes, R. M. (2001). *Rational choice in an uncertain world*. Thousand Oaks, CA: Sage.

Herbrich, R., Graepel, T., & Obermayer, K. (1999). *Regression models for ordinal data: A machine learning approach*. TechReport TR99-3. Berlin, Germany: Technical University of Berlin. Retrieved from http://research.microsoft.com/apps/pubs/default.aspx?id=65632

Herbrich, R., Graepel, T., Bolmann, P., & Obermayer, K. (1998). Supervised learning of preferences relations. In *Proceedings Fachgruppentreffen Maschinelles Lernen*, (pp. 43-47). Retrieved from http://research.microsoft.com/apps/pubs/default.aspx?id=65615

Hsu, J., & Meyer, A. (1972). *Modern control principles and applications*. New York, NY: McGraw-Hill.

Huber, G. P. (1980). *Managerial decision-making*. New York, NY: Scott, Foresman, and Company.

Huber, G. P., & McDaniel, R. (1986). The decision-making paradigm of organizational design. *Management Science, 32*(5), 572–589. doi:10.1287/mnsc.32.5.572

Humphreys, P. C. (2007). The decision hedgehog: Group communication and decision support. In Adam, F., & Humphreys, P. (Eds.), *Encyclopedia of Decision-Making and Decision Support Technologies* (pp. 148–154). Hershey, PA: IGI Global.

Humphreys, P. C. (2007). Decision support systems and representation levels in the decision spine. In Adam, F., & Humphreys, P. (Eds.), *Encyclopedia of Decision-Making and Decision Support Technologies* (pp. 225–231). Hershey, PA: IGI Global.

Hutson, V., & Pim, J. (1980). *Applications of functional analysis and operator theory*. London, UK: Elsevier.

Jakimavicius, M. (2008). *Multi-criteria assessment of urban areas transports systems development according to sustainability*. (Doctoral Dissertation). Vilnius-Technika. Retrieved from http://vddb.laba.lt/fedora/get/LT-eLABa-0001:E.02~2008~D_20090119_094516-74428/DS.005.1.02.ETD

Jonson, N., & Leone, F. (1977). *Statistics and experimental design*. New York, NY: John Wiley & Sons.

Kahneman, D., & Tversky, A. (1979). Prospect theory: An analysis of decision under risk. *Econometrica, 47*, 263–291. doi:10.2307/1914185

Kay, J. (2001). Learner control. *User Modeling and User-Adapted Interaction, 11*(1-2), 111–127. doi:10.1023/A:1011194803800

Keeney, R. (1988). Value-driven expert systems for decision support. *Decision Support Systems, 4*(4), 405–412. doi:10.1016/0167-9236(88)90003-6

Keeney, R., & Raiffa, H. (1993). *Decision with multiple objectives: Preferences and value trade-offs* (2nd ed.). Cambridge, UK: Cambridge University Press. doi:10.1109/TSMC.1979.4310245

Keller, J. M. (1983). Motivational design of instruction. In Reigeluth, C. M. (Ed.), *Instructional-Design Theories and Models: An Overview of their Current Status*. Hillsdale, NJ: Lawrence Erlbaum Associates.

Kelly, G. A. (1955). *The psychology of personal constructs*. New York, NY: Norton.

Kelly, J. (1955). *General topology*. New York, NY: Ishi Press.

Kepner, C. H., & Tregoe, B. B. (1981). *The new rational manager*. Princeton, NJ: Princeton Research Press.

Kersten, G. E., & Noronha, S. J. (1999). WWW-based negotiation support: Design, implementation, and use. *Decision Support Systems, 25*(2), 135–154. doi:10.1016/S0167-9236(99)00012-3

Kivinen, J., Smola, A., & Williamson, R. (2004). Online learning with kernels. *IEEE Transactions on Signal Processing*, *52*(8), 2165–2176. doi:10.1109/TSP.2004.830991

Klein, D., & Shortliffe, E. (1994). A framework for explaining decision-theoretic advice. *Artificial Intelligence*, *67*(2), 201–243. doi:10.1016/0004-3702(94)90053-1

Klein, M. R., & Methlie, L. (1990). *Expert systems: A DSS approach*. Reading, MA: Addison Wesley.

Knight, F. H. (1921). *Risk, uncertainty and profit*. New York, NY: Houghton Mifflin.

Koper, R., & Oliver, B. (2004). Representing the learning design of units of learning. *Journal of Educational Technology & Society*, *7*(3), 97–111.

Krantz, D. H., Luce, R. D., Suppes, P., & Tversky, A. (1971). *Foundations of measurement* (*Vol. 1*). New York, NY: Academic Press.

Krantz, D. H., Luce, R. D., Suppes, P., & Tversky, A. (1989). *Foundations of measurement* (*Vol. 2*). New York, NY: Academic Press.

Krantz, D. H., Luce, R. D., Suppes, P., & Tversky, A. (1990). *Foundations of measurement* (*Vol. 3*). New York, NY: Academic Press.

Krotov, V., & Gurman, V. (1973). *Methods and problems in the optimal control*. Moscow, Russia: Nauka.

Lang, S. (1965). *Algebra*. Reading, MA: Addison-Wesley.

Larichev, O., & Olson, D. (2010). *Multiple criteria analysis in strategic sitting problems* (2nd ed.). London, UK: Springer.

Lave, J., & Wenger, E. (1998). *Communities of practice: Learning, meaning, and identity*. Cambridge, UK: Cambridge University Press.

Leontiev, A. N. (1978). *Activity, consciousness, personality*. Englewood Cliffs, NJ: Prentice Hall.

Levy, M., Pliskin, N., & Ravid, G. (2010). Studying decision processes via a knowledge management lens: The Columbia space shuttle case. *Decision Support Systems*, *48*(4), 559–567. doi:10.1016/j.dss.2009.11.006

Linstone, H. A. (1999). *Decision-making for technology executives: Using multiple perspectives to improve performance*. Boston, MA: Artech House. doi:10.1109/TEM.2000.865908

Litvak, B. (1982). *Expert information, analysis and methods*. Moscow, Russia: Nauka.

Ljakova, K. (2005). Knowledge base implementation in biotechnological processes. In *Proceedings of the International Symposium Bioprocess System*. Sofia, Bulgaria: Bioprocess System.

Ljakova, K., & Pavlov, Y. (2002). Modeling and prognoses in birdproduction decision support. *Automatica & Informatics*, *1*, 42–48.

Ljakova, K., & Pavlov, Y. (2003). Methodological and information provision for some aspects in poultry farming. *Ecological Engineering and Environment Protection*, *2*, 52–59.

Machina, M. (1982). Expected utility analysis without the independence axiom. *Econometrica*, *50*(2), 277–323. doi:10.2307/1912631

Machina, M. (1987). Choice under uncertainty: Problems solved and unsolved. *The Journal of Economic Perspectives*, *1*, 121–154. doi:10.1257/jep.1.1.121

MacLane, S., & Birkhoff, G. (1974). *Algebra*. New York, NY: The Macmillan Company.

Makarov, I., et al. (1987). *Theory of choice and decision-making*. Moscow, Russia: Mir.

Manos, B., Ciani, A., Bournaris, T., Vassiliadou, I., & Papathanasiou, J. (2004). A taxonomy survey of decision support systems in agriculture. *Agricultural Economics Research*, *5*(2), 80–94.

Markowitz, H. M. (1952). Portfolio selection. *The Journal of Finance*, *7*, 77–91.

Marx, M. H., & Cronan-Hillix, W. A. (1987). *Systems and theories in psychology*. New York, NY: McGraw-Hill.

Meisel, S. (2011). *Anticipatory optimization for dynamic decision-making*. New York, NY: Springer Science + Business Media.

Meisel, S., & Mattfeld, D. (2007). Synergies of data mining and operations research. In *Proceedings of the 40th Hawaii International Conference on System Sciences,* (pp. 54-58). Los Almos, CA: IEEE Press.

Mengov, G. (2010). *Decision making under risk and uncertainty*. Sofia, Bulgaria: Publishing house JANET 45.

Montgomery, D. (1980). *Design and analysis of experiments*. New York, NY: John Wiley & Sons.

Mora, M., Cervantes, F., Forgionne, G., & Gelman, O. (2008). On frameworks and architectures of intelligent decision-making support systems. In Adam, F., & Humphreys, P. (Eds.), *Encyclopedia of Decision-Making and Decision Support Technologies* (pp. 680–690). Hershey, PA: IGI Global. doi:10.4018/978-1-59904-843-7.ch076

Morgan, G. (2006). *Images of organization*. Thousand Oaks, CA: Sage Publications, Inc.

Muller, K., Mika, S., Ratsch, G., Tsuda, K., & Scholkopf, B. (2001). An introduction to kernel-based learning algorithms. *IEEE Transactions on Neural Networks, 12*(2), 181–205. doi:10.1109/72.914517

Neeleman, R. (2002). *Biomass performance: Monitoring and control in biopharmaceutical production*. (Thesis). Wageningen University. Wageningen, The Netherlands. Retrieved from http://edepot.wur.nl/121354

Nielsen, J. (1993). *Usability engineering*. Boston, MA: Academic Press.

Novikof, A. (1963). A convergence proof for perceptrons. In *Symposium on Mathematical Theory of Automation*, (vol. 5, p. 12). Brookline.

Paradice, D., & Davice, R. A. (2008). DSS and multiple perspectives of complex problems. In Adam, F., & Humphreys, P. (Eds.), *Encyclopedia of Decision-Making and Decision Support Technologies* (pp. 286–295). Hershey, PA: IGI Global. doi:10.4018/978-1-59904-843-7.ch033

Parthasarathy, K. R. (1978). *Introduction to probability and measure*. Berlin, Germany: Springer-Verlag.

Pavlov, Y. (2004). Value based decisions and correction of ambiguous expert preferences: An expected utility approach. In *Proceedings of the International Conference BioPS 2004*. Sofia, Bulgaria: BioPS.

Pavlov, Y. (2005). Subjective preferences, values and decisions: Stochastic approximation approach. *Comptes rendus de L'Academie bulgare des Sciences, 58*(4), 367-372.

Pavlov, Y. (2010). Normative utility and prescriptive analytical presentation: A stochastic approximation approach. In *Social Welfare Enhancement in EU, 13th International Conference*. Sofia, Bulgaria: Government Publication. Retrieved from http://www.uni-sofia.bg/index.../SofiaConferenceProgramme2010.pdf

Pavlov, Y. (2011). Preferences based stochastic value and utility function evaluation. In *Proceeding of Conferences InSITE 2011,* (pp. 403-411). Novi Sad, Serbia: InSITE.

Pavlov, Y., & Ljakova, K. (2001). Expert information and decision support in complex bird production systems. In *Proceedings of the International Symposium and Young Scientists' School Bioprocess Systems*. Sofia, Bulgaria: Bioprocess Systems.

Pavlov, Y., & Ljakova, K. (2003). Machine learning and expert utility assessment. In *Proceedings of the CompSysTech 2003*. Retrieved from http://ecet.ecs.ru.acad.bg/cst/Docs/proceedings/S3A/IIIA-5.pdf

Pavlov, Y. (1989). Recurrent algorithm for value function construction. *Proceedings of Bulgarian Academy of Sciences, 7*, 41–42.

Pavlov, Y. (2001). Exact linearization of a non linear biotechnological model/Brunovsky model. *Proceedings of Bulgarian Academy of Sciences, 10*, 25–30. Retrieved from http://adsabs.harvard.edu/abs/2001CRABS.54j.25P

Pavlov, Y. (2005). Subjective preferences, values and decisions: Stochastic approximation approach. *Proceedings of Bulgarian Academy of Sciences, 58*(4), 367–372.

Pavlov, Y. (2007). Brunovsky normal form of monod kinetics and growth rate control of a fed-batch cultivation process. *Online Journal Bioautomation, 8*, 13–26.

Pavlov, Y. (2008). Equivalent form of Wang -Yerusalimsky kinetic model and optimal growth rate control of fed-batch cultivation processes. *Bioautomation, 11*, 1–13.

Pavlov, Y. (2010). Preferences, utility function and control design of complex process. *Proceedings in Manufacturing Systems, 5*(4), 225–231.

Pavlov, Y., Grancharov, D., & Momchev, V. (1996). Economical and ecological utility oriented analysis of the process of anaerobic digestion of waste waters. *European Journal of Operational Research*, *88*(2), 251–256. doi:10.1016/0377-2217(94)00193-6

Pavlov, Y., & Ljakova, K. (2004). Equivalent models and exact linearization by the optimal control of monod kinetics models. *Bioautomation*, *1*, 42–56.

Pavlov, Y., & Lyakova, K. (2001). Fuzzy logic and utility theory: An expert system based comparison. [Sofia, Bulgaria: BIOPS.]. *Proceedings of the*, *BIOPS-2001*, 37–40.

Pavlov, Y., & Tzonkov, St. (1999). An algorithm for constructing of utility functions. *Proceedings of the Bulgarian Academy of Sciences*, *52*(1-2), 21–24.

Pavlov, Y., & Vassilev, K. (1992). Recurrent construction of utility functions. *Comptes Rendus de l'Academie Bulgare de Science*, *3*, 5–8.

Perry, M. (2003). Distributed cognition. In Carroll, J. M. (Ed.), *HCI Models, Theories, and Frameworks: Toward an Interdisciplinary Science* (pp. 193–223). San Francisco, CA: Morgan Kaufmann. doi:10.1016/B978-155860808-5/50008-3

Pfanzagl, J. (1971). *Theory of measurement*. Wurzburg, Germany: Physical-Verlag.

Philips-Wren, G. E. (2008). Inteligent agents in decision support systems. In Adam, F., & Humphreys, P. (Eds.), *Encyclopedia of Decision-Making and Decision Support Technologies* (pp. 505–512). Hershey, PA: IGI Global. doi:10.4018/978-1-59904-843-7.ch058

Pinson, S. (1987). A multi-attribute approach to knowledge representation for loan granting. In *Proceedings of the 9th Joint Conference on Artificial Intelligence (IJCAI 1987)*, (pp. 588-591). IJCAI.

Popchev, I. (1986). *Technology for multi-criteria evaluation and decision-making*. Sofia, Bulgaria: BISA.

Power, D. J. (2002). *Decision support systems: Concepts and resources for managers*. Westport, CT: Quorum Books.

Power, D. J. (2008). Decision support system concept. In Adam, F., & Humphreys, P. (Eds.), *Encyclopedia of Decision-Making and Decision Support Technologies* (pp. 232–235). Hershey, PA: IGI Global. doi:10.4018/978-1-59904-843-7.ch027

Pratt, J. (1964). Risk aversion in the small and in the large. *Econometrica*, *32*, 122–136. doi:10.2307/1913738

Raiffa, H. (1968). *Decision analysis*. Reading, MA: Addison-Wesley.

Raiffa, H. (1968). *Decision analysis: Introductory lectures on choices under uncertainty*. Reading, MA: Addison-Wesley. doi:10.2307/2987280

Rastrigin, L. (1981). *Adaptation of complex systems*. Riga, Russia: Zinatne.

Rastrigin, L., & Ponomarev, Y. (1986). *Extrapolation methods for design and control*. Moscow, Russia: Mashinostroene.

Robbins, H., & Monro, S. (1951). A stochastic approximation method. *Annals of Mathematical Statistics*, *22*(3), 400–407. doi:10.1214/aoms/1177729586

Roberts, F. (1976). *Discrete mathematical models with application to social, biological and environmental problems*. Englewood Cliffs, NJ: Prentice-Hall, Inc.

Rogers, P., & Blenko, M. (2006). Who has the D? How clear decision roles enhance organizational performance. *Harvard Business Review*, *84*(1), 53–61.

Rudin, W. (1974). *Principles of mathematical analysis* (3rd ed.). Retrieved from http://dangtuanhiep.files.wordpress.com/2008/09/principles_of_mathematical_analysis_walter_rudin.pdf

Saaty, T. L., & Vargas, L. G. (1984). The legitimacy of rank reversal. *Omega*, *12*(5), 513–516. doi:10.1016/0305-0483(84)90052-5

Sage, A. (1981). Behavioural and organizational considerations in the design of information systems and processes for planning and decision support. *IEEE Transactions on Systems, Man, and Cybernetics*, *11*(9), 640–678. doi:10.1109/TSMC.1981.4308761

Sauter, V. L. (1999). Intuitive decision-making. *Communications of the ACM*, *42*(6), 109–115. doi:10.1145/303849.303869

Savage, J. (1954). *The foundations of statistics*. New York, NY: Wiley.

Shiryaev, A. N. (1989). *Probability*. New York, NY: Springer.

Shmeidler, D. (1989). Subjective probability and expected utility without additivity. *Econometrica*, *57*(3), 571–587. doi:10.2307/1911053

Simon, H. (1986). *Research briefings 1986: Report of the research briefing panel on decision-making and problem solving*. Washington, DC: National Academy of Sciences.

Simon, H. A. (1997). *Administrative behaviour: A study of decision-making process in administrative organizations* (4th ed.). New York, NY: Free Press.

Smith, G. (1988). Towards a heuristic theory of problem structuring. *Management Science*, *35*(12), 1489–1506. doi:10.1287/mnsc.34.12.1489

Sobol, I. (1973). *Monte-Carlo numerical methods*. Moscow, Russia: Nauka.

Sobol, I. (1979). On the systematic search in a hypercube. *SIAM Journal on Numerical Analysis*, *16*, 790–793. doi:10.1137/0716058

Stanoulov, N., Pavlov, Y., & Tonev, M. (1985). Multi-criteria routes selection in urban conditions. In I. Popchev & P. Petrov (Eds.), *Optimization, Decision-Making, Microprocessors systems, Sofia 1983- Eight Bulgarian-Polish Symposium*. Sofia, Bulgaria: Publishing House of Bulgarian Academy of Sciences.

Stephanidis, C. (2001). Adaptive techniques for universal access. *User Modeling and User-Adapted Interaction*, *11*(1-2), 159–179. doi:10.1023/A:1011144232235

Strang, G. (1976). *Linear algebra and its applications*. New York, NY: Elsevier.

Technical Documentation. (1995). *ISABROWN breed chickens*. ISABROWN.

Tenekedjiev, K. (2004). *Quantitative decision analyze: Utility theory and subjective statistics*. Sofia, Bulgaria: Marin Drinov.

Terzieva, V., Pavlov, Y., & Andreev, R. (2007). E-learning usability: A learner-adapted approach based on the evaluation of learner's preferences. In *E-Learning III and the Knowledge Society*. Brussels, Belgium: Academic Press.

Trenogin, V. (1985). *Functional Analysis* (2nd ed.). Moscow, Russia: Nauka.

Tversky, A., & Kahneman, D. (1991). Loss aversion in riskless choice: A reference dependent model. *The Quarterly Journal of Economics*, *106*, 1039–1061. doi:10.2307/2937956

Tversky, A., & Kahneman, D. (1992). Advances in prospect theory: Cumulative representation of uncertainty. *Journal of Risk and Uncertainty*, *5*(4), 297–323. doi:10.1007/BF00122574

Tzonkov, S., & Hitzmann, B. (2006). *Functional state approach to fermentation process modeling*. Sofia, Bulgaria: Prof. Marin Drinov Academic Publishing House.

Vapnik, V. (2006). *Estimation of dependences based on empirical data* (2nd ed.). London, UK: Springer.

Vári, A., & Vecsenyi, J. (1984). Selecting decision support methods in organizations. *Journal of Applied Systems Analysis*, *11*, 23–36.

Vilkas, E. (1990). *Optimum in games and decisions*. Moscow, Russia: Nauka.

Vygotsky, L. S. (1978). *Mind and society: The development of higher mental processes*. Cambridge, MA: Harvard University Press.

Wang, T., Moore, C., & Birdwell, D. (1987). Application of a robust multivariable controller to non-linear bacterial growth systems. In *Proceedings of the 10th IFAC Congress on Automatic Control,* (vol. 39). Munich, Germany: IFAC. Retrieved from http://www.ifac-control.org/events/congresses

Watt, D. (2004). Value based decision models of management for complex systems. In *Proceedings-2004 IEEE International, Engineering Management Conference*, (Vol. 3, pp. 1278 – 1283). IEEE Press.

Weber, E. (1994). From subjective probabilities to decision weights: The effect of asymmetric loss functions on the evaluation of uncertain outcomes and events. *Psychological Bulletin, 115*(2), 228–242. doi:10.1037/0033-2909.115.2.228

Westcott, M. (1968). *Toward a contemporary psychology of intuition: A historical and empirical inquiry*. New York, NY: Holt, Rinehart & Winston, Inc.

Zaichenko, P. (1988). *Operations research*. Kiev, Russia: Vishta Shkola.

Zarate, P. (2008). Cooperative decision support systems. In Adam, F., & Humphreys, P. (Eds.), *Encyclopedia of Decision-Making and Decision Support Technologies* (pp. 109–115). Hershey, PA: IGI Global. doi:10.4018/978-1-59904-843-7.ch013

Zeleznikow, J. (2002). Risk, negotiation and argumentation—A decision support system based approach. *Law Probability and Risk, 1*(1), 37–48. doi:10.1093/lpr/1.1.37

Zukerman, I., & Albrecht, D. W. (2001). Predicative statistical models for user modeling. *User Modeling and User-Adapted Interaction, 11*(1-2), 5–18. doi:10.1023/A:1011175525451

Zukerman, I., & Litman, D. (2001). Natural language processing and user modeling: Synergies and limitations. *User Modeling and User-Adapted Interaction, 11*(1-2), 129–158. doi:10.1023/A:1011174108613

About the Authors

Rumen Andreev is Head of the Department of Communication Systems and Services in the Institute of Information and Communication Technologies, Bulgarian Academy of Sciences. He received his MSc degree in Automation and holds a PhD in Computer-Aided Design from Technical University of Sofia. He has both research and practical expertise in computer graphics, computer-aided design, human-computer interactions, and enhancement of resources for collaborative, cooperative work, and personalized e-learning. His research has been published in the monograph *Graphics Systems: Architecture and Realization* (Elsevier) and in international journals including the *Computers and Graphics*, *Computer Graphics Forum*, *Interacting with Computers*, *Communication and Cognition*. He is a member of the Union of Automation and Informatics, UAI-Bulgaria.

Yuri Pavlov is an Associate Professor in the Institute of Information and Communication Technologies in the Bulgarian Academy of Sciences, Bulgaria. He has received DUES from Paris VI – France, MSc degree in Automation from the Technical University of Sofia, and holds PhDs from the Bulgarian Academy of Sciences. His research has been published in international journals: *Proceedings of the Bulgarian Academy of Sciences, International Online Journal Bioautomation, European Journal of OR, E-learning III and the Knowledge Society – Belgium, Proceedings in Manufacturing Systems – Romania*. He has research and practical expertise in innovative and creative decision-making, optimal control, and control design of complex systems. He is also a member of the International Institute of Informatics and Systemics (IIIS).

Index

Q

R

S

T

U

V

W